Spectrum and Network Measurements

SciTech Publishing, Inc.

New Age

Spectrum and Network Measurements

Robert A. Witte
Agilent Technologies

PUBLISHING FOR ONE WORLD

NEW AGE INTERNATIONAL (P) LIMITED, PUBLISHERS
New Delhi • Bangalore • Chennai • Cochin • Guwahati • Hyderabad
Jalandhar • Kolkata • Lucknow • Mumbai • Ranchi
Visit us at www.newagepublishers.com

First Indian Sub-Continent Edition : 2010
by New Age International (P) Ltd., Publishers

This special low-priced edition is for sale in **India, Bangladesh, Bhutan, Myanmar (Burma), Sri Lanka, Nepal, Pakistan, Maldives and countries geographically located under Middle Eastern region** only.

Originally published by *SciTech Publishing, Inc., USA*

Branches:

- No. 37/10, 8th Cross (Near Hanuman Temple), Azad Nagar, Chamrajpet, **Bangalore**-560 018. Tel.: (080) 2675 6823, Telefax: 2675 6820, E-mail: bangalore@newagepublishers.com
- 26, Damodaran Street, T. Nagar, **Chennai**-600 017. Tel.: (044) 24353401, Telefax: 24351463, E-mail: chennai@newagepublishers.com
- CC-39/1016, Carrier Station Road, Ernakulam South, **Cochin**-682 016. Tel.: (0484) 2377004, Telefax: 4051303. E-mail: cochin@newagepublishers.com
- Hemsen Complex, Mohd. Shah Road, Paltan Bazar, Near Starline Hotel, **Guwahati**-781 008. Tel.: (0361) 2513881. Telefax: 2543669, E-mail: guwahati@newagepublishers.com
- No. 105, 1st Floor, Madhiray Kaveri Tower, 3-2-19, Azam Jahi Road, Nimboliadda, **Hyderabad**-500 027. Tel.: (040) 24652456, Telefax: 24652457, E-mail:hyderabad@newagepublishers.com
- RDB Chambers (Formerly Lotus Cinema)106A, 1st Floor, S.N. Banerjee Road, **Kolkata**-700 014. Tel.: (033) 22273773, Telefax: 22275247, E-mail:kolkata@newagepublishers.com
- 16-A, Jopling Road, **Lucknow**-226 001. Tel.: (0522) 2209578, 4045297, Telefax: 2204098, E-mail: lucknow@newagepublishers.com
- 142C, Victor House, Ground Floor, N.M. Joshi Marg, Lower Parel, **Mumbai**-400 013. Tel.: (022) 24927869. Telefax: 24915415, E-mail: mumbai@newagepublishers.com
- 22, Golden House, Daryaganj, **New Delhi**-110 002. Tel.: (011) 23262370, 23262368, Telefax: 43551305. E-mail: sales@newagepublishers.com

Printed in India at Glorious Printers, Delhi.
C-10-06-4703

ISBN: 978-81-224-3097-4
Original Edition Price: $ 75.00
Indian Edition Price: Rs. 495.00

PUBLISHING FOR ONE WORLD
NEW AGE INTERNATIONAL (P) LIMITED, PUBLISHERS
4835/24, Ansari Road, Daryaganj, New Delhi-110002
Visit us at **www.newagepublishers.com**

Contents

Preface to the Noble Publishing Edition

Welcome to the latest printing of Spectrum and Network Measurements. This book covers the basic theory and practice of measuring the spectrum of a signal and the transfer characteristics of networks.

The concept of wireless communication has been around for decades, evolving from spark gap transmitters to handheld digital mobile phones. Spark gap transmissions relied on Morse code (the original digital format), occupied wide spectrum bandwidth and were relatively inefficient. Over time, communication systems adopted AM and FM analog modulation techniques to implement AM and FM broadcast radio, two-way radio and early cellular telephones. More recently, digital formats have emerged as the most efficient and versatile modulation schemes. It has been fascinating to witness the explosion in wireless communications devices, and it is not over yet.

The increasing usage of wireless devices has caused a corresponding high demand for engineers and technicians who understand RF and microwave circuits and systems. This resurgence in RF and microwave design proceeds and contributes more to the wave of digital technology that swept through the electronics industry in the 1980s. A merging of wireless and digital technology is producing an unprecedented level of connectivity in our society like never before.

Despite these recent changes in technology, the fundamentals of signals propagating through circuits and through the air have not changed. The basic theory of "signals and systems" and the measurements that accompany it still apply. Concepts such as Fourier analysis, transmission lines, intermodulation distortion, signal-to-noise ratio and *S*-parameters represent a critical foundation for this new era of wireless development. The purpose of this book is to enable the reader to understand the basic theory that is the foundation of signals and systems, relate the basic theory to measured results and apply it in creating new RF and microwave designs.

Portions of Chapter 4 (Sections 4.21 through 4.32) were contributed by the R&D department of the Lake Stevens Instrument Division of Hewlett-Packard Company (now Agilent Technologies).

My thanks to Gary Breed, Dennis Ford and Martina Voigt of Noble Publishing for making possible the publishing of this book.

This version of the book is dedicated to the memory of William Hewlett and David Packard, who together created one of the world's premier technology leaders: the Hewlett-Packard Company. Test and Measurement was the foundation of the company and has been recently carried over into a new company called Agilent Technologies. Bill and Dave were talented engineers, technology pioneers, solid businessmen, creators of innovative management practices, philanthropists and two very down-to-earth guys.

–Bob Witte

Preface

This book is about the theory and practice of spectrum and network measurements in electronic systems. It is intended for readers who have a background in electrical engineering and use spectrum analyzers and network analyzers to characterize electronic signals or systems in the frequency domain.

Although some of the internal functions of spectrum analyzers and network analyzers are discussed, the real emphasis of the book is on the theory and practice of frequency domain measurements. Enough theory is provided so that the reader can understand how a particular measurement is made, what the possible sources of error are, and the significance of the results. Many numerical examples are given to aid the reader in understanding the material and to help relate theory and practice.

Spectrum and network analyzers are widely used in electronic design, manufacturing, and service. Spectrum measurements show the frequency components that are present in a particular signal. Some typical spectrum measurement applications are found in radar systems, radio receivers, mobile radio, cable television, electronic surveillance, electromagnetic interference and control, oscillator characterization, and audio measurements. Network measurements are used to determine how an electric circuit behaves in response to a range of frequencies. Network measurements are used to characterize devices such as amplifiers, filters, cables, attenuators, and components. Network measurements often imply a full network analyzer system complete with an *s*-parameter test set, but many measurements (especially transmission measurements) can be made with simpler equipment.

The reader will probably find many familiar topics throughout the book since electrical engineers usually have been exposed to such things as decibels, Fourier theory, two-port networks, and so on. However, the usual exposure to these topics involves multiple textbooks perhaps encountered through multiple courses. Most often, these sources of information are not oriented toward actual measurements. One aim of this book is to consolidate the pertinent theory into one comprehensive treatment of frequency domain measurements.

This book is primarily intended for the person making measurements in the frequency range below 1 GHz. The same basic theory is applicable to higher frequencies, but the intricacies of the microwave world such as striplines, klystrons, waveguides, and flow graphs are not explored in this book. Still, microwave engineers will find the material in this book helpful in their work.

The influence of emerging digital technology is apparent in the field of frequency domain measurements. Digital technology first appeared as display storage which provided an effective way to eliminate display flicker and to perform measurement functions such as peak hold, normalization, and error correction for network measurements. This was accompanied by microprocessor control of the instrument, which automated some of the operating and control tasks. Finally, digital signal processing has appeared in the form of averaging, digital filtering, and fast Fourier transforms. These digital enhancements to spectrum and network measurements are covered throughout this book.

Most of the book is oriented toward the workhorses of the industry—the swept spectrum analyzer and the swept network analyzer. However, a chapter is also devoted to spectrum analyzers which use the fast Fourier transform to determine the spectral content of a signal. These analyzers have the advantage of measurement speed and excellent frequency resolution but are currently somewhat limited in frequency range. The hybrid analyzer, which uses a combination of swept and FFT techniques, provides FFT measurements at higher frequencies.

The book can certainly be read cover to cover, but it is also organized into independent chapters and subchapters. This allows the reader to read selectively and enhances the usefulness of the book as a reference.

Chapter 1 is an introduction to spectrum and network measurements. Decibels are an often used and misused concept so Chapter 2 is devoted to that topic. Fourier theory, the theoretical basis for spectrum analysis, is summarized in Chapter 3. The two main types of spectrum analyzers (FFT analyzers and swept analyzers) are discussed in Chapters 4 and 5. Chapters 6 through 9 cover several important measurement applications: modulated signals, signal distortion, noise, and pulsed waveforms. Averaging and filtering are covered together in Chapter 10.

Chapters 11 and 12 cover transmission lines and measurement connection techniques. Chapter 13 introduces two-port network theory, which is the basis for network analysis. Chapters 14 through 16 cover network analyzers and their use in making transmission and reflection measurements. Chapter 17 ends the book with a discussion of instrument performance and specifications.

Many people contribute to the undertaking of writing a book such as this one, either directly or indirectly. The professors, engineers and technicians with whom I have been associated in my career deserve some credit since they have influenced me through the exchange of ideas. My special thanks goes to the following engineers at Hewlett-Packard who, in a variety of ways, helped me with this book: Jerry Daniels, Glenn Engel, Bryan Hoog, Roy Mason, Harry Plate, Bill Spaulding, Joe Tarantino, and Ken Wyatt.

—Robert A. Witte

Spectrum and Network Measurements

1

Introduction to Spectrum and Network Measurements

1.1 SIGNALS AND SYSTEMS

An electrical system normally has one or more input ports and one or more output ports. Figure 1-1 shows a system with a single input and a single output. Electrical devices such as filters, attenuators, and amplifiers fall into this category. Shown at the input is a signal, $x(t)$ and at the output a signal, $y(t)$.

A more complex system (a phase lock loop) is shown in Figure 1-2. Although there is still only one input and one output, there are several blocks or subsections of the system, each having its own input and output. (Each block of the system may be considered as another system.) A design engineer thinks in terms of the individual blocks and signals while designing the system. Measurement instrumentation is used in the design phase when the engineer checks the performance of the individual blocks and signals. The signals and system blocks may also be measured in production, and later, when the system is maintained in the field.

Network measurements characterize the circuit blocks of the system while spectrum measurements characterize the signals present. For example, the phase noise or spurious frequency components of the output, $y(t)$ might be a critical parameter in the performance of the system. These characteristics are measured using a spectrum analyzer. Similarly, the transfer characteristics of the low-pass (loop) filter might be of interest, which is measured with a network analyzer.

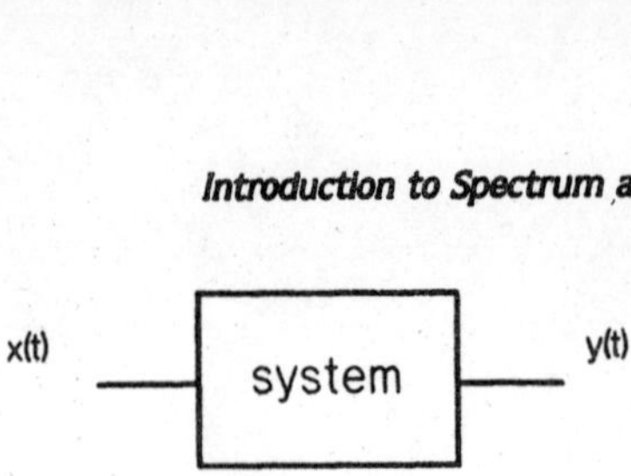

Figure 1-1. A simple system having one input, $x(t)$, and one output, $y(t)$.

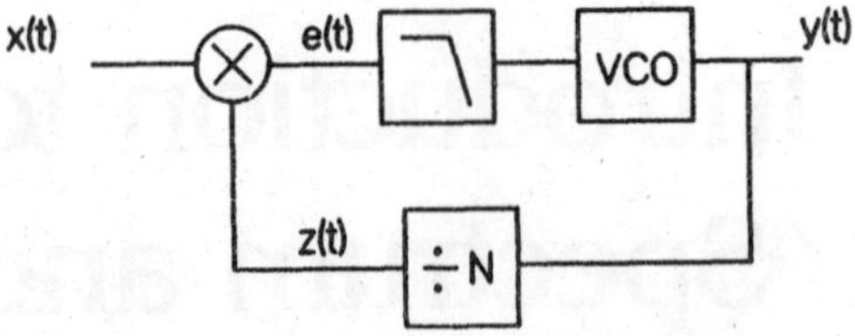

Figure 1-2. A phase lock loop is a complex system with multiple blocks and multiple signals.

1.2 TIME DOMAIN AND FREQUENCY DOMAIN RELATIONSHIPS

The most obvious way to describe an electrical signal is its *time domain representation* (i.e., the voltage or current as a function of time) as shown in Figure 1-3a. An oscilloscope displays the time domain representation of a signal. The system blocks can be characterized in the time domain by measuring the step response, pulse response, or the response at the output due to some other input signal.

Another way to describe a signal is using its *frequency domain representation* (i.e., the amplitude of the signal as a function of frequency) as shown in Figure 1-3b. Fourier theory relates the time domain and frequency domain representations.

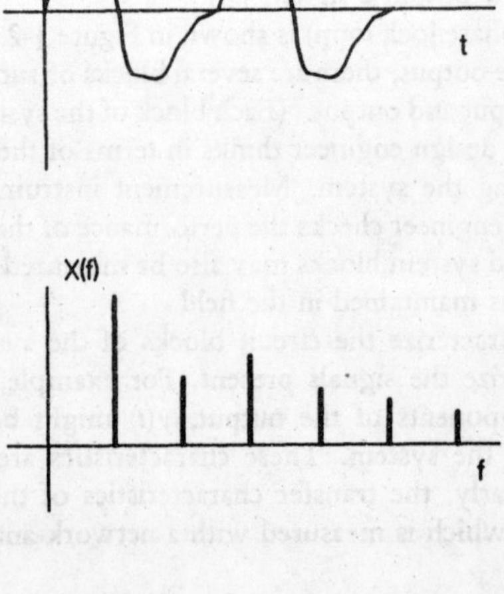

Figure 1-3. (a) A signal shown as a function of time. (b) A signal shown as a function of frequency.

Appropriate use of the Fourier series, Fourier transform, and the discrete Fourier transform (DFT) allow the transformation of a time domain function, $x(t)$, into a frequency domain function, $X(f)$.

The spectrum analyzer's frequency-selective circuitry measures the amplitude of a signal over the frequency range of interest. Thus, the spectrum analyzer is to the frequency domain as the oscilloscope is to the time domain. Network measurements require a stimulus (usually supplied by the measuring instrument) at the input to the system. This stimulus must cover a wide range of frequencies in order for the output signal to adequately represent the frequency domain performance of the system. Most often, a sine wave source which is automatically swept through the frequency range of interest provides the stimulus, but a broadband noise source can also be used.

1.3 SYSTEM TRANSFER FUNCTION

The stimulus signal, $X(f)$, is applied to the input of a system and the output, $Y(f)$, is measured (Figure 1-4). The transfer function is ratio of the output over the input, both as a function of frequency.

$$\text{transfer function:}\quad H(f) = \frac{Y(f)}{X(f)}$$

This implies a simple model of the system. That is, the input signal and the transfer function completely determine the output signal, with no loading effects present. Two-port parameters (discussed in Chapter 13) provide for a more complete model of a system.

1.4 ADVANTAGES OF USING FREQUENCY DOMAIN MEASUREMENTS

Why use frequency domain measurement techniques? The answer varies, but frequency domain instrumentation has several distinct advantages.

Narrowband frequency domain measurements provide greater sensitivity than time domain measurements. Since the measurement bandwidth can be narrowed almost arbitrarily, frequency domain analyzers can greatly reduce the amount of noise present in the measurement. Similarly, narrowband measurements can remove large interfering signals at undesired frequencies. Consider the measurement of harmonic distortion of a near-perfect sine wave. A spectrum analyzer can ignore the large fundamental frequency when measuring the harmonic level. A time do-

Figure 1-4. The transfer function of a system describes its behavior in the frequency domain.

main measurement with an oscilloscope must simultaneously measure the fundamental and the much smaller harmonics in the signal. Harmonic distortion measurements with an oscilloscope are limited to a few percent, while spectrum analyzers routinely allow 0.01% distortion measurements.

Some systems are inherently frequency domain oriented. For instance, the frequency division multiplexing (FDM) systems used in telecommunications systems operate by sandwiching together multiple signals in the frequency domain. Radio stations are also frequency domain multiplexed, with each station in a given geographical area occupying a particular frequency band. A radio receiver is inherently a frequency domain device, since it is essentially a frequency selective detector.

Even systems that are not usually thought of as being inherently frequency domain in nature may still require frequency domain measurements. For instance, stray capacitance and resistive losses in a digital circuit connection may limit the bandwidth of the circuit and the speed of a digital pulse. A network analyzer can determine the bandwidth of the circuit by measuring its transfer function in the frequency domain.

Multiple signals are usually easier to separate in the frequency domain than in the time domain. For instance, suppose the output of a switching power supply contains significant levels of the 60 Hz line frequency (and its harmonics) and the switching frequency of the power supply. Whichever of these is the largest will be discernable by a time domain measurement. Usually, though, if there are multiple frequencies present, it will be difficult to view them with an oscilloscope. A spectrum analyzer, on the other hand, can separate these frequency components and measure them accurately.

1.5 SPECTRUM MEASUREMENTS

A signal is characterized using a spectrum analyzer as shown in Figure 1-5. The measurement is usually as simple as connecting the analyzer to the source of the signal. However, the loading effects and other sources of measurement error may need to be considered. The frequency domain representation of the signal will appear on the analyzer's display (an example is shown in Figure 1-6).

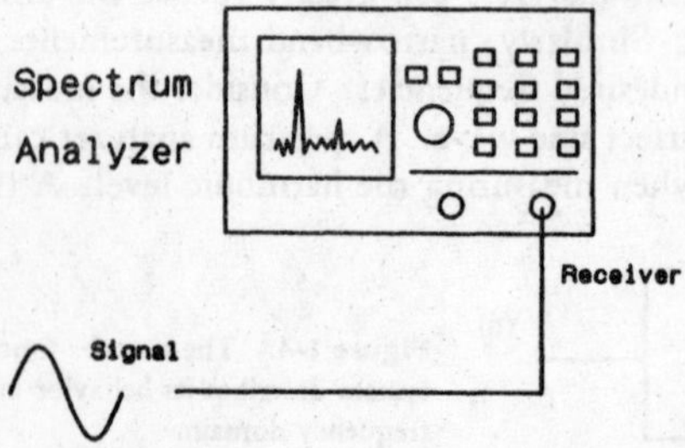

Figure 1-5. A spectrum measurement is performed by applying the signal to be analyzed to the input of a spectrum analyzer.

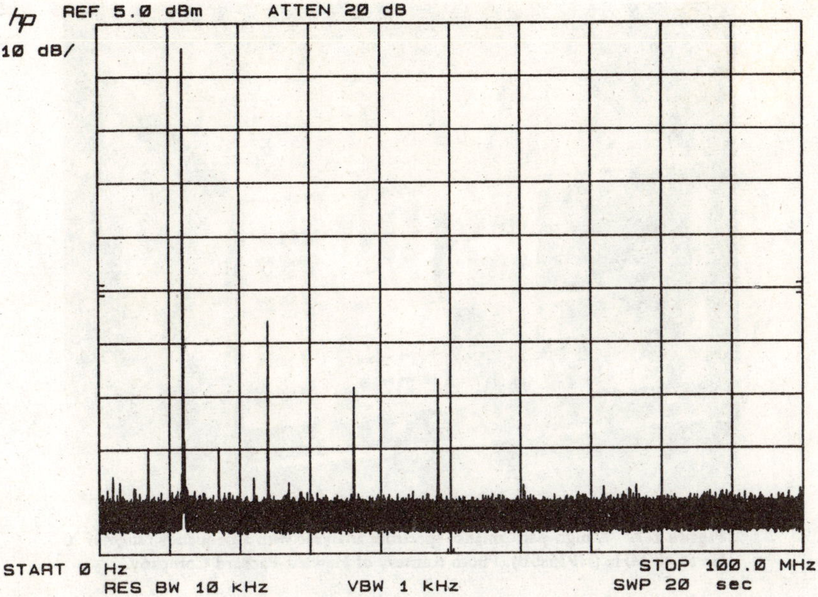

Figure 1-6. A typical spectrum analyzer measurement showing the harmonic content of a signal generator's output.

The complexity of the measurement varies according to the application. In a simple case, the spectrum analyzer may be used to measure the amplitude and frequency of a signal spectral line. More often, the spectral content of the signal includes multiple responses such as harmonics, modulation sidebands, and spurious responses. Noise levels can also be measured (if the measured noise is greater than the analyzer's noise), and the noise level can be displayed as a function of frequency.

The standard vertical scale on a spectrum analyzer is logarithmic and marked in decibels. This allows a large dynamic range to be displayed on a reasonable-sized screen. Many analyzers also provide a linear vertical scale for users that prefer to work in terms of volts. The horizontal scale is, of course, frequency. It is usually a linear frequency scale, but there are applications where a logarithmic frequency scale is used.

1.6 SPECTRUM ANALYZERS

Spectrum analyzers are available in a wide range of configurations with a corresponding wide range of performance. Frequency range is the most obvious parameter to use to categorize spectrum analyzers. Different measurement technologies are

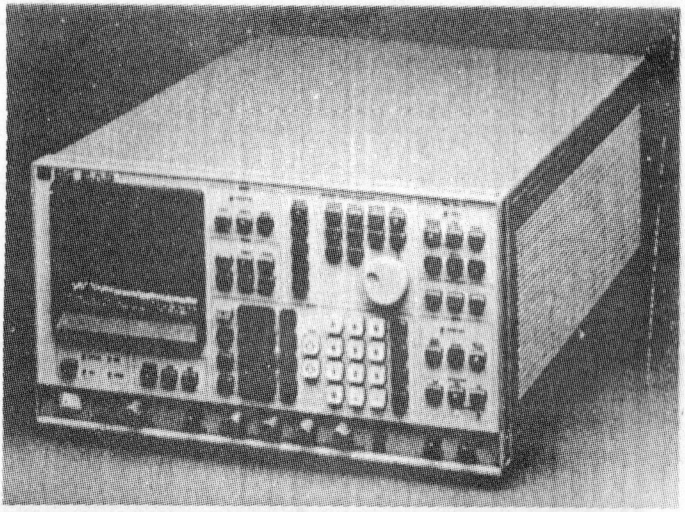

Figure 1-7. A high-performance spectrum analyzer with a frequency range of 20 Hz to 40 MHz (HP3585B). Photo courtesy of Hewlett-Packard Company.

most effective in different frequency bands. At low frequencies, analyzers using the fast Fourier transform (FFT) provide good performance from near DC to a few hundred kilohertz. The next group of analyzers reach the high-frequency (HF) region, extending from a low end of 10 Hz to a high-frequency limit of 100 MHz or so. The radio frequency/microwave analyzers fill out the high end with a low-frequency limit of 100 kHz and a high-frequency limit anywhere from 1 to 100 GHz.

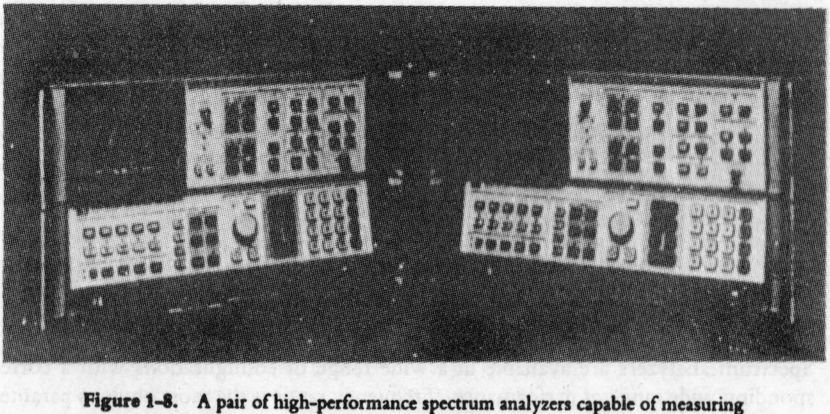

Figure 1-8. A pair of high-performance spectrum analyzers capable of measuring microwave frequencies (HP8566B and HP8568B). Photo courtesy of Hewlett-Packard Company.

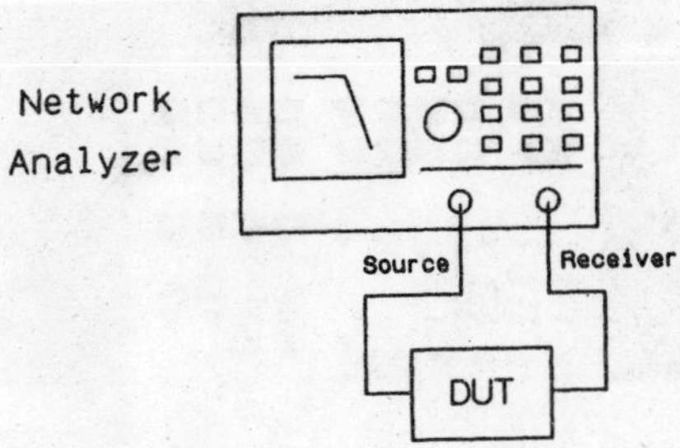

Figure 1-9. A network analyzer provides a signal source to the device under test (DUT) and measures the response at the device's output.

Figure 1-7 shows a high performance 40 MHz spectrum analyzer while Figure 1-8 shows a pair of analyzers which operate above 1 GHz. Besides frequency range, other factors such as cost, dynamic range, sensitivity, accuracy, and feature set vary from analyzer to analyzer.

1.7 NETWORK MEASUREMENTS

A network is characterized in the frequency domain by connecting the source of a network analyzer to the input of the network and the analyzer's receiver to the

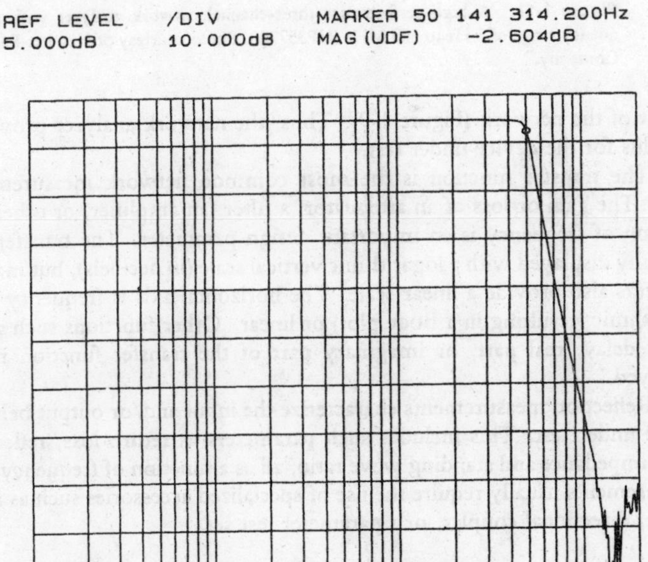

Figure 1-10. A typical network measurement showing the transmission characteristics of a low-pass filter.

Figure 1-11. A high-performance three-channel network analyzer with a frequency range of 5 Hz to 200 MHz (HP3577A). Photo courtesy of Hewlett-Packard Company.

output of the network (Figure 1-9). Thus, the network analyzer provides its own stimulus for the device under test.

The transfer function is the most common network measurement (Figure 1-10). The gain or loss of an attenuator, a filter, an amplifier, or other circuit as a function of frequency is an important design parameter. The transfer function is normally displayed with a logarithmic vertical scale (in decibels), but many network analyzers also provide a linear scale. The horizontal axis is frequency and may be logarithmic (resulting in a Bode plot) or linear. Other functions such as the phase, group delay, real part, or imaginary part of the transfer function may also be displayed.

Reflection measurements characterize the input and/or output behavior of the device under test. This includes such parameters as return loss, reflection coefficient, impedance and standing wave ratio, all as a function of frequency. Reflection measurements usually require the use of specialized accessories such as a directional bridge, directional coupler, or *s* parameter test set.

1.8 NETWORK ANALYZERS

Network analyzers are available in two main varieties: *scalar* and *vector*. Scalar analyzers provide only magnitude information (no phase information) and are therefore less expensive for equivalent-frequency range. Scalar analyzers usually operate in the microwave region, since this is where their cost effectiveness is most needed. Vector analyzers are available in all frequency ranges, since phase information is required for many measurement applications. A typical vector network analyzer is shown in Figure 1-11.

1.9 SPECTRUM/NETWORK ANALYZERS

Recently some manufacturers have combined the functions of the spectrum analyzer and network analyzer in one instrument (Figure 1-12). This hybrid approach is a natural one, since the same instrument user often needs to perform both spectrum and network measurements. Furthermore, the block diagrams and technologies used in the two types of instruments are similar enough that significant economy can be achieved.

Spectrum/network analyzers usually have vector network capability. One can usually think of the instrument as a "spectrum analyzer with phase." This leaves the market seemingly a bit confused, with instrument offerings that range from pure spectrum analyzers (with no source or tracking generator) to pure network analyzers

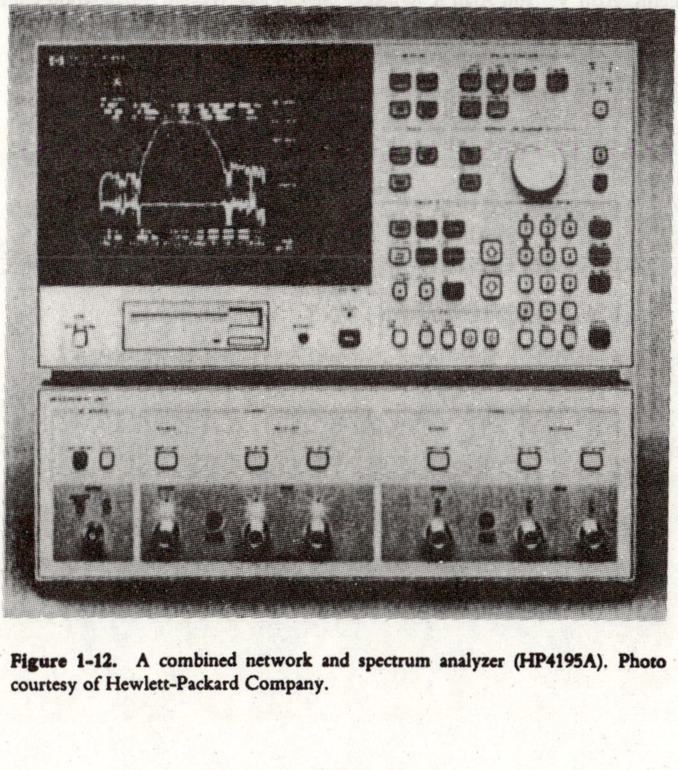

Figure 1-12. A combined network and spectrum analyzer (HP4195A). Photo courtesy of Hewlett-Packard Company.

(no spectrum capability). In between, there are spectrum analyzers that have tracking generators that allow scalar network measurements and spectrum analyzers that have phase detectors built in for simple vector network measurements.

Why doesn't a network analyzer inherently have the ability to also make spectrum measurements? The answer varies, but often a network analyzer design will capitalize on the fact that the frequency of the stimulus signal is known (since it is supplied by the network analyzer). This allows the use of a simpler receiver block diagram than one which must reject images and other off-carrier frequency components. This is unlike the spectrum case, where the frequency of the measured signal has to be assumed to be totally arbitrary. Thus, a network analyzer can have a simpler and less expensive block diagram.

REFERENCES

1. McGillem, Clare D., and George R. Cooper. *Continuous and Discrete Signal and System Analysis*. New York: Holt, Rinehart and Winston, Inc., 1974.
2. Oliver, Bernard M., and John M. Cage. *Electronic Measurements and Instrumentation*. New York: McGraw-Hill Book Company, 1971.
3. Oppenheim, Alan V., and Alan S. Willsky. *Signals and Systems*. Englewood Cliffs, NJ: Prentice-Hall, Inc., 1983.
4. Schwartz, Mischa. *Information, Transmission, Modulation, and Noise*, 3rd ed. New York: McGraw-Hill Book Company, 1980.

2

Decibels

Decibels are used to specify ratios of powers and voltages in a logarithmic fashion. Absolute levels can also be specified via suitable reference values. Decibels are commonly used for gain and loss calculations in electronic systems.

Most, if not all, spectrum and network analyzers display measurement results with their displays calibrated in decibels. The popularity of the decibel in such applications is due to its ability to compress logarithmically widely varying signal levels. For example, a 1 volt signal and a 10 microvolt signal can both be represented on a display with 100 dB of range. To show these two signals simultaneously with reasonable clarity on a linear scale is impossible.

Decibels also are useful for gain and loss calculations, where multiplication operations are transformed into (easier) additions.

2.1 DEFINITION OF THE DECIBEL

The definition of the decibel (dB) is in terms of a power ratio. Two powers, P_1 and P_2, can be related in dB by the following equation:

$$A_{(dB)} = 10 \log(P_2/P_1) \tag{2-1}$$

where log indicates the base 10 logarithm.

As shown, P_2 is expressed relative to P_1. Reversing P_1 and P_2 changes the sign of the result in decibels.

If the powers P_1 and P_2 resulted from a pair of voltages across a pair of resistors then

$$A_{(dB)} = 10 \log \frac{(V_2^2/R_2)}{(V_1^2/R_1)} \tag{2-2}$$

$$A_{(dB)} = 10 \log(V_2/V_1)^2 + 10 \log(R_1/R_2) \tag{2-3}$$

$$A_{(dB)} = 20 \log(V_2/V_1) + 10 \log(R_1/R_2) \tag{2-4}$$

The first term is the voltage form of the decibel equation, and the second term accounts for differences in the two resistances. If the two resistances are equal, this equation can be further simplified.

$$A_{(dB)} = 20 \log(V_2/V_1) \tag{2-5}$$

The last equation has taken on a life of its own and is often used as the defining equation for the decibel. Strictly speaking, the decibel is defined in terms of power only. If the resistances associated with each of the powers (voltages) are equal, then the power equation and the voltage equation are consistent. If the voltage equation is used when the resistances are not equal, incorrect results will occur.

Despite this potential problem, the voltage equation is widely used in situations where the two resistances are not equal. For instance, the voltage equation is often used to specify the voltage gain of operational amplifier circuits. In these circuits, the input impedance is usually very high and the output impedance is usually low. The voltage form of the decibel equation can be used successfully in such a case as long as power gain is not inferred from it.

Example 2.1

Calculate the ratio of P_2 and P_1 and express in decibels. $P_1 = 2$ watts, $P_2 = 12$ watts. Exchange P_1 and P_2 and recalculate.

The linear ratio is $A = P_2/P_1 = 12/2 = 6$. Expressed in decibels

$$A_{(dB)} = 10 \log(P_2/P_1) = 10 \log(12/2) = 7.78 \text{ dB}$$

With P_1 and P_2 reversed,

$$A_{(dB)} = 10 \log(P_1/P_2) = 10 \log(2/12) = -7.78 \text{ dB}$$

Solving the decibel equations for the power or voltage ratio results in

$$P_2/P_1 = 10^{(A_{(dB)}/10)} \tag{2-6}$$

$$V_2/V_1 = 10^{(A_{(dB)}/20)} \tag{2-7}$$

Example 2.2

The voltage gain of a circuit (the ratio of the output voltage to the input voltage) is 25 dB. If the output voltage is 5 volts, what is the input voltage?

$$V_2/V_1 = 10^{(25/20)} = 17.78$$

$$V_1 = V_2/17.78 = 5/17.78 = 0.281 \text{ volts}$$

2.2 CARDINAL VALUES

It is worth summarizing some of the common cardinal values for decibels.

Voltage ratio	Power ratio	Decibels
1	1	0 dB
1.414	2	3 dB
2	4	6 dB
3.16	10	10 dB
10	100	20 dB

Some mathematical identities can be used to develop some rules for working with decibels. Although precise calculations would be best accomplished using a scientific calculator, these rules can provide a more intuitive, working knowledge of decibels.

Rule 1. Changing the sign of the decibel value, corresponds to the taking the reciprocal of the linear ratio.

If

$$A_{(dB)} = 10 \log(A)$$

then

$$-A_{(dB)} = 10 \log(1/A)$$

Rule 2. Adding two decibel values is equivalent to multiplying their corresponding linear ratios.

$$10 \log(A_1) + 10 \log(A_2) = 10 \log(A_1 \times A_2)$$

Example 2.3

Use the table of cardinal values and the foregoing rules to express the following linear ratios in terms of decibels: (a) $V_2/V_1 = 20$; (b) $P_2/P_1 = 0.5$; (c) $V_2/V_1 = 40$.

a. $V_2/V_1 = 10 \times 2$. From the table of cardinal values, the ratios of 10 and 2 can be converted to decibel values of 20 dB and 6 dB. Using rule 2, $A_{(dB)} = 20$ dB + 6 dB = 26 dB.

b. The reciprocal of P_2/P_1 is 2, which in decibels is 3 dB. Thus, by rule 1, the decibel value is −3 dB.

c. $V_2/V_1 = 10 \times 2 \times 2$. From the table of cardinal values, these can be converted to decibel values of 20 dB, 6 dB, and 6 dB. Summing these provides the complete result, $A_{(dB)} = 32$ dB.

2.3 ABSOLUTE DECIBEL VALUES

The original definition of the decibel allows only for representing ratios of two values in decibel form. By providing a reference value (either a voltage or a power), decibels can be used to refer to absolute voltage or power values.

$$P_{(dB)} = 10 \log(P/P_{REF}) \tag{2-8}$$

$$V_{(dB)} = 20 \log(V/V_{REF}) \tag{2-9}$$

dBm

The most common power reference for spectrum and network measurements is 1 milliwatt, resulting in dBm.

$$P_{(dBm)} = 10 \log(P/0.001) \tag{2-10}$$

Note that this definition does not depend on the impedance which dissipates the power.

It is convenient to develop the voltage form of the equation, since many power measurements are actually calibrated voltage measurements. In order to do so, it is necessary to specify the impedance level involved since it relates the voltage and power levels. Therefore, these equations are valid only for the specified impedances.

The voltage reference produces 1 milliwatt of power in a resistor of the appropriate impedance. For $R = 50\ \Omega$

$$V_{REF} = \sqrt{PR} = \sqrt{0.001 \times 50} = 0.2236 \text{ volts} \tag{2-11}$$

$$P_{(dBm)} = 20 \log(V_{RMS}/0.2236) \quad 50\ \Omega \tag{2-12}$$

For $R = 75\ \Omega$,

$$P_{(dBm)} = 20 \log(V_{RMS}/0.2739) \quad 75\ \Omega \tag{2-13}$$

Note that the reference voltage and the voltage to be converted to dBm are both in RMS volts. The symbol $P_{(dBm)}$ was used even though the decibel equation used voltage, in order to emphasize that dBm is defined in terms of power. The voltage form of the equation is valid only for one particular impedance while the power form of the equation is independent of impedance. A particular dBm value will always indicate the same power level, but will correspond to different voltages for different impedances.

Example 2.4

Express the following voltages and powers in terms of dBm: (a) $P = 25\ \mu W$; (b) $V_{RMS} = 1$ volt, 50 Ω impedance; (c) $V_{RMS} = 1$ volt, 75 Ω impedance.

a. $P_{(dBm)} = 10 \log(25 \times 10^{-6}/0.001) = -16.0$ dBm

b. $P_{(dBm)} = 20 \log(1/0.224) = 13.0$ dBm

c. $P_{(dBm)} = 20 \log(1/0.274) = 11.24$ dBm

dBV

The most common voltage reference is 1 volt (RMS), resulting in dBV.

$$V_{(\mathrm{dBV})} = 20 \log(V_{\mathrm{RMS}}/1) = 20 \log(V_{\mathrm{RMS}}) \qquad (2\text{-}14)$$

Measurements in dBV are based on voltage only. A particular dBV value will always have a corresponding voltage value, independent of the impedance present. This means that a constant dBV value will supply differing amounts of power to different impedances. This runs counter to the previous assertion that the decibel is strictly defined in terms of power.

Figure 2-1 allows convenient conversion from volts to dBV and dBm.

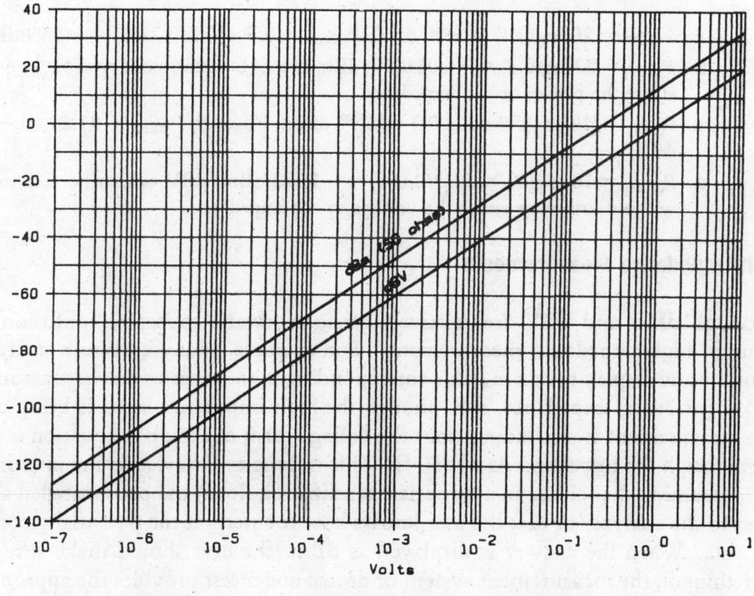

Figure 2-1. Plot of dBm and dBV versus RMS voltage.

dBm/dBV Conversions

In order to convert between dBm and dBV, it is necessary to specify the impedance. Both dBm and dBV can be computed using the voltage form of the decibel equation, but with different voltage references. Due to the logarithmic nature of decibels, for any particular impedance dBm and dBV differ by a constant.

$$P_{(\mathrm{dBm})} = V_{(\mathrm{dBV})} + 10 \log[1/(0.001 \times R)] \qquad (2\text{-}15)$$

For $R = 50\ \Omega$ and $75\ \Omega$

$$P_{(dBm)} = V_{(dBV)} + 13.01 \quad \text{for } 50\ \Omega \tag{2-16}$$

$$P_{(dBm)} = V_{(dBV)} + 11.25 \quad \text{for } 75\ \Omega \tag{2-17}$$

The equations showing a power on the left side and a voltage on the right side may seem inconsistent at first. However, they are the logarithmic equivalent of $P = V^2/R$ and serve to emphasize the differing nature of dBm (which is based on power) and dBV (which is based on voltage).

Example 2.5

Convert the following measured values to dBV and dBm, as possible, within the limitations of the information provided: (a) 0.1 volt RMS across 50 Ω; (b) 0.5 volt RMS, unknown impedance; (c) 5 mW, into 75 Ω; (d) 30 μW, unknown impedance.

a. $V_{(dBV)} = 20 \log(0.1) = -20$ dBV; $P_{(dBm)} = 20 \log(0.1/0.2236) = -6.99$ dBm.

b. $V_{(dBV)} = 20 \log(0.5) = -6.02$ dBV; dBm cannot be determined without knowing either the power or the impedance.

c. $P_{(dBm)} = 10 \log(0.005/0.001) = 6.99$ dBm; $V_{(dBV)} = P_{(dBm)} - 11.25 = -4.26$ dBV.

d. $P_{(dBm)} = 10 \log(30 \times 10^{-6}/0.001) = -15.23$ dBm; dBV cannot be determined without knowing either the voltage or the impedance.

High-Impedance Measurements

Although dBm and dBV are relatively straightforward concepts, confusion can occur in high-impedance measurements. For example, many spectrum analyzers compute and display the dBm value corresponding to the measured *voltage* assuming a 50 Ω (or other) impedance, even though the high-impedance input is being used. The measurement is misleading since the voltage form of the dBm equation is used even though the impedance is not 50 Ω. This is done as a service to the instrument user, assuming that either an appropriate termination (load) has been installed at the input to the analyzer or that the user knows how to interpret the potentially confusing data. When the answer is displayed as dBm, the user should make sure that something in the measurement system or device under test provides the appropriate load impedance. Some analyzers with high-impedance inputs allow the user to specify the impedance to be used in computing dBm.

2.4 GAIN AND LOSS CALCULATIONS

The power gain of a system is the ratio of the output power to the input power.[1]

$$G_P = P_2/P_1 \tag{2-18}$$

[1] The definition of power gain is sometimes further refined into several different definitions: operating power gain, transducer power gain, available power gain, and insertion power gain. See Ralph S. Carson, *High Frequency Amplifiers* (New York: John Wiley & Sons, Inc., 1975).

where

$$P_2 = \text{output power}$$

$$P_1 = \text{input power}$$

Power gain is often specified in terms of decibels.

$$G_{P(\mathrm{dB})} = 10 \log(P_2/P_1) \tag{2-19}$$

If P_2 is greater than P_1, the system exhibits actual power gain. The ratio P_2/P_1 is greater than unity and is positive when expressed in decibel form. If P_2 is less than P_1, the system has a power gain of less than unity and actually exhibits a loss. When expressed in decibels, the gain is negative. If P_1 and P_2 are equal, the gain is 1, or in dB, 0 dB.

In ratio form,

$$\text{loss} = P_1/P_2 = 1/G_P \tag{2-20}$$

Using decibels,

$$\text{loss}_{(\mathrm{dB})} = 10 \log(P_1/P_2) = 10 \log(1/G_P) \tag{2-21}$$

A loss is the negative of the corresponding gain, when both are expressed in decibels. For example, a loss of 10 dB is the same as a gain of −10 dB.

Voltage Gain

Gain calculations are not limited to power gain. Voltage gain is often used. Again, the warnings apply about using voltage ratios expressed in decibels when the two impedances involved are not equal.

$$G_V = V_2/V_1 \tag{2-22}$$

where

$$V_2 = \text{output voltage}$$

$$V_1 = \text{input voltage}$$

In decibels, the voltage gain is

$$G_{V(\mathrm{dB})} = 20 \log(V_2/V_1) \tag{2-23}$$

Example 2.6

Compute the gain and loss (both ratio and dB) for a circuit having an input power of 0.40 mW and an output power of 0.25 mW.

The power gain

$$G_P = 0.25/0.40 = 0.625$$

The power loss

$$\text{loss} = 1/G_P = 1.6$$

In decibels

$$G_{(dB)} = 10 \log(0.625) = -2.04 \text{ dB}$$

$$\text{loss(dB)} = 2.04 \text{ dB}$$

Multiple Blocks

When multiple circuits are cascaded together, decibels are often used to simplify the calculations involved. The electronic system shown in Figure 2-2 has three individual blocks, each with its own gain. The total gain of this system can be computed using the following equation:

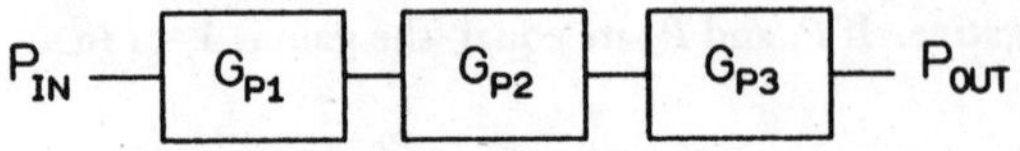

Figure 2-2. Decibels can be used to simplify gain calculations of multiple blocks.

$$G_T = P_{OUT}/P_{IN} = G_{P1}\, G_{P2}\, G_{P3} \tag{2-24}$$

In terms of decibels,

$$G_{T(dB)} = 10 \log(G_{P1}\, G_{P2}\, G_{P3}) \tag{2-25}$$

$$G_{T(dB)} = 10 \log(G_{P1}) + 10 \log(G_{P2}) + 10 \log(G_{P3}) \tag{2-26}$$

$$G_{T(dB)} = G_{1(dB)} + G_{2(dB)} + G_{3(dB)} \tag{2-27}$$

Thus, when decibels are used for gain calculations, multiplication operations are transformed into additions. Similarly, the output power can be expressed in absolute decibels (such as dBm).

$$G_T = P_{OUT}/P_{IN} \tag{2-28}$$

$$G_{T(dB)} = P_{OUT(dB)} - P_{IN(dB)} \tag{2-29}$$

$$P_{OUT(dB)} = G_{T(dB)} + P_{IN(dB)} \tag{2-30}$$

Expanding the total gain into its individual components,

$$P_{OUT(dB)} = G_{1(dB)} + G_{2(dB)} + G_{3(dB)} + P_{IN(dB)} \tag{2-31}$$

Example 2.7

Compute the total system gain in dB for the system shown in Figure 2-3. If the input power is 150 μW, what is the output power in dBm?

The total system gain in dB is $G_{(dB)} = -3 \text{ dB} + 20 \text{ dB} + 5 \text{ dB} = 22 \text{ dB}$. The input power is 150 μW, which is

$$10 \log(150 \times 10^{-6}/0.001) = -8.24 \text{ dBm}$$

$$P_{OUT(dB)} = -8.24 \text{ dBm} + 22 \text{ dB} = 13.76 \text{ dBm}$$

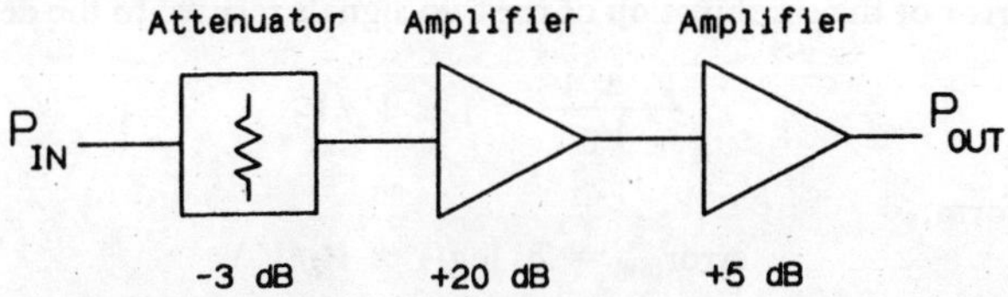

Figure 2-3. A simple system with gain and loss (see Example 2.7).

2.5 DECIBELS AND PERCENT

Often, decibels are used to compare the relative sizes of two signals on a spectrum or network analyzer. The smaller of the two signals can be described as being a certain number of dB down from the larger signal, which acts as a reference. Modulation measurements and harmonic distortion measurements are often stated this way.

It may also be desirable to express the size of the smaller signal as a percent of the larger one. Since

$$A_{(dB)} = 10 \log(P_2/P_1) \tag{2-32}$$

$$P_2/P_1 = 10^{A_{(dB)}/10} \tag{2-33}$$

(P_2/P_1 is just the ratio of the two signal powers, but not expressed in percent. It is understood that the ratio must be multiplied by 100 to get percent.)

Similarly, for voltage

$$V_2/V_1 = 10^{A_{(dB)}/20} \tag{2-34}$$

Example 2.8

A signal is 42 dB smaller than another signal. What percent of the second signal is the first signal (in terms of voltage)?

$$A_{(dB)} = -42 \text{ dB},\ V_2/V_1 = 10^{(-42/20)} = 0.00794, \text{ which is } 0.794\%.$$

2.6 ERROR EXPRESSED IN DECIBELS

The smaller of the two signals may actually be a source of error in a measurement. For example, the smaller signal may be a spurious response occurring at the same frequency as the (desired) larger signal. Depending on how the two signals add together, an error is produced in the measurement. In most analyzer measurements, the smaller signal may add constructively or destructively (or somewhere in between), depending on the relative phases of the signals. Adding the signals together gives a maximum bound on the error and subtracting them gives a minimum bound.

The error of the combination of the two signals relative to the desired signal is

$$\frac{V_1 \pm V_2}{V_1} = 1 \pm V_2/V_1 \tag{2-35}$$

In decibel form,

$$\text{error}_{(dB)} = 20 \log(1 \pm V_2/V_1) \tag{2-36}$$

This is the error induced in V_1 (expressed in dB), due to the presence of V_2. If V_2 is zero, then $\text{error}_{(dB)}$ is 0 dB, indicating that the decibel value of V_1 has no error in it.

Example 2.9

If the smaller signal in Example 2.8 introduces an error into the large signal, express this error in dB. If the large signal is −20 dBV, what is the measured signal (including the error)?

$$\text{error}_{(dB)} = 20 \log(1 \pm 0.00794) = \pm 0.069 \text{ dB}$$

If the error is positive,

$$V_{(dBV)} = -20 \text{ dBV} + 0.069 \text{ dB} = -19.931 \text{ dBV}$$

If the error is negative,

$$V_{(dBV)} = -20 \text{ dBV} - 0.069 \text{ dB} = -20.069 \text{ dBV}$$

TABLE 2-1 ERROR DUE TO INTERFERING SIGNAL

Interfering signal level relative to desired signal			Error introduced into desired signal	
dB	Power %	Voltage %	Error (dB) (positive)	Error (dB) (negative)
0	100.00	100.00	3.0103	−∞
−1	79.43	89.13	2.5390	−6.8683
−2	63.10	79.43	2.1244	−4.3292
−3	50.12	70.79	1.7643	−3.0206
−4	39.81	63.10	1.4554	−2.2048
−5	31.62	56.23	1.1933	−1.6509
−6	25.12	50.12	0.9732	−1.2563
−7	19.95	44.67	0.7901	−0.9665
−8	15.85	39.81	0.6389	−0.7494
−9	12.59	35.48	0.5150	−0.5844
−10	10.00	31.62	0.4139	−0.4576
−20	1.00	10.00	0.0432	−0.0436
−30	0.10	3.16	0.004341	−0.004345
−40	0.010	1.00	0.000434	−0.000434
−50	0.0010	0.32	0.000043	−0.000043
−60	0.00010	0.10	0.00000434	−0.00000434
−70	0.000010	0.032	0.00000043	−0.00000043
−80	0.0000010	0.010	0.0000000434	−0.0000000434
−90	0.00000010	0.0032	0.0000000043	−0.0000000043
−100	0.000000010	0.0010	0.00000000043	−0.00000000043

Table 2-1 is a list of useful decibel relationships. The first column lists the difference between two signal levels, expressed in decibels. The corresponding percent (either power or voltage) is shown in the next two columns, and the worst case errors introduced into the larger signal by the smaller signal (according to equation 2-36) are listed in the last two columns.

Example 2.10

Using Table 2-1, predict the error introduced into a −10 dBm signal by an interfering signal which is 20 dB lower in power level.

The interfering signal is −20 dB relative to the desired signal. From Table 2-1, an error of +0.0432 dB or −0.0436 dB will be introduced, depending on the phase of the interfering signal. So the measured signal level is somewhere between −9.9568 dBm and −10.0436 dBm.

3

Fourier Theory

The most familiar way of representing signals is in the time domain (i.e., a voltage or current represented as functions of time). An alternative representation which is extremely powerful and is inherent in spectrum and network measurements is the frequency domain representation, which describes the signal or system in terms of its frequency content (i.e., how much energy is present at each particular frequency). The frequency domain is related to the time domain by a body of knowledge generally known as Fourier theory, named for Jean Baptiste Joseph Fourier (1768–1830). This includes the series representation know as the *Fourier series* and the transform techniques known as the *Fourier transform*. Discrete (digitized) signals can be transformed into the frequency domain using the *discrete Fourier transform*.

3.1 PERIODICITY

A signal or function is *periodic* if it meets the following criterion:

$$x(t) = x(t + T) \quad \text{for all } t \tag{3-1}$$

where T is the period of the function.

In other words, a periodic function can be shifted in time by exactly one period and the resulting new function will look the same as the original one. A periodic function of time repeats itself every T seconds (Figure 3-1).

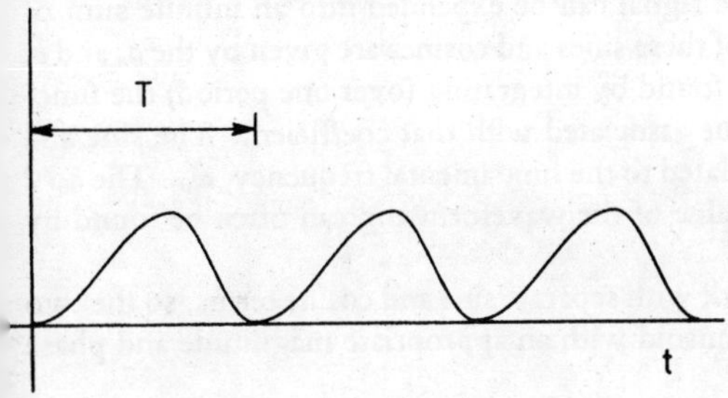

Figure 3-1. A periodic signal repeats every T seconds.

3.2 FOURIER SERIES

Most periodic signals can be represented by a series expansion of sines and cosines. There are some mathematical limitations on the represented signal, but physically realizable signals meet these constraints.[1]

The Fourier series representation of a periodic function has the form[2]

$$x(t) = \frac{a_0}{2} + \sum_{n=1}^{\infty} (a_n \cos 2\pi n f_0 t + b_n \sin 2\pi n f_0 t) \tag{3-2}$$

where

$$a_n = \frac{2}{T} \int_{-T/2}^{T/2} x(t) \cos 2\pi n f_0 t \, dt \tag{3-3}$$

$$b_n = \frac{2}{T} \int_{-T/2}^{T/2} x(t) \sin 2\pi n f_0 t \, dt \tag{3-4}$$

where

f_0 = fundamental frequency in hertz

T = period of the signal

T and f_0 are related by

$$f_0 = \frac{1}{T} \tag{3-5}$$

The frequency in rad/sec (ω_0) is

$$\omega_0 = 2\pi f_0 \tag{3-6}$$

---

[1] We will take a less than rigorous mathematical approach.

[2] The reader should be aware that there are several different ways of defining the Fourier series, with subtle differences in formulation.

Using the Fourier series, a periodic signal can be expanded into an infinite sum of sines and cosines. The weighting of these sines and cosines are given by the a_n and b_n coefficients. These coefficients are found by integrating (over one period) the function multiplied by the sine or cosine associated with that coefficient. The sine and cosine terms are all harmonically related to the fundamental frequency, ω_0. The $a_0/2$ term is simply the average (DC) value of the waveform and can often be found by inspection.

It is often inconvenient to work with separate sine and cosine terms, so the two terms can be combined into one sinusoid with an appropriate magnitude and phase angle.

$$x(t) = \frac{a_0}{2} + \sum_{n=1}^{\infty} \sqrt{a_n^2 + b_n^2} \cos (2\pi n f_0 t + \theta_n) \tag{3-7}$$

where

$$\theta_n = \tan^{-1}(-b_n/a_n)$$

Alternatively, the a_n and b_n terms can be combined into a complex coefficient which gives the complex form of the Fourier series. Instead of sines and cosines, a complex exponential is used:

$$x(t) = \sum_{n=-\infty}^{\infty} c_n e^{j2\pi n f_0 t} \tag{3-8}$$

where

$$c_n = \frac{1}{T} \int_{-T/2}^{T/2} x(t)\, e^{-j2\pi n f_0 t}\, dt \tag{3-9}$$

The two Fourier series representations are related by

$$c_n = (a_n - jb_n)/2 \tag{3-10}$$

The complex coefficient can also be expressed in magnitude/phase format:

$$c_n = |c_n| \angle \theta_n \tag{3-11}$$

Note that the complex form of the Fourier series is usually shown with n ranging from negative infinity to positive infinity while the original form restricts n to positive values. The complex form is chosen in anticipation of the Fourier transform, which includes negative frequencies. The factor of 2 which appears in equation 3-10 accounts for the presence of twice as many terms (both positive and negative) in the complex form. Frequency domain representations that include only positive frequencies are called *single sided;* those that include both positive and negative frequencies are called *double sided.*

3.3 FOURIER SERIES OF A SQUARE WAVE

As an example of the significance and utility of the Fourier series, the coefficients of a square wave will be determined. In addition, the square wave is a common signal in electrical systems (Figure 3-2a).

$$a_n = \frac{2}{T}\int_{-T/2}^{T/2} x(t)\cos 2\pi n f_0 t\,dt$$

$$= \frac{2}{T}\int_{-T/2}^{0} (-1)\cos 2\pi n f_0 t\,dt + \frac{2}{T}\int_{0}^{T/2} (1)\cos 2\pi n f_0 t\,dt$$

$$= \frac{2}{T}\left[-\frac{1}{2\pi n f_0}\sin(2\pi n f_0 t)\Big|_{-T/2}^{0} + \frac{1}{2\pi n f_0}\sin(2\pi n f_0 t)\Big|_{0}^{T/2}\right]$$

$$= 0$$

$$b_n = \frac{2}{T}\int_{-T/2}^{T/2} x(t)\sin(2\pi n f_0 t)\,dt$$

$$= \frac{2}{T}\int_{-T/2}^{0} (-1)\sin(2\pi n f_0 t)\,dt + \frac{2}{T}\int_{0}^{T/2} (1)\sin(2\pi n f_0 t)\,dt$$

$$= \frac{2}{T}\left[\frac{1}{2\pi n f_0}\cos(2\pi n f_0 t)\Big|_{-T/2}^{0} - \frac{1}{2\pi n f_0}\cos(2\pi n f_0 t)\Big|_{0}^{T/2}\right]$$

$$= \frac{1}{n\pi}(2 - 2\cos n\pi)$$

$$= \frac{4}{n\pi} \quad \text{for } n \text{ odd}$$

$$= 0 \quad \text{for } n \text{ even}$$

The Fourier series for the square wave is

$$x(t) = \frac{4}{\pi}\sin(2\pi f_0 t) + \frac{4}{3\pi}\sin(6\pi f_0 t) + \frac{4}{5\pi}\sin(10\pi f_0 t) + \cdots \tag{3-12}$$

Thus, the ideal square wave has only odd harmonics. With the particular phase chosen for the square wave, the a_n (cosine) terms are all zero, while the odd b_n (sine) terms remain nonzero. If the phase of the square wave were changed relative to $t = 0$, the a_n terms could be nonzero but only for the odd harmonics. Similarly, at just the right phase the b_n terms could become zero.

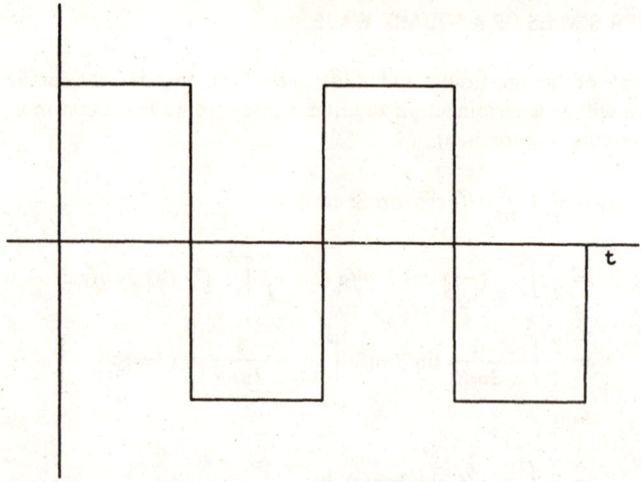

Figure 3-2a. The square wave.

Examining the square wave and its harmonics graphically is usually of some help. Figure 3-2b shows the first three harmonics of the square wave. Figures 3-2c through 3-2f show a square wave constructed from a finite number of its harmonics. Note how the harmonics, which are sinusoids, tend to fill in the square wave as

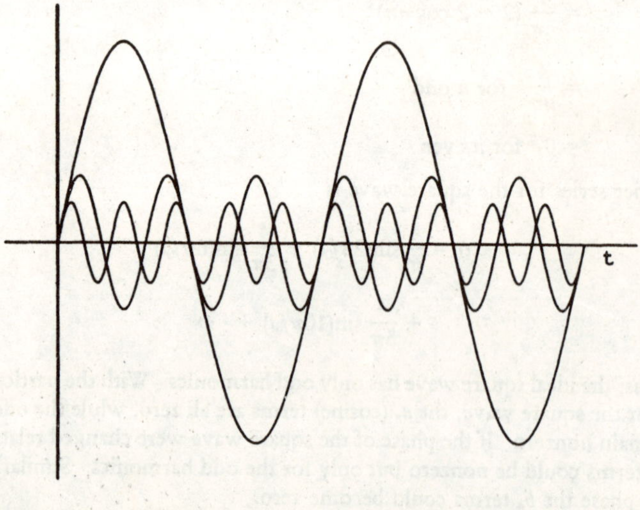

Figure 3-2b. The fundamental, third harmonic, and fifth harmonic of the square wave.

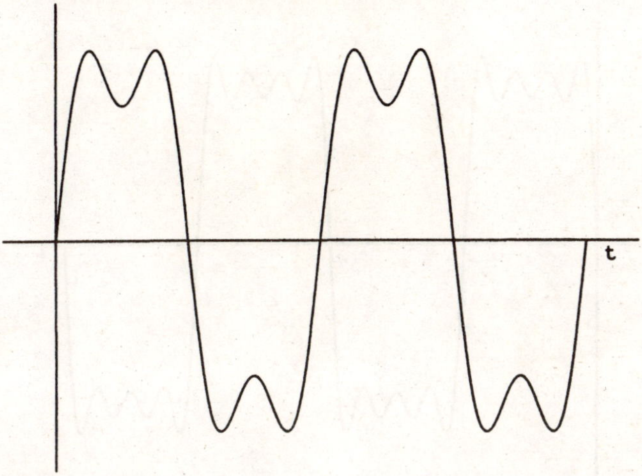

Figure 3-2c. The square wave with only the fundamental and third harmonic included.

each additional harmonic is added to the plot. It takes an infinite number of harmonics to produce a perfect square wave, but in practice the higher harmonics are often ignored since their amplitudes are small compared to the fundamental.

Note how the harmonics (which are sine terms in this case) are just the right phase to fill in the square wave. Had the square wave been shifted to the left by 90

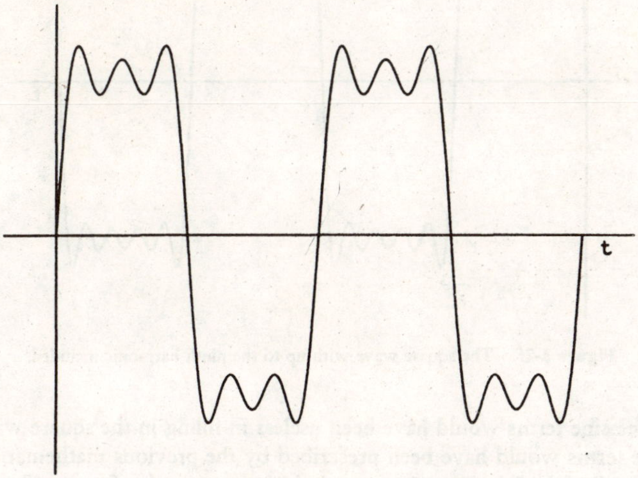

Figure 3-2d. The square wave with up to the fifth harmonic included.

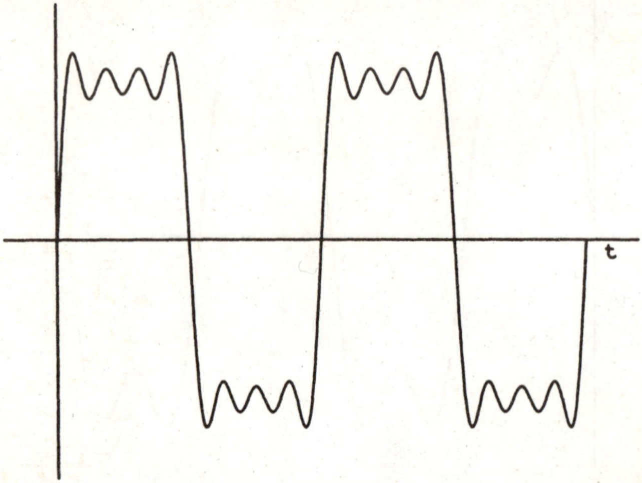

Figure 3-2e. The square wave with up to the seventh harmonic included.

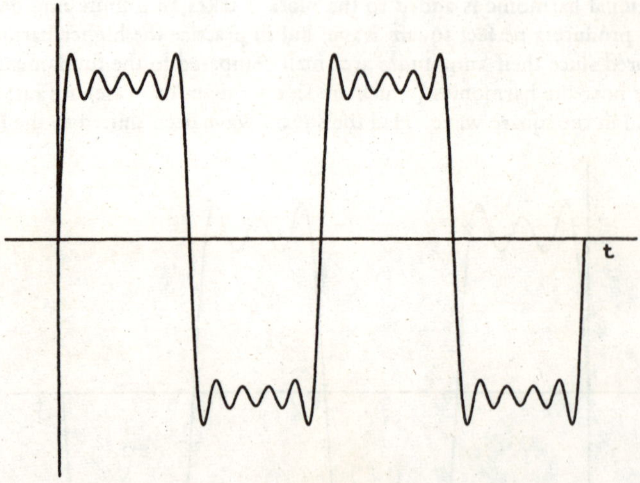

Figure 3-2f. The square wave with up to the ninth harmonic included.

degrees, the sine terms would have been useless in filling in the square wave shape and cosine terms would have been prescribed by the previous mathematics. If the complex form of the Fourier series was used, the magnitude of c_n would remain the same with changes in the waveform's phase, but the phase of c_n would change.

Although the Fourier series is a mathematical technique, an intuitive feel can be acquired by looking at the waveform graphically.

The square wave can be shown in the frequency domain by plotting the amplitude of each harmonic on a volts versus frequency type of plot (Figure 3-3). Since each harmonic appears as a single vertical line they are called *spectral lines* and such a frequency domain plot is called a *line spectra*. The Fourier series will always result in a frequency domain representation which has only line spectra, since the series form includes only the fundamental and harmonic frequencies. This contrasts with frequency domain spectrums which are continuous and will be encountered later.

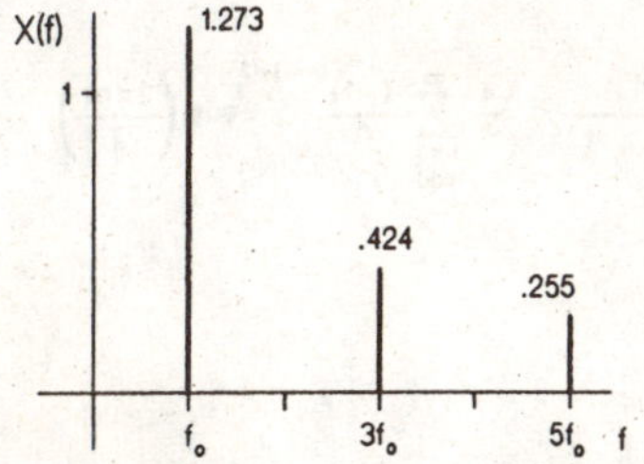

Figure 3-3. The frequency domain representation of a square wave showing the fundamental, third harmonic, and fifth harmonic.

Practical Considerations

The amplitude of each odd harmonic of the square wave is $1/n$ times the fundamental (where n is the harmonic number). When examined on a spectrum analyzer, we can expect to see each harmonic reduced by this factor. In decibels, the nth harmonic will be 20 $\log(1/n)$ decibels relative to the fundamental.

Ideally, the even harmonics are nonexistent. This is true if the square wave being measured is perfect. If the waveform is not perfectly symmetrical or has other forms of distortion, the even harmonics will be present. This is true of most practical measurements. Modern spectrum analyzers easily detect such imperfections in a square wave even though the square wave may look quite good when measured with an oscilloscope.

By the nature of the mathematical formula used, the coefficients a_n and b_n in the Fourier series represent the zero-to-peak value of the particular harmonic. Spectrum analyzers, however, are normally calibrated to measure the RMS value of a spectral line, usually expressed in dBm or RMS volts. Thus, to correlate the Fourier series representation and the typical measured result it is necessary to multiply the Fourier series coefficient by $1/\sqrt{2}$ and, if desired, convert to dBm.

3.4 FOURIER SERIES OF OTHER WAVEFORMS

The Fourier series representation of other periodic waveforms can be determined using the techniques given. For convenience, the Fourier series representations of some common waveforms are tabulated in Table 3-1.

TABLE 3-1 FOURIER SERIES OF WAVEFORMS

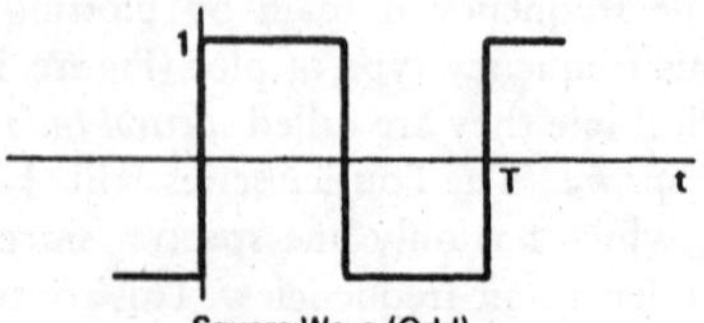

Square Wave (Odd)

$$\frac{4}{\pi}\sum_{\substack{n=1\\ \text{odd}}}^{\infty}\frac{1}{n}\sin\left(\frac{2\pi nt}{T}\right)$$

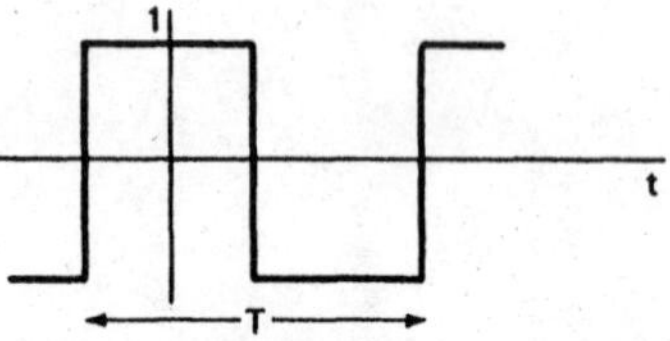

Square Wave (Even)

$$\frac{4}{\pi}\sum_{\substack{n=1\\ \text{odd}}}^{\infty}\frac{(-1)^{(n-1)/2}}{n}\cos\left(\frac{2\pi nt}{T}\right)$$

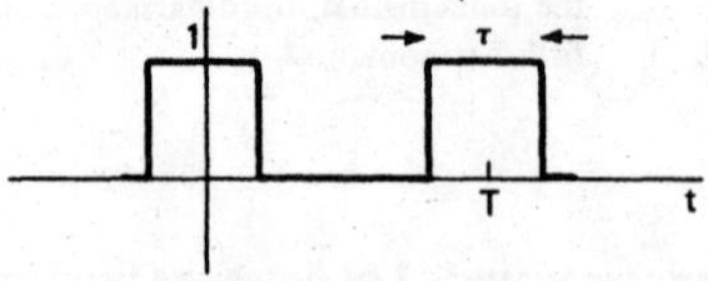

Pulse Train

$$\frac{\tau}{T}+\frac{2\tau}{T}\sum_{n=1}^{\infty}\frac{\sin\left(\frac{\pi n\tau}{T}\right)}{\frac{\pi n\tau}{T}}\cos\left(\frac{2\pi nt}{T}\right)$$

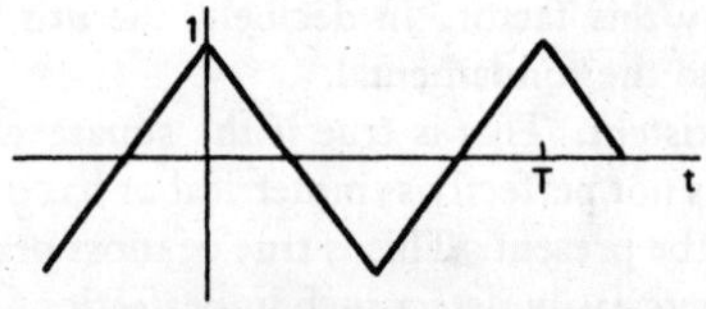

Triangle Wave

$$\frac{8}{\pi^2}\sum_{\substack{n=1\\ \text{odd}}}^{\infty}\frac{1}{n^2}\cos\left(\frac{2\pi nt}{T}\right)$$

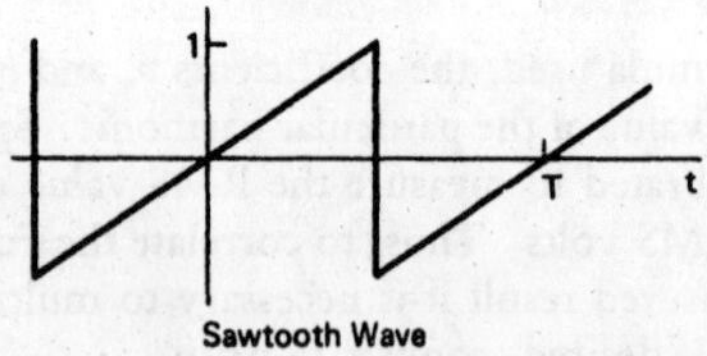

Sawtooth Wave

$$\frac{2}{\pi}\sum_{\substack{n=1\\ \text{odd}}}^{\infty}\frac{(-1)^{n+1}}{n}\sin\left(\frac{2\pi nt}{T}\right)$$

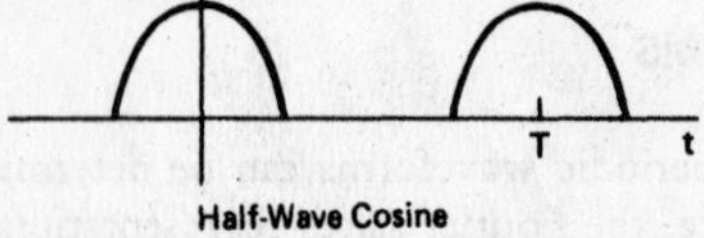

Half-Wave Cosine

$$\frac{1}{\pi}+\frac{1}{2}\cos\left(\frac{2\pi t}{T}\right)-\frac{2}{\pi}\sum_{\substack{n=2\\ \text{even}}}^{\infty}\frac{(-1)^{n/2}}{n^2-1}\cos\left(\frac{2\pi nt}{T}\right)$$

Example 3.1

Determine the amplitude and frequency of the fundamental of the waveform shown in Figure 3-4. If the signal is a voltage present across 50 Ω, what is the power level in dBm of the fundamental? Determine the amplitude of the second harmonic and express it in decibels relative to the fundamental.

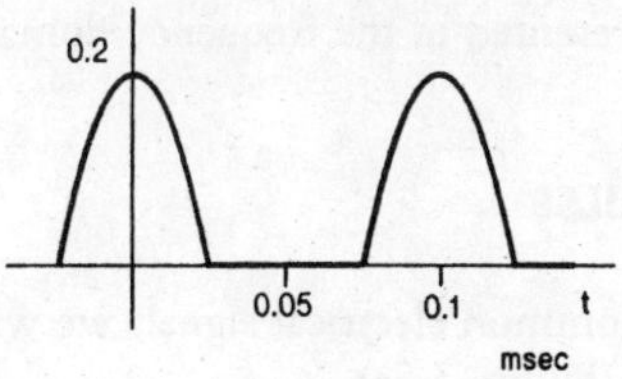

Figure 3-4. The half-wave rectified sine wave is a periodic signal.

From Table 3-1, the first few terms of the Fourier series of the half cosine wave are

$$x(t) = \frac{1}{\pi} + \frac{1}{2}\cos(2\pi t/T) + \frac{2}{3\pi}\cos(4\pi t/T)$$

The waveform shown in Figure 3-4 has a peak voltage of 0.2 volts, so the Fourier series is multiplied by 0.2.

$$x(t) = 0.2\left[\frac{1}{\pi} + \frac{1}{2}\cos(2\pi t/T) + \frac{2}{3\pi}\cos(4\pi t/T)\right]$$

The frequency of the fundamental is $1/T = 1/(0.1 \text{ msec}) = 10$ kHz.

The amplitude of the fundamental is 0.2 (1/2) = 0.1 volts zero to peak. Converting this value to RMS gives 0.707 × 0.1 = 0.0707 volts. In dBm (50 Ω), this is 20 log (0.0707/0.223) = −9.98 dBm.

The amplitude of the second harmonic is $0.2\,(2/3\pi) = 0.0424$ volts zero to peak, or 0.030 volts RMS. Expressed as decibels relative to the fundamental, the second harmonic is 20 log (0.030/0.0707) = −7.45 dB.

3.5 FOURIER TRANSFORM

Although the Fourier series representation of a signal is very powerful, it is limited to periodic signals. Signals that are not periodic may be represented in the frequency domain by the Fourier transform. The Fourier transform of a time domain signal $x(t)$ is

$$X(f) = \int_{-\infty}^{\infty} x(t)\, e^{-j2\pi ft}\, dt \tag{3-13}$$

where

$X(f)$ = frequency domain representation of the signal

$x(t)$ = time domain representation of the signal

f = frequency

The Fourier transform transforms a time domain signal into a continuous frequency domain representation. Recall that the Fourier series representation, by definition, contains only the fundamental frequency and its harmonics. Not only are these discrete frequencies, but they are harmonically related. The Fourier transform can not only represent discrete frequencies, but more often is used to represent continuous distributions in the frequency domain. Thus, a one time event (such as a pulse) in the time domain can also be represented in the frequency domain.

3.6 FOURIER TRANSFORM OF A PULSE

As an example and because it is a common electrical signal, we will determine the Fourier transform of a single pulse (Figure 3-5a).

$$\begin{aligned} X(f) &= \int_{-\infty}^{\infty} x(t)\, e^{-j2\pi ft}\, dt \\ &= \int_{-T/2}^{T/2} e^{-j2\pi ft}\, dt = \left. \frac{e^{-j2\pi ft}}{-j2\pi f} \right|_{-T/2}^{T/2} \\ &= \frac{e^{j2\pi f\frac{T}{2}} - e^{-j2\pi f\frac{T}{2}}}{j^{2\pi f}} = T\,\frac{\sin[2\pi f(T/2)]}{2\pi f(T/2)} \end{aligned} \qquad (3\text{-}14)$$

The frequency domain representation for a pulse is of the form $(\sin x)/x$ (Figure 3.5b). Notice that the function is continuous and extends over the entire frequency axis, both positive and negative. Thus, the bandwidth which a perfect pulse occupies is infinite. However, the amplitude of the frequency content tends to decrease with increasing frequency, and, in practice, a finite bandwidth can be assumed.

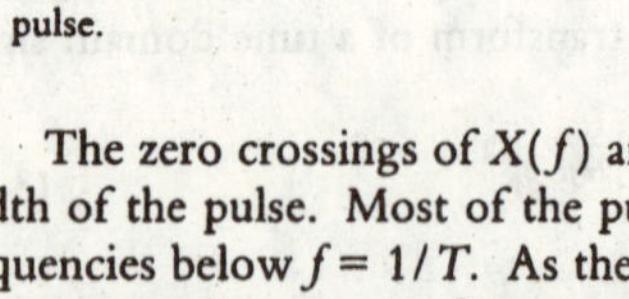

Figure 3-5a. A single time domain pulse.

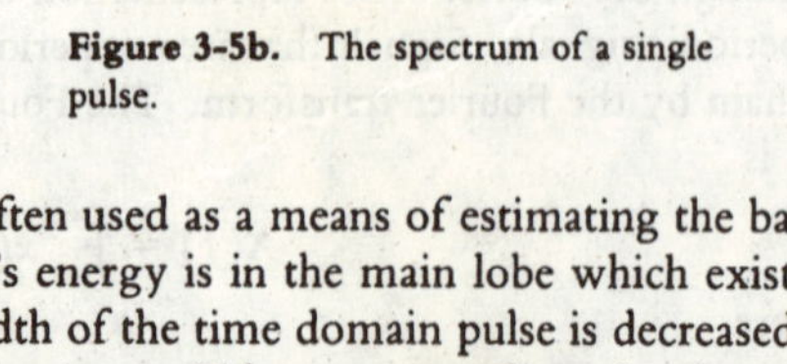

Figure 3-5b. The spectrum of a single pulse.

The zero crossings of $X(f)$ are often used as a means of estimating the bandwidth of the pulse. Most of the pulse's energy is in the main lobe which exists at frequencies below $f = 1/T$. As the width of the time domain pulse is decreased, T becomes smaller. In the frequency domain, as T becomes smaller, the first zero crossing moves out to a higher frequency. Therefore, the narrower the pulse, the wider the bandwidth in the frequency domain. This should make sense intuitively, since a narrower pulse requires a faster change in instantaneous voltage. A faster change in voltage implies more high-frequency content. This is true of signals in general—the faster the voltage changes in the time domain, the wider the bandwidth in the frequency domain.

3.7 INVERSE FOURIER TRANSFORM

The *inverse Fourier transform* converts the frequency domain representation (obtained by the Fourier transform) back into the time domain representation. The inverse transform is given by

$$x(t) = \int_{-\infty}^{\infty} X(f)\, e^{j2\pi ft}\, df \tag{3-15}$$

Thus, Fourier theory provides a means of transforming a time domain signal into the frequency and (just as important) provides a means of getting the frequency domain representation back into the time domain.

The time domain and frequency domain representations of a signal are known as *transform pairs*. They are unique in that each time domain representation has only one frequency domain representation and vice versa. A table of common Fourier transform pairs has been included in Table 3-2.

3.8 FOURIER TRANSFORM RELATIONSHIPS

Many mathematical operations in the time domain have a corresponding operation in the frequency domain. These relationships are often used to reduce the difficulty of finding a transform of a particular function. These relationships also lend insight into how the time and frequency domain relate. Table 3-3 is a compilation of commonly used Fourier transform relationships.

3.9 DISCRETE FOURIER TRANSFORM

The Fourier transform is mostly an analysis tool, a powerful means of understanding how signals behave in a system. It is not directly used in a measurement system to produce the frequency domain representation of a signal. The discrete Fourier transform (DFT) is a discrete version of the Fourier transform. It allows a sampled time domain signal to be transformed into a sampled frequency domain form. Digitizing a real-world signal in the time domain and performing a DFT produces the frequency domain representation of the signal. Thus, the DFT goes beyond being just an analysis tool. It can be used in a spectrum or network analyzer to directly compute the desired result.

TABLE 3-2 FOURIER TRANSFORM PAIRS

	$x(t)$		$X(f)$	
Unit Impulse	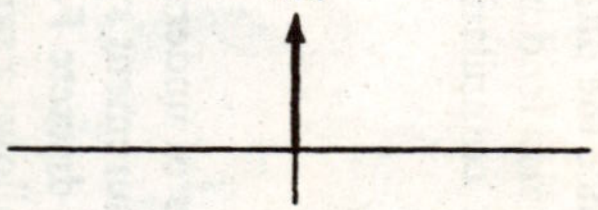	$\delta(t)$	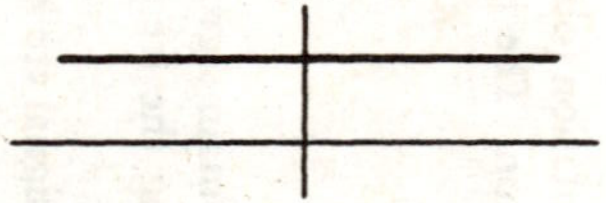	1
Constant	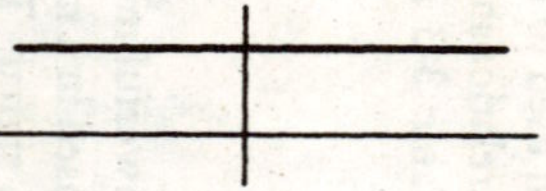	1		$\delta(f)$
Unit Step	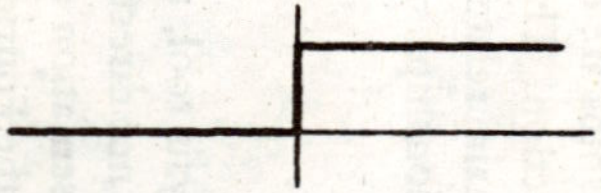	$u(t)$		$\frac{1}{2}\,\delta(f) + \frac{1}{j2\pi f}$
Pulse	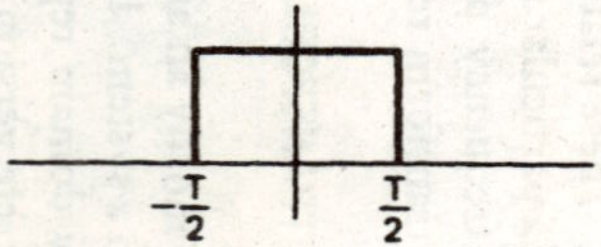	$u\left(t+\frac{T}{2}\right) - u\left(t-\frac{T}{2}\right)$		$\frac{T\sin(\pi Tf)}{\pi Tf}$

Function	Time waveform	$x(t)$	Transform	$X(f)$
Expotential	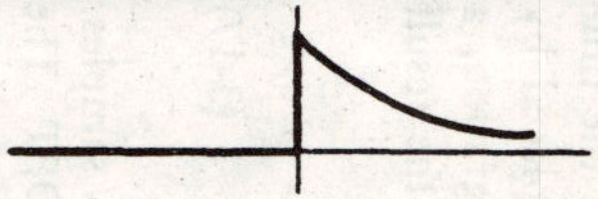	$e^{-at}u(t)$, $a > 0$		$\frac{1}{a+j2\pi f}$
Cosine	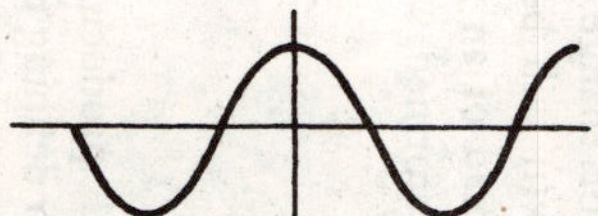	$\cos 2\pi f_0 t$	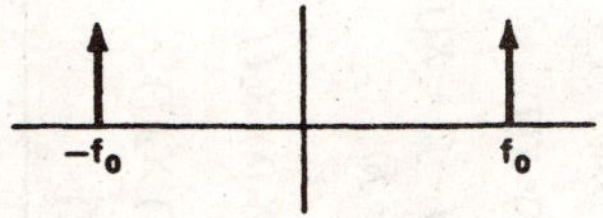	$\frac{1}{2}[\delta(f-f_0)+\delta(f+f_0)]$
Complex Sinusoid	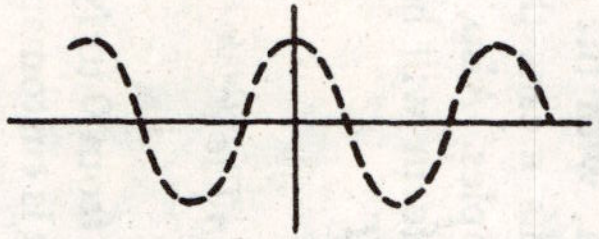	$e^{j2\pi f_0 t}$	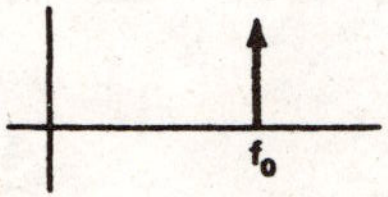	$\delta(f-f_0)$
Damped Cosine	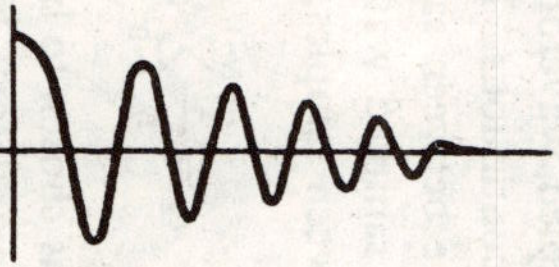	$e^{-at}\cos(2\pi f_0 t)u(t)$	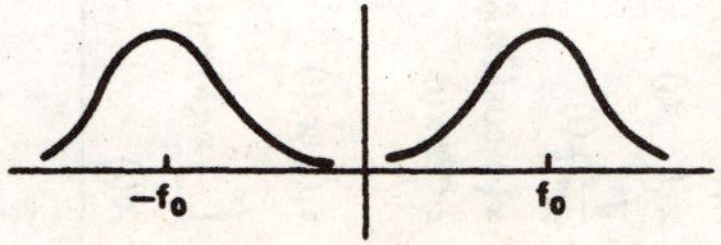	$\frac{a+j2\pi f_0}{(a+j2\pi f)^2+(2\pi f_0)^2}$
Sampling Function	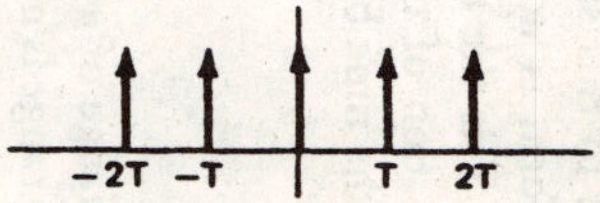	$\sum_{n=-\infty}^{\infty}\delta(t-nT)$	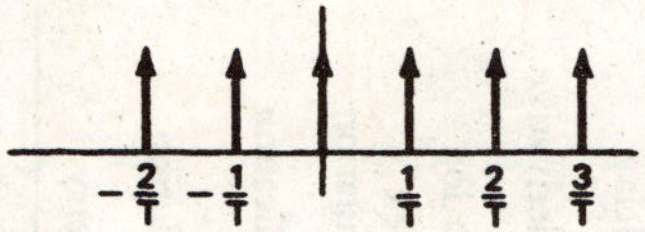	$\frac{1}{T}\sum_{n=-\infty}^{\infty}\delta\left(f-\frac{n}{T}\right)$

TABLE 3-2 (*concluded*)

TABLE 3-3 PROPERTIES OF THE FOURIER TRANSFORM

	$x(t)$	$X(f)$
Magnitude scaling	$Ax(t)$	$AX(f)$
Time scaling	$x(at)$	$\frac{1}{\lvert a\rvert} X\left(\frac{f}{a}\right)$
Linearity	$x_1(t) + x_2(t)$	$X_1(f) + X_2(f)$
Time delay	$x(t - t_0)$	$e^{-j2\pi f t_0} X(f)$
Time derivative	$\frac{d^n}{dt^n} x(t)$	$(j2\pi f)^n X(f)$
Modulation	$x(t) \cos(2\pi f_0 t)$	$\frac{1}{2}[X(f - f_0) + X(f + f_0)]$
Complex modulation	$e^{j2\pi f_0 t} x(t)$	$X(f - f_0)$
Multiplication	$x_1(t)\, x_2(t)$	$\int_{-\infty}^{\infty} X_1(\lambda)\, X_2(f - \lambda)\, d\lambda$
Convolution	$\int_{-\infty}^{\infty} x_1(\lambda) x_2(t - \lambda)\, d\lambda$	$X_1(f)\, X_2(f)$
Symmetry	$X(t)$	$x(-f)$

We had previously introduced the complex form of the Fourier series. It is rewritten here with a slight change of variable (the period T has become t_p and harmonic number n is replaced by k).

$$c_k = \frac{1}{t_p} \int_{-t_p/2}^{t_p/2} x(t) e^{-j2\pi k f_0 t}\, dt \tag{3-16}$$

Consider the periodic waveform shown in Figure 3-6a. Suppose that a sampled version of one period of this waveform is available (Figure 3-6b). The Fourier series can be applied to this sampled waveform, with the minor change that the time domain waveform is not continuous. This means that $x(t)$ will be replaced by $x(nT)$, where T is the time between samples. Also, instead of an integration, a discrete summation of the sampled waveform will be performed with the result multiplied by the time between samples, T.

$$c_k = \frac{T}{t_p} \sum_{n=0}^{N-1} x(nT) e^{-j2\pi k f_0 n T} \tag{3-17}$$

Note that the range of n was chosen to be from 0 to $N - 1$, producing N samples. This particular range is not mandatory, but is customary for defining the DFT. The fundamental frequency, f_0 is also the spacing between the discrete frequency points. We will rename f_0 to be F and attempt to provide consistent notation. Finally, the DFT is usually defined to be N times the complex Fourier series coefficient.[3]

[3] This is only a scale factor and does not affect the frequency content of the DFT. In instrumentation use, the DFT must have appropriate scale factors added to properly calibrate the instrument.

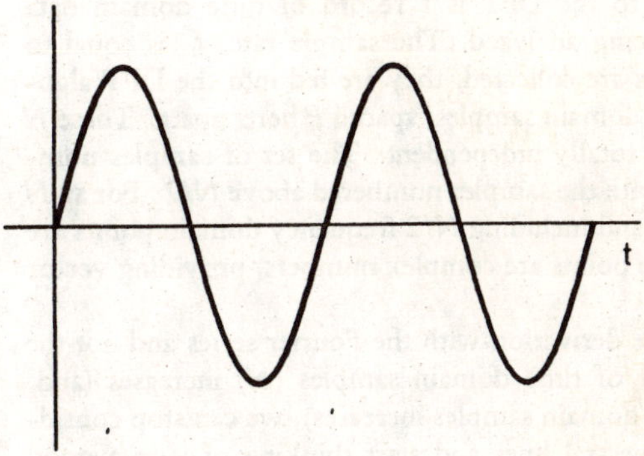

Figure 3-6a. A periodic signal to be sampled.

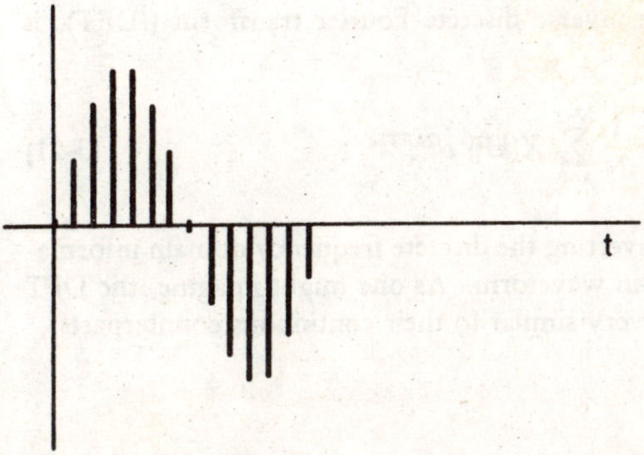

Figure 3-6b. The sampled version of one period of the signal.

$$X(kF) = N\,c_k \tag{3-18}$$

$$X(kF) = \frac{NT}{t_p}\sum_{n=0}^{N-1} x(nT)e^{-j2\pi kFnT} \tag{3-19}$$

Since the number of samples (N) times the sample time (T) equals the period (t_p), the equation simplifies to give the common form of the discrete Fourier transform.

$$X(kF) = \sum_{n=0}^{N-1} x(nT)\,e^{-j2\pi kFnT} \tag{3-20}$$

where

N = number of samples

F = spacing of the frequency domain samples

T = sample period in the time domain

In instrumentation use, the input to the DFT is a record of time domain data obtained by sampling the signal being analyzed. The sample rate, f_s, is equal to $1/T$. After N time domain samples are collected, they are fed into the DFT algorithm which produces N frequency domain samples, spaced F hertz apart. These N frequency domain samples are not totally independent. The set of samples numbered less than $N/2$ are redundant with the samples numbered above $N/2$. For an N point DFT, only the samples up to and including $N/2$ frequency domain points are normally retained. In general, these points are complex numbers, providing vector information.

Remember that we started the derivation with the Fourier series and not the Fourier transform. As the number of time domain samples (N) increases (and, therefore, the number of frequency domain samples increases), we can stop considering the DFT as a small set of spectral lines and start thinking of it as a good approximation to the continuous Fourier transform.

The inverse of the DFT, the inverse discrete Fourier transform (IDFT), is given by

$$x(nT) = \frac{1}{N} \sum_{k=0}^{N-1} X(kF)\, e^{j2\pi FTkn} \tag{3-21}$$

The IDFT provides a means for converting the discrete frequency domain information back into a discrete time domain waveform. As one might imagine, the DFT and IDFT have properties that are very similar to their continuous counterparts.

3.10 LIMITATIONS OF THE DFT

The DFT is only an approximation to the Fourier transform. It differs from the continuous Fourier transform in several important ways.

Obviously, due to the quantized nature of the DFT, it is valid at only certain frequencies. The frequency resolution of the DFT can be increased by using a larger number of samples.

The theory behind the DFT implicitly assumes that the waveform was periodic. Whether this is the case or not, the mathematics of the DFT will treat the sampled waveform as if it repeats. This causes a phenomenon known as leakage which is an important limitation of the DFT, but one which can be minimized by proper use of time domain windowing. Leakage is discussed further in Chapter 4.

Since the DFT is performed with digital arithmetic, it is subject to the limitations imposed by the particular algorithm chosen. In particular, finite arithmetic effects due to the number of bits used can limit the dynamic range and noise performance of the DFT.

3.11 FAST FOURIER TRANSFORM

The *fast Fourier transform* (FFT) is a very quick and efficient algorithm for implementing a DFT. The original basis for the FFT was developed by J. W. Cooley and J. W. Tukey in 1965.[4] Although it is often implied that there is just one FFT, in reality, there is an entire class of algorithms commonly referred to as the FFT. An FFT algorithm gains a significant speed advantage over the DFT by carefully selecting and organizing intermediate results. Ignoring finite arithmetic effects, the results are the same whether an FFT or a DFT is used.

The number of computations required for a DFT is on the order of N^2, where N is the number of samples, or record length. The FFT, on the other hand, requires $N \log_2 N$ computations ($\log_2$ indicates the base 2 logarithm). The most common FFT algorithms require N to be a power of 2. A typical record length in a spectrum analyzer might be 2^{10} or 1024. This means a DFT would require over 1 million computations while an FFT would require only 10,240 computations. Assuming all computations take the same amount of time, the FFT could be computed in less than 1% of the DFT computation time. Clearly, this is a substantial time savings and explains why the FFT dominates in instrumentation use.

Examining the details of how and why an FFT is implemented is beyond the scope of this book.[5] For our purposes, we will consider the FFT to be simply an efficient implementation of a DFT.

3.12 RELATING THEORY TO MEASUREMENTS

When the instrument user attempts to relate Fourier theory to an actual measurement, some notable differences will appear. The major differences are summarized here.

1. The spectrum analyzer normally shows a one-sided spectrum, while the Fourier transform and perhaps the Fourier series (depending on which form is used) show a two-sided spectrum.
2. The frequency resolution (resolution bandwidth) of the spectrum analyzer determines the width and shape of discrete spectral lines. Ideally, the lines are infinitely thin, but they appear with a finite width due to the resolution bandwidth of the analyzer.
3. Other distortion and noise effects generated internal to the spectrum analyzer will affect the measurement. For example, the noise floor of the analyzer may

[4] J. W. Cooley and J. W. Tukey, "An Algorithm for the Machine Calculation of Complex Fourier Series," *Math. Computation*, Vol. 19 (1965).

[5] For more information, see Oppenheim and Schafer (1975).

obscure low-level frequency components or distortion products may appear as additional spectral lines.

In particular, it can be a problem relating the amplitude predicted by Fourier theory to the measured amplitude. In an attempt to reconcile theory and measurement, let us consider a simple, but instructive case—the cosine. We apply both the Fourier series and the Fourier transform to this signal and then compare the results with a practical spectrum measurement.

Consider the time domain waveform,

$$v(t) = V_0 \cos 2\pi f_0 t \tag{3-22}$$

The RMS value, as measured by an RMS-reading voltmeter, would be 0.707 times the zero-to-peak value. This value should agree with the spectrum analyzer measurement:

$$V_{RMS} = 0.707\ V_0 \tag{3-23}$$

The Fourier series for this voltage waveform can easily be found by inspection.

$$v(t) = a_1 \cos 2\pi f_0 t \tag{3-24}$$

where

$$a_1 = V_0$$

This implies a single spectral line at f_0, with a zero-to-peak amplitude of V_0.

From Table 3-2, the Fourier transform of the waveform is

$$F(f) = \frac{V_0}{2}[\delta(f - f_0) + \delta(f + f_0)] \tag{3-25}$$

Since the Fourier transform is a two-sided representation, with both positive and negative frequencies, the frequency domain representation indicates two impulse functions, one at $+f_0$ and the other at $-f_0$. The amplitudes of each of these impulse functions is $V_0/2$. This amplitude is doubled in order to convert the double-sided amplitude to the equivalent single-sided amplitude. Thus, the zero-to-peak amplitude equal to V_0 agrees with the Fourier series analysis and if multiplied by 0.707 to obtain the RMS value, agrees with the voltmeter reading and a spectrum analyzer reading.

3.13 FINITE MEASUREMENT TIME

The discussion of the Fourier series and the Fourier transform both involved integrals which cover all time, that is, from $-\infty$ to $+\infty$. Therefore, to ascertain correctly the frequency domain representation of a signal, the time domain function must be known for all time. For theoretical analysis, this does not present a problem. However, practical measurements occur in a finite time. Normally, the spectrum analyzer user simply performs the measurement over some convenient time interval and assumes that the time interval chosen adequately represents the signal. Mathematically speaking, the signal is assumed to be *stationary*.[6]

The characteristics of many signals are constant over time in which case such an assumption is justified. By definition, a periodic signal repeats over and over again for all time, producing a constant spectrum. Some other signals change quite rapidly and should not be assumed to have constant spectrums. As an example consider a radio transmitter. If the modulating signal is a person's voice, the spectrum of the signal will change quickly and unpredictably as the radio operator speaks different words. A measurement taken at any particular time will not represent the signal over all time. However, if a constant audio tone modulates the radio signal, the spectrum is constant.

When measuring a signal's spectrum, we should consider the possibility that the signal's spectral content may be varying. If this variation is slow compared with the duration of the measurement, it is not of concern. However, if the signal varies fast enough, the spectrum analyzer may not produce the desired result. The measurement duration may need to be reduced, but this is not always possible. The choice of spectrum analyzer type may also be important, since FFT analyzers are usually better at characterizing fast-changing signals. (See Chapter 5 for a comparison of swept and FFT analyzers.)

REFERENCES

1. Brigham, E. Oran. *The Fast Fourier Transform and Its Applications.* Englewood Cliffs, NJ: Prentice Hall, Inc., 1988.
2. Hayt, William H., and Jack E. Kemmerly. *Engineering Circuit Analysis,* 2nd ed. New York: McGraw-Hill Book Company, 1971.
3. Hewlett-Packard Company. "Fundamentals of Signal Analysis," Application Note 243, Publication Number 5952-8898, Palo Alto, CA: 1981.
4. Irwin, J. David. *Basic Engineering Circuit Analysis.* New York: Macmillan Publishing Company, 1984.
5. McGillem, Clare D., and George R. Cooper. *Continuous and Discrete Signal and System Analysis.* New York: Holt, Rinehart and Winston, Inc., 1974.
6. Oppenheim, Alan V., and Ronald W. Schafer. *Digital Signal Processing.* Englewood Cliffs, NJ: Prentice-Hall, Inc. 1975.
7. Oppenheim, Alan V., and Alan S. Willsky. *Signals and Systems.* Englewood Cliffs, NJ: Prentice-Hall, 1983.
8. Ramirez, Robert W. *The FFT, Fundamentals and Concepts.* Englewood Cliffs, NJ: Prentice Hall, Inc., 1985.

---

[6] A signal is stationary if its statistical nature does not change with time, which implies that its spectrum is constant. For a more rigorous discussion, see Oppenheim and Willsky (1983).

9. Schwartz, Mischa. *Information, Transmission, Modulation, and Noise,* 3rd ed. New York: McGraw-Hill Book Company, 1980.

10. Schwartz, Mischa, and Leonard Shaw. *Signal Processing.* New York: McGraw-Hill Book Company, 1975.

11. Stanley, William D., Gary R. Dougherty, and Ray Dougherty. *Digital Signal Processing,* 2nd ed. Reston, VA: Reston Publishing Company, Inc., 1984.

4

Fast Fourier Transform Analyzers

The fast Fourier transform (FFT) can be used to implement a spectrum or network analyzer by digitizing the input waveform and performing an FFT on the time domain signal to get the frequency domain representation. What seems to be a simple measurement technique often turns out to be much more complicated in practice. Given reasonable computational power (usually in the form of a microprocessor or custom integrated circuit), the FFT-based analyzer can provide significant speed improvement over the more traditional swept analyzer. FFT analyzers usually have limited bandwidth (less than a few hundred kilohertz), due to the lack of fast, high-resolution analog-to-digital converters. The FFT analyzer is also referred to as the *dynamic signal analyzer*.

4.1 THE BANK-OF-FILTERS ANALYZER

The bank-of-filters technique is not common in general electronic instrumentation but has been used in some applications, such as low-frequency audio meters ($\frac{1}{3}$-octave spectrum analyzers). This technique is included here to provide a theoretical base for discussing more practical spectrum analyzer block diagrams.

One simple approach to implementing a spectrum analyzer is to connect a bank of electronic filters together, each with its own output device (Figure 4-1). For a small number of filters, this technique has the advantage of simplicity. Also, this measurement technique is quite fast and can result in a real-time measurement system.

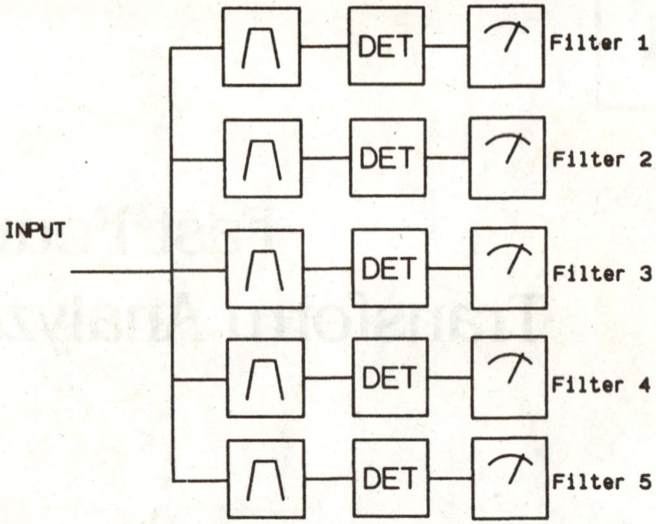

Figure 4-1. The bank-of-filters spectrum analyzer uses a set of filters to determine the frequency content of a signal.

Each of the electronic filters is a bandpass filter tuned to a different center frequency. The bandwidths and center frequencies of the filters are aligned as shown in Figure 4-2 to provide complete coverage of the frequency range of interest

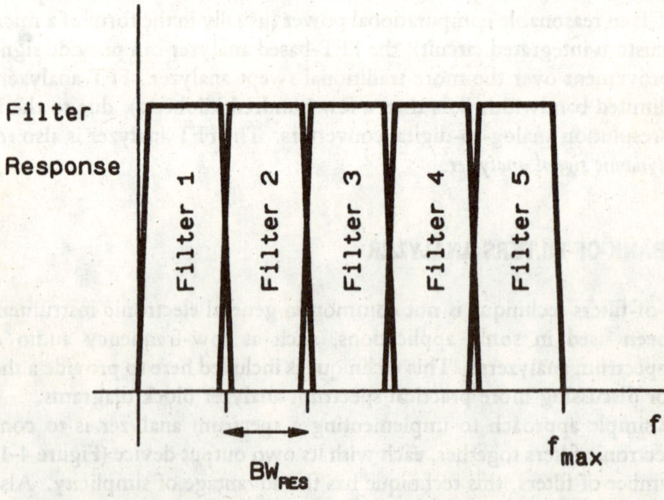

Figure 4-2. The filters are adjacent in the frequency domain, aligned for minimum overlap.

with minimal overlap of filter shapes. Ideally, infinitely steep "brick wall" filters would be used to provide zero overlap between filter passbands. The outputs of the filters are connected to detectors which convert the AC (sine wave) signal into a DC level which is displayed by a meter. Alternatively, the detector outputs could be multiplexed together and displayed on a graphics device such as a cathode ray tube (CRT).

4.2 FREQUENCY RESOLUTION

Each filter is designed to pass only one small range of frequencies onto the detector. Thus, each filter/detector/meter combination displays the energy present over that particular range of frequencies. If two frequencies are present within the same filter, both of them will affect the meter reading. (The exact meter reading will depend on the type of detector used.) The analyzer cannot resolve two signals in the same filter; thus the filter bandwidth, BW_{RES}, determines the frequency resolution of the analyzer. BW_{RES} is called the *resolution bandwidth.*

For example, consider Figure 4-3. Frequency components f_1 and f_2 appear in the passband of the same filter. Therefore, they cannot be resolved (in frequency). Frequency components f_3 and f_4, on the other hand, do not appear within the same filter and each will be measured individually. The frequency of f_3 and f_4 are known to the extent that they are within the passband of their respective filters. Thus, their frequencies are known to within BW_{RES}.

Assuming extremely sharp filters (with steep skirts) precisely positioned such that the edges of their passbands just touch, the resolution bandwidth of a bank-of-filters analyzer is given by

$$BW_{RES} = f_{max}/M \tag{4-1}$$

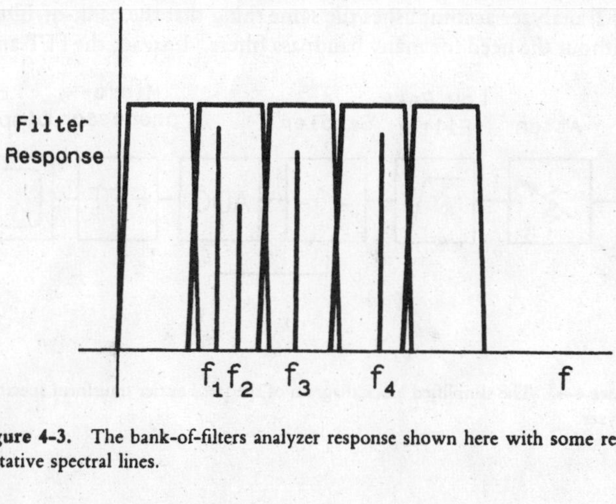

Figure 4-3. The bank-of-filters analyzer response shown here with some representative spectral lines.

where

f_{max} = maximum frequency of the analyzer

M = number of filters

This equation can be used to explain the major limitation of the bank-of-filters analyzer. Suppose that a spectrum analyzer with a frequency resolution (resolution bandwidth) of 100 Hz must cover the 0 to 100 kHz frequency range. The number of filters required is

$$M = f_{max}/BW_{RES} = 100 \text{ kHz}/100 = 1000 \qquad (4\text{-}2)$$

Not only would this be a large number of filters to implement, building a steep-walled 100 Hz wide filter with a center frequency near 100 kHz would be very difficult. For this reason, the bank-of-filters analyzer is used mainly where a much wider resolution bandwidth is acceptable.

4.3 THE FFT ANALYZER

As discussed in Chapter 3, the fast Fourier transform can be used to determine the frequency domain representation (spectrum) of a time domain signal. The signal must be digitized in the time domain; then the FFT algorithm is executed to find the spectrum. Figure 4-4 shows a simplified block diagram of an FFT analyzer. The input signal is first passed through a variable attenuator to provide various measurement ranges. Then the signal is low-pass filtered to remove undesirable high-frequency content which is beyond the frequency range of the instrument. The waveform is sampled and converted to digital form by the combination of the sampler circuit and the analog-to-digital converter. The microprocessor (or other digital circuitry) receives the sampled waveform, computes the spectrum of the waveform using the FFT, and writes the results on the display.

The FFT analyzer accomplishes the same thing that the bank-of-filters analyzer does, but without the need for many bandpass filters. Instead, the FFT analyzer uses

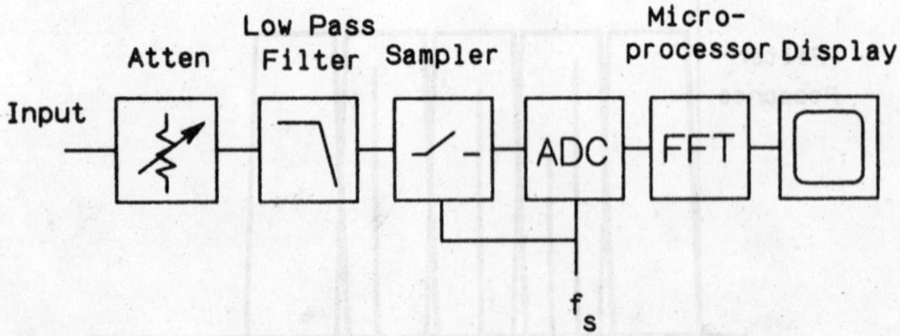

Figure 4-4. The simplified block diagram of the fast Fourier transform spectrum analyzer.

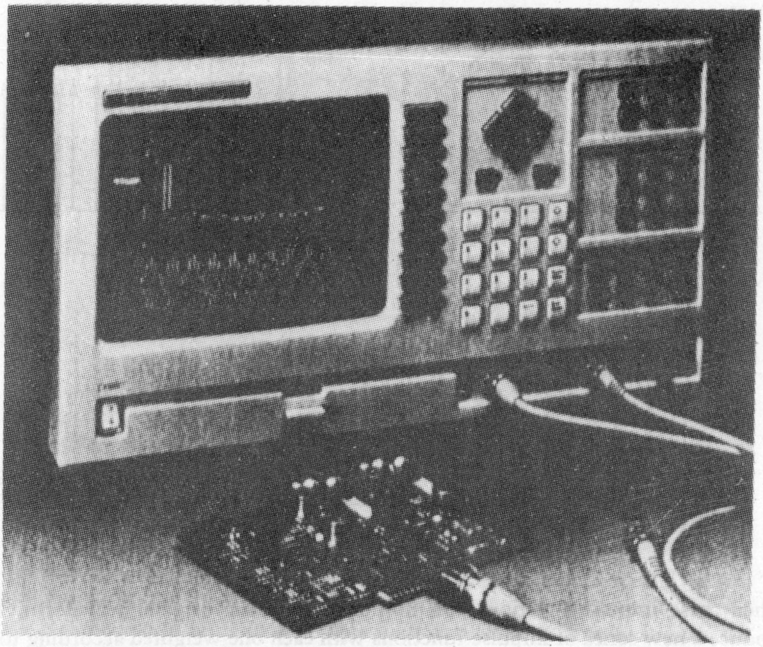

Figure 4-5. A typical fast Fourier transform spectrum analyzer. Photo courtesy of Hewlett-Packard Company.

digital signal processing to implement the equivalent of many individual filters. Thus, when considering the operation of the FFT analyzer, it is appropriate to think in terms of a bank of parallel filters, each filtering a portion of the frequency spectrum. A typical FFT spectrum analyzer is shown in Figure 4-5.

Conceptually, the FFT approach is simple and straightforward—digitize the signal and compute the spectrum. In practice, there are some effects which must be accounted for in order for the measurement to be meaningful.

4.4 SAMPLED WAVEFORM

In a sampled system, the time domain waveform (Figure 4-6a) is effectively multiplied by the sample function (Figure 4-6b) to produce the sampled waveform (Figure 4-6c). The sampling function is shown as a series of impulse functions, spaced at $T = 1/f_s$, where f_s is the sample rate of the system.

$$s(t) = \sum_{n=-\infty}^{\infty} \delta(t - nT) \qquad (4\text{-}3)$$

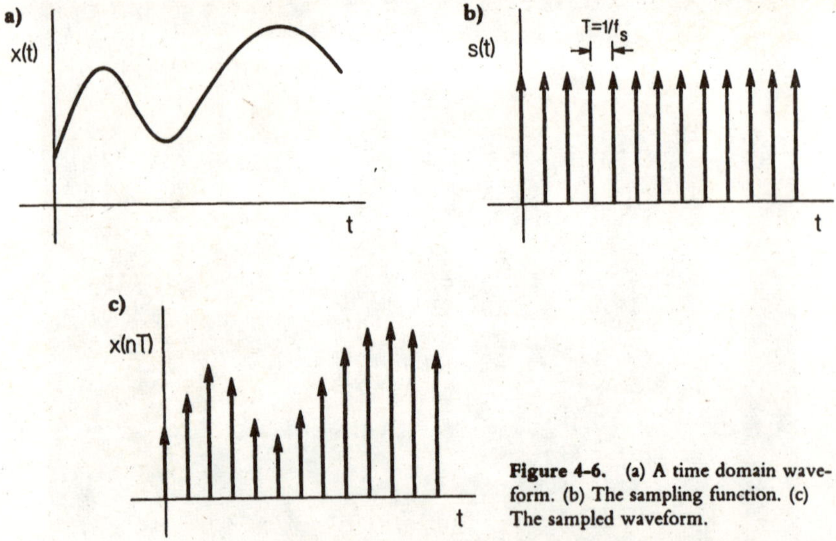

Figure 4-6. (a) A time domain waveform. (b) The sampling function. (c) The sampled waveform.

When these impulse functions are multiplied with the original waveform, they produce a new series of impulse functions with each one weighted according to the original waveform.

$$x(nT) = \sum_{n=-\infty}^{\infty} x(t)\, \delta(t - nT) \tag{4-4}$$

The sampled analog waveform is converted into a sequence of digital numbers using an analog-to-digital converter (ADC). The output of the ADC is an array or record of numbers representing the sampled waveform. The sampled and digitized version of the waveform still retains the shape and information content of the unsampled waveform, if the sample rate is sufficiently high.

4.5 SAMPLING THEOREM

The waveform must be sampled often enough to produce a digitized time record that faithfully represents the original waveform. The *sampling theorem* states that a baseband signal must be sampled at a rate greater than twice the highest frequency present in the signal. The minimum acceptable sample rate is called the *Nyquist rate*. Thus,

$$f_s > 2\, f_{max} \tag{4-5}$$

where

$$f_s = \text{sample rate}$$

$$f_{max} = \text{highest frequency present}$$

Figure 4-7a shows the frequency spectrum, $X(f)$, of a signal, $x(t)$, with a maximum frequency of f_{max}. The frequency spectrum of the sampling function, as given by Table 3-1, is an infinite number of impulse functions spaced every f_s in frequency (Figure 4-7b). The spectrum of the sampled waveform can be derived by convolving[1] $X(f)$ with $S(f)$, which results in the original spectrum $X(f)$ appearing centered around each impulse function of $S(f)$ (Figure 4-7c).

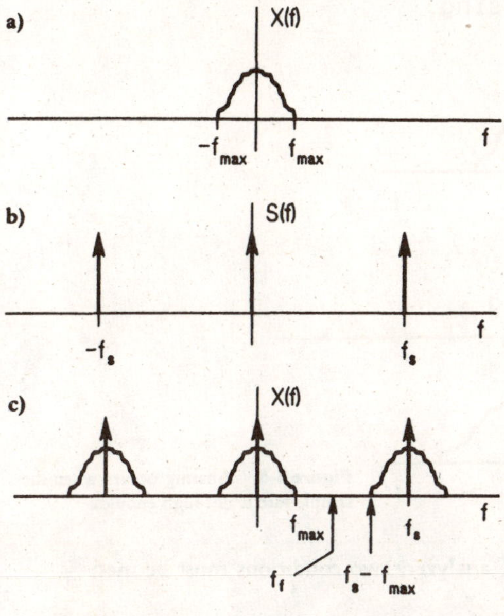

Figure 4-7. (a) The spectrum of the unsampled waveform. (b) The spectrum of the sampling function. (c) The spectrum of the sampled waveform.

This type of spectrum is always found in sampled systems—the baseband signal is repeated at integer multiples of the sample frequency. Notice that the spectrum between 0 and f_s is symmetrical about $f_s/2$, which is also called the *folding frequency*, f_f. The original signal can be recovered by applying a low-pass filter with a cutoff frequency of f_f, as long as the frequency content centered around f_s does not encroach on the baseband signal. Stated mathematically, the following condition must be met:

$$f_s - f_{max} > f_f \tag{4-6}$$

which is just a restatement of the sampling theorem since

$$f_s - f_{max} > f_s/2 \tag{4-7}$$

[1] For a discussion of the fine points of convolution, see McGillem and Cooper (1974).

$$f_s/2 > f_{max} \tag{4-8}$$

$$f_s > 2\, f_{max} \tag{4-9}$$

Figure 4-8 shows the spectra of two sampled signals, one where the sampling theorem is met and another which violates the sampling theorem. Notice that when the sampling theorem is violated, unwanted frequency components show up below f_f. This phenomenon is known as *aliasing,* since these undesirable frequency components appear under the alias of another (baseband) frequency.

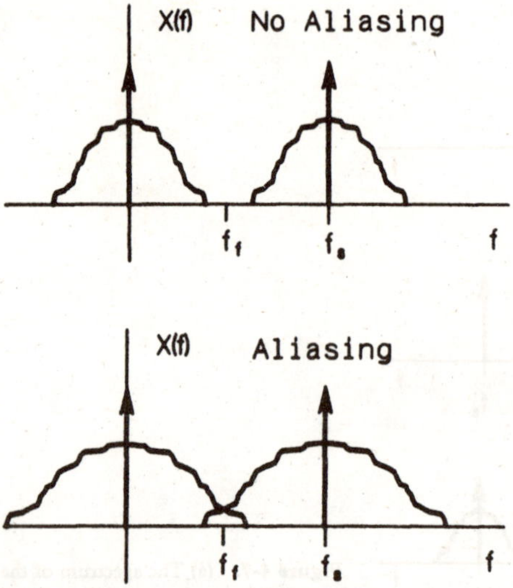

Figure 4-8. Aliasing occurs when the sample rate is not high enough.

To prevent aliasing in an FFT analyzer, two conditions must be met:

1. The input signal must be band limited. In other words, there must exist an f_{max} above which no frequency components are present.[2] This is usually accomplished by inserting an anti-alias filter in the signal path. (This is the low-pass filter shown in Figure 4-4.)
2. The input signal must be sampled at a rate that satisfies the sampling theorem.

The sampling frequency required by the sampling theorem is the minimum theoretical value that can reconstruct the signal properly. In practice, it is necessary to use a sampling frequency somewhat higher than this value. Figure 4-9 shows the frequency response of a practical low-pass filter. The filter will have a finite slope above its cutoff frequency, f_{max}. The mirrored response of the filter above the

---

[2] In practice, frequency components above f_{max} are allowed to exist, but must be sufficiently attenuated so that they do not affect the measurement.

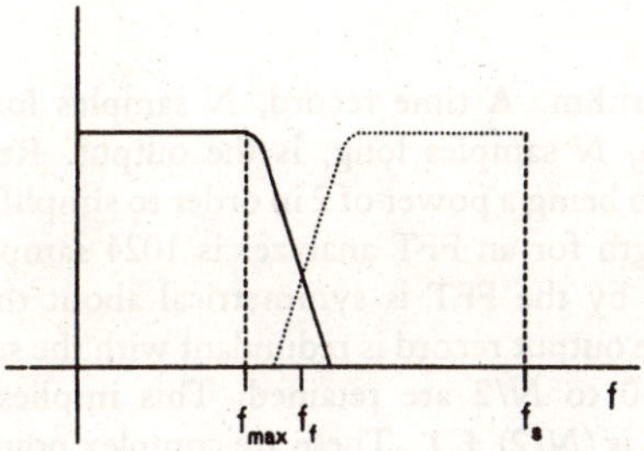

Figure 4-9. The response of the anti-alias filter requires that the sample rate be somewhat higher than the sampling theorem states.

folding frequency is also shown. This response represents the possible alias frequencies which could result with the anti-alias filter in place. The system is designed so that the folding frequency (and the sampling frequency) are large enough that the anti-alias response has room to roll off. Thus, f_{max}, the highest frequency that the analyzer will measure, must be significantly less than f_f. For practical filter implementations, f_s is typically 2.5 times f_{max}.

As shown, aliasing can be explained in the frequency domain, but it is also helpful to consider it briefly in the time domain. Figure 4-10 shows a set of sample points which fit two different waveforms. One of the waveforms has a frequency which violates the sampling theorem; the other does not. (The higher-frequency waveform violates the sampling theorem, of course.) Unless an anti-alias filter removes the unwanted alias frequency, the two sampled sine waves will be indistinguishable when processed digitally.

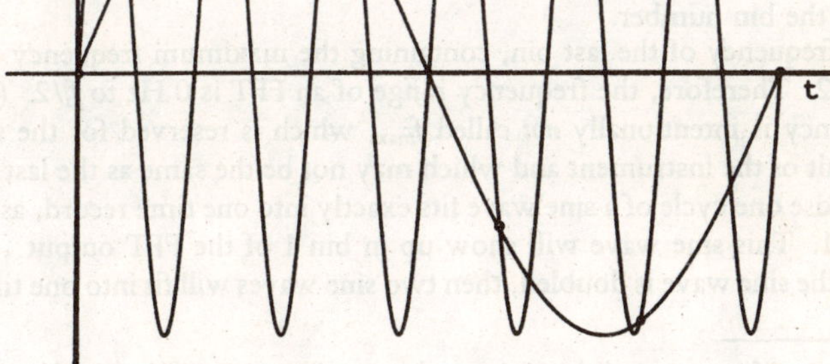

Figure 4-10. Aliasing in the time domain.

4.6 FFT PROPERTIES

The FFT is a record-oriented algorithm. A time record, N samples long, is the input, and the frequency spectrum, N samples long, is the output. Recall from Chapter 3 that N is often restricted to being a power of 2 in order to simplify the FFT computation. A typical record length for an FFT analyzer is 1024 sample points. The frequency spectrum produced by the FFT is symmetrical about the folding frequency. Thus, the first half of the output record is redundant with the second half and the sample points numbered 0 to $N/2$ are retained. This implies that the effective length of the output record is $(N/2) + 1$. These are complex points (real + j imaginary) which contain both magnitude and phase information.

Practically speaking, the output of the FFT is $(N/2) + 1$ points, extending from 0 Hz to f_f. Not all of these points are usually displayed though, since the anti-alias filter begins to roll off before f_f. A common configuration is 1024 samples in the time record, producing 513 unique complex frequency domain points, with 401 of these actually displayed.

The $N/2$ (or so) frequency domain points are often referred to as *bins* and are usually numbered from 0 to $N/2$ (e.g., 0 to 512 for $N = 1024$). These bins are equivalent to the individual filter/detector outputs in the bank-of-filters analyzer. Bin 0 contains the DC level present in the input signal and is also referred to as the *DC bin*. The bins are spaced equally in frequency, with the frequency step, f_{step} being the reciprocal of the time record length.[3]

$$f_{step} = 1/\text{length of time record} \tag{4-10}$$

The length of the time record can be determined from the sample rate and the number of sample points in the time record.

$$f_{step} = f_s/N \tag{4-11}$$

The frequency associated with each bin is given by

$$f_n = n\, f_s/N \tag{4-12}$$

where n is the bin number.

The frequency of the last bin, containing the maximum frequency out of the FFT, is $f_S/2$. Therefore, the frequency range of an FFT is 0 Hz to $f_s/2$. (Note that this frequency is intentionally *not* called f_{max}, which is reserved for the upper-frequency limit of the instrument and which may not be the same as the last FFT bin.)

Suppose one cycle of a sine wave fits exactly into one time record, as shown in Figure 4-11. This sine wave will show up in bin 1 of the FFT output. If the frequency of the sine wave is doubled, then two sine waves will fit into one time record

[3] The term "frequency step" does not mean that some frequencies will be missed by the FFT. The output of the FFT is equivalent to the bank-of-filters analyzer, with contiguous bandpass filters centered at each bin.

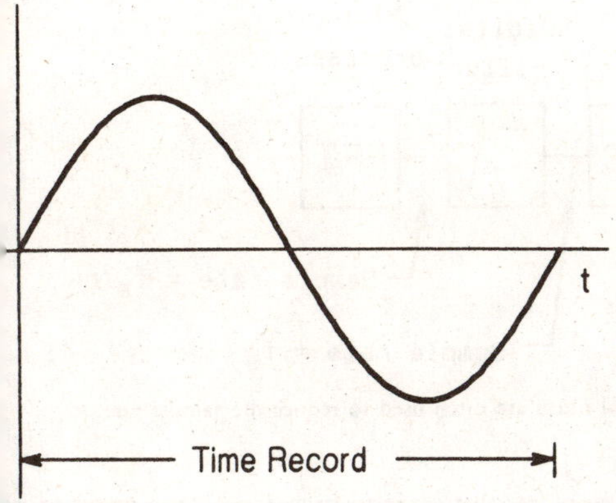

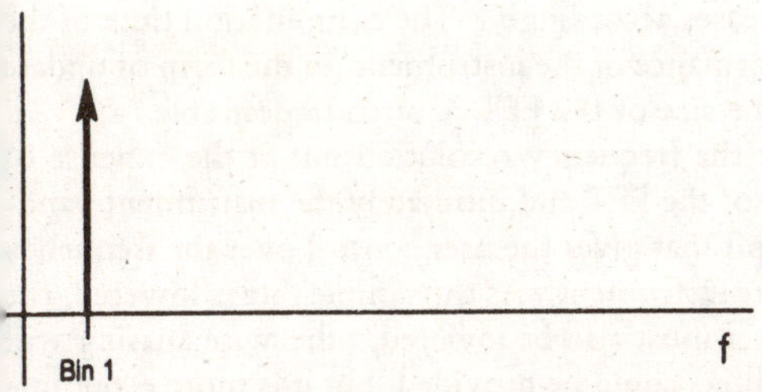

Figure 4-11. (a) A sine wave which exactly fills one time record. (b) The spectral line shows up in bin 1 of the FFT output.

and their energy will appear in bin 2. Tripling the original sine wave frequency will cause a frequency domain response in bin 3, and so forth.

4.7 CONTROLLING THE FREQUENCY SPAN

The FFT is inherently a baseband transform. In other words, the frequency range of the FFT always starts at 0 Hz and extends to some maximum frequency, $f_S/2$. This can be a significant limitation in measurement situations where a small frequency band (not starting at DC) needs to be analyzed.

For example, suppose an FFT analyzer has a sample rate, f_s = 256 kHz. The frequency range of the FFT would be 0 Hz to 128 kHz ($f_s/2$). If N = 1024, the frequency resolution would be f_s/N = 250 Hz. Spectral lines closer than 250 Hz could not be resolved.[4]

[4] This is an approximation since the frequency resolution will also depend on the window function, discussed in Section 4.9.

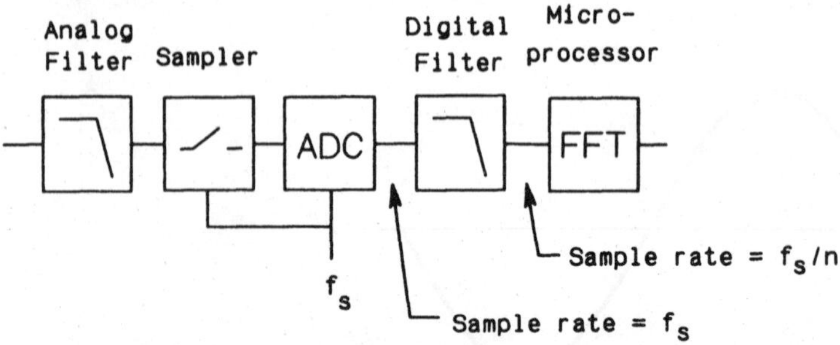

Figure 4-12. Decimating digital filters are often used to reduce the sample rate into the FFT.

One way to increase the frequency resolution is to increase N, the number of samples in the time record, which also increases the number of bins in the FFT output. Unfortunately, this increases the size of the arrays that the FFT has to deal with and the computation time increases accordingly. The computation time of the FFT algorithm often limits the performance of the instrument (in the form of update rate to the display), so increasing the size of the FFT is often undesirable.

Reducing f_s will also improve the frequency resolution but at the expense of reducing the upper-frequency limit of the FFT and ultimately the instrument bandwidth. This is a worthwhile trade-off that gives the user control over the frequency resolution and frequency range of the instrument. As the sample rate is lowered, the cutoff frequency of the anti-alias filter must also be lowered, otherwise aliasing will occur. Selectable analog anti-alias filters could be provided, but it is more economical to implement the additional filters digitally. A *decimating digital filter* simultaneously decreases the bandwidth of the signal and decreases the sample rate (Figure 4-12). The sample rate into the digital filter is f_s, while the sample rate out of the filter is f_s/n, where n is the *decimation factor,* which is an integer number. Similarly, the bandwidth at the input of the filter is BW and the bandwidth at the output of the filter is BW/n.

The digital filter provides alias protection while reducing the sample rate so that the FFT frequency resolution is increased. The analog anti-alias filter is still required, since the digital filter is itself a sampled system which must be protected from aliasing. The analog filter protects the instrument at its widest frequency span, with operation at f_s. The digital filters fall in behind the protection of the analog filter and are used when narrower spans are selected by the user.

4.8 BAND SELECTABLE ANALYSIS

By varying the sample rate, the frequency span of the analyzer can be controlled but the start frequency of the span is always at DC. The frequency resolution of the measurement can be improved arbitrarily but at the expense of a lower maximum

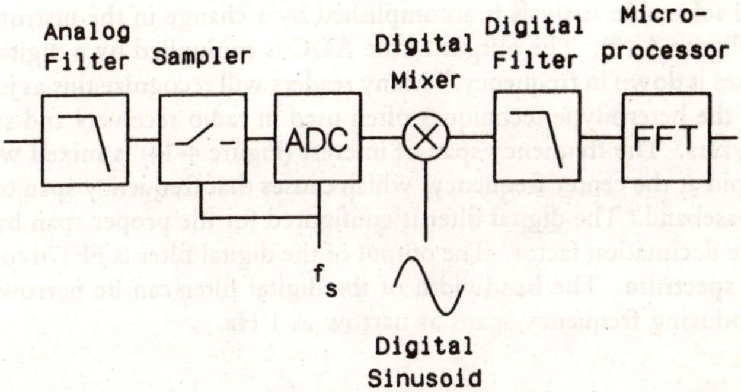

Figure 4-13. A digital mixer provides band selectable analysis in an FFT analyzer.

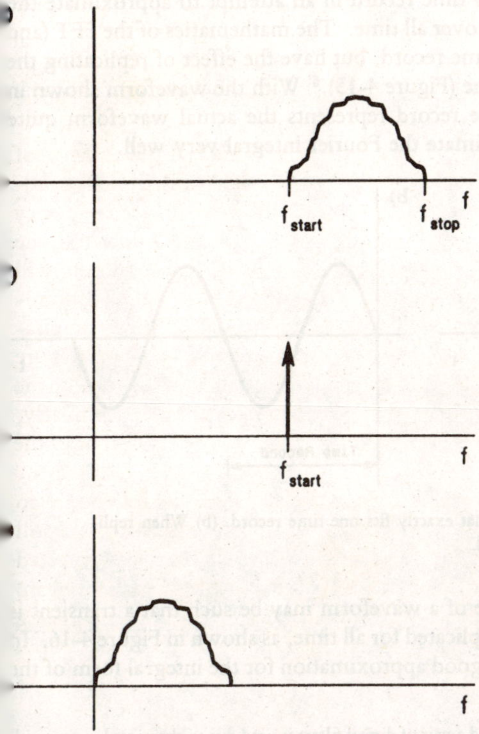

Figure 4-14. (a) The spectrum of the signal to be measured. (b) The spectrum of the digital oscillator. (c) The frequency translated version of the original spectrum.

frequency. *Band selectable analysis* (also known as *zoom* operation) allows the user to reduce the frequency span while maintaining a constant center frequency. In other words, the displayed frequency range is not limited to starting at DC. This is useful because very narrow spans away from DC can be analyzed.

Band selectable analysis is accomplished by a change in the instrument block diagram (Figure 4-13). The output of the ADC is multiplied by a digital sinusoid, which mixes it down in frequency.[5] Many readers will recognize this as just a digital version of the heterodyne techniques often used in radio receivers and swept spectrum analyzers. The frequency span of interest (Figure 4-14) is mixed with a complex sinusoid at the center frequency, which causes that frequency span to be mixed down to baseband. The digital filter is configured for the proper span by using the appropriate decimation factor. The output of the digital filter is FFT'd to obtain the frequency spectrum. The bandwidth of the digital filter can be narrowed significantly, producing frequency spans as narrow as 1 Hz.

4.9 LEAKAGE

The FFT operates on a finite length time record in an attempt to approximate the Fourier transform, which integrates over all time. The mathematics of the FFT (and DFT) operate on the finite length time record, but have the effect of replicating the finite length time record over all time (Figure 4-15).[6] With the waveform shown in Figure 4-15b, the finite length time record represents the actual waveform quite well, so the FFT result will approximate the Fourier integral very well.

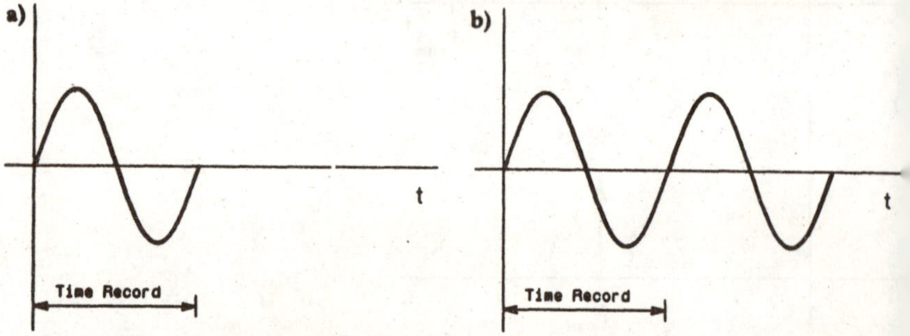

Figure 4-15. (a) A waveform that exactly fits one time record. (b) When replicated, no transients are introduced.

However, the shape and phase of a waveform may be such that a transient is introduced when the waveform is replicated for all time, as shown in Figure 4-16. In this case, the FFT spectrum is not a good approximation for the integral form of the

[5] Normally, a pair of digital mixers and a pair of digital filters is used due to the complex sinusoid, and a complex FFT is required, but the operations are shown simplified here.

[6] The FFT has the *effect* of replicating the time record. This is a consequence of the mathematics, and there is no need for the algorithm to actually produce the replicated time record.

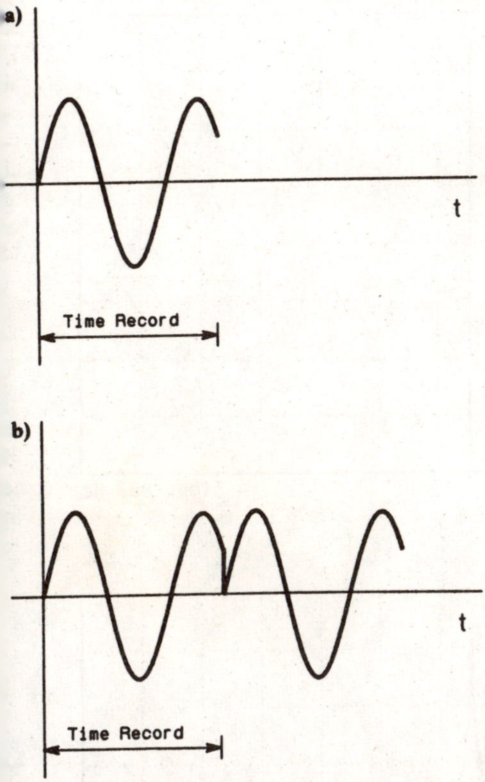

Figure 4-16. (a) A waveform that does not exactly fit into one time record. (b) When replicated, severe transients are introduced, causing leakage in the frequency domain.

Fourier transform. Since the instrument user often does not have control over how the waveform fits into the time record, in general, it must be assumed that a discontinuity may exist. This effect, known as *leakage,* is very apparent in the frequency domain. Instead of the spectral line appearing thin and slender, it spreads out over a wide frequency range (Figure 4-17).

The usual solution to the problem of leakage is to force the waveform to zero at the ends of the time record; then they will always be the same and no transient will exist when the time record is replicated. This is accomplished by multiplying the time record by a *window* function. Of course, the shape of the window is important as it will affect the data, and it must not introduce a transient of its own. Many different window functions have been developed for particular digital signal processing applications. The ones common to spectrum analyzers will be examined here.

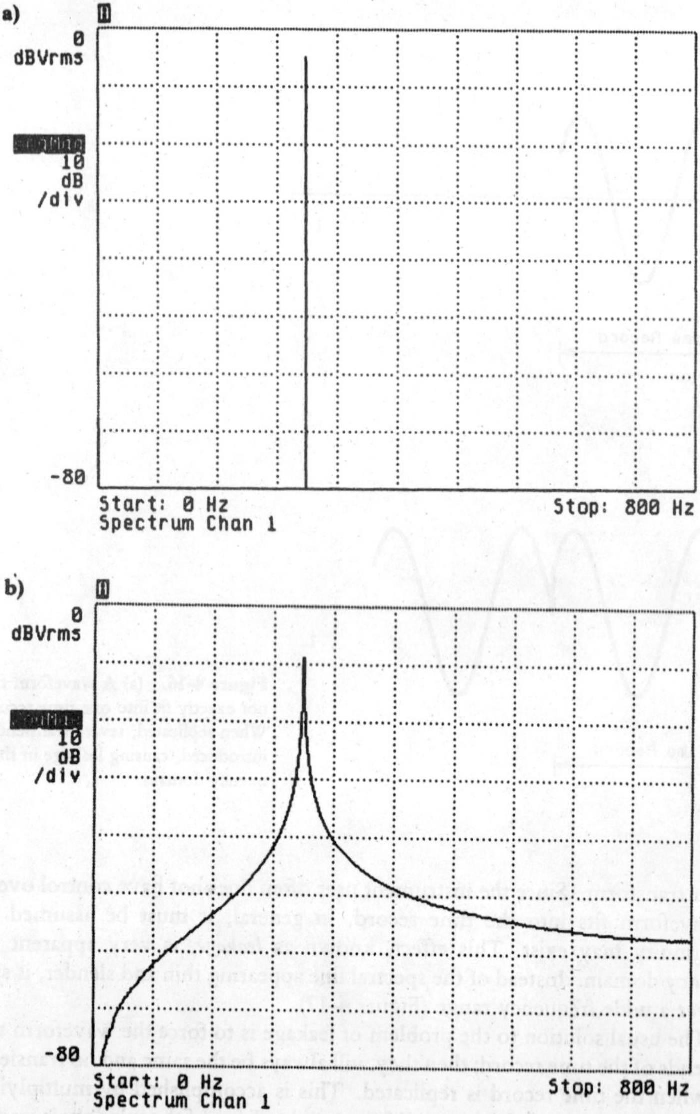

Figure 4-17. (a) Measurement of a spectral line with no leakage. (b) Measurement of a spectral line with leakage.

4.10 HANNING WINDOW

Also known as the Hann window, the *Hanning window* is one of the most common windows in digital signal processing. The time record samples are weighted by the following function:

$$w_n = \frac{1}{2}\{1 - \cos[2\pi n/(N-1)]\} \tag{4-13}$$

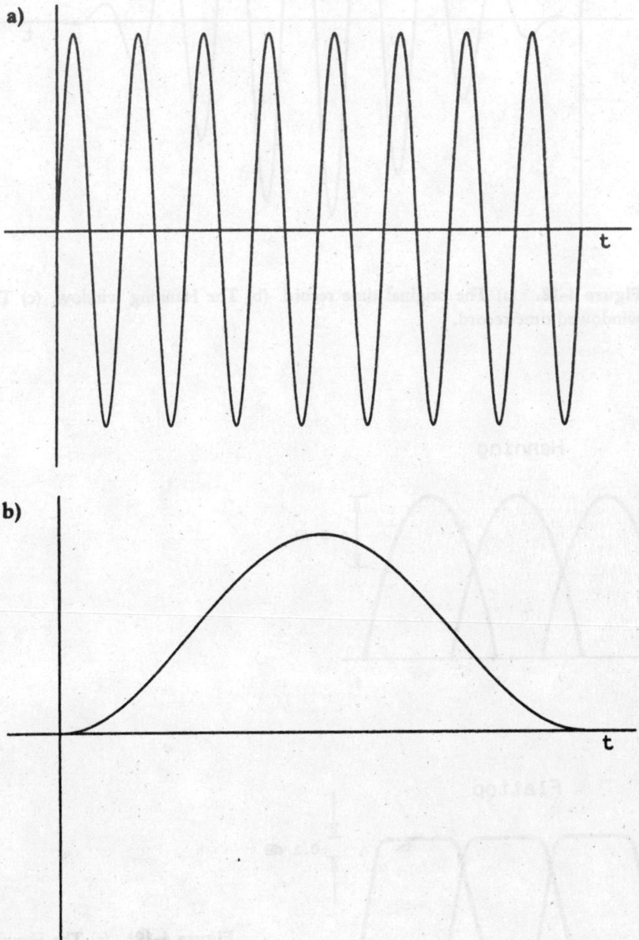

Figure 4-18. Continued on next page.

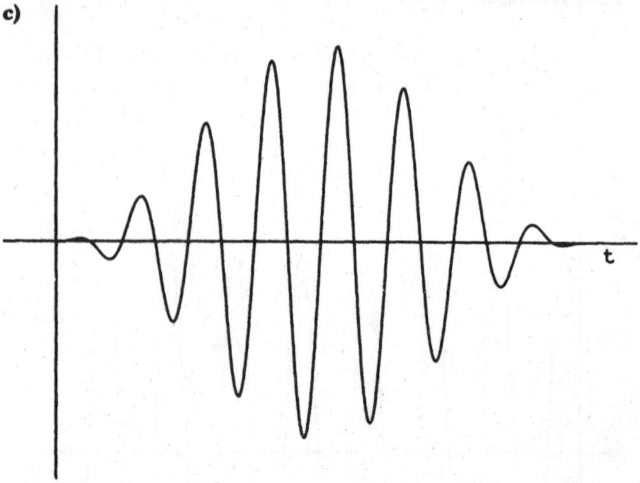

Figure 4-18. (a) The original time record. (b) The Hanning window. (c) The windowed time record.

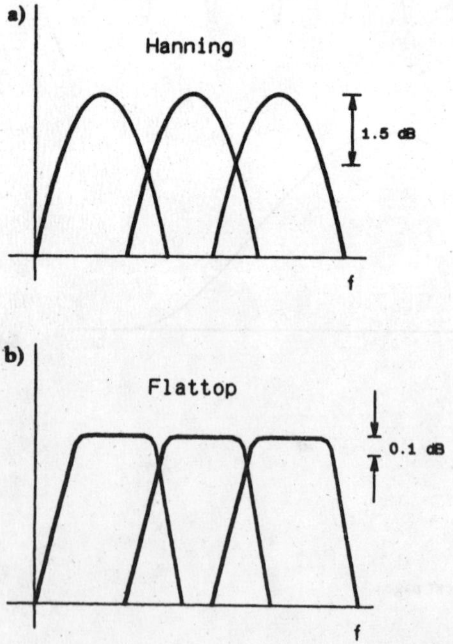

Figure 4-19. (a) The Hanning window introduces a maximum amplitude error of 1.5 dB. (b) The flattop window introduces a maximum amplitude error of 0.1 dB.

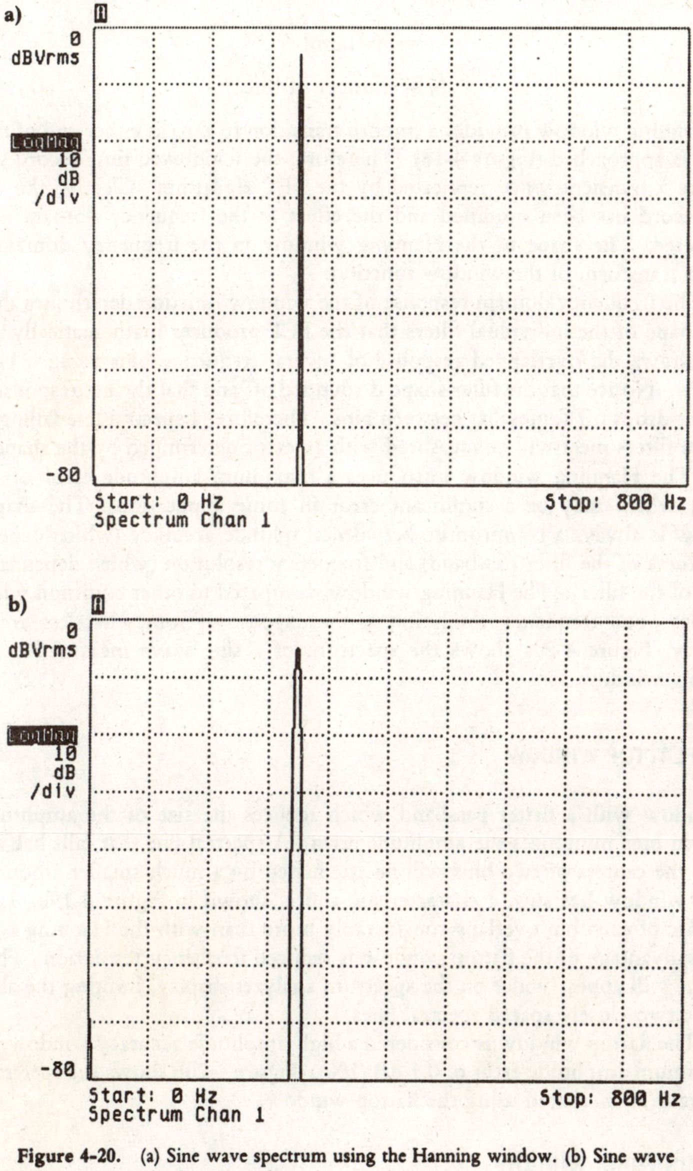

Figure 4-20. (a) Sine wave spectrum using the Hanning window. (b) Sine wave spectrum using the flattop window.

where

$$n = \text{bin number}$$

$$N = \text{number of bins}$$

The Hanning window provides a smooth transition to zero as either end of the time record is approached (Figure 4-18). Therefore, the windowed time record will not produce a transient when replicated by the FFT algorithm. Clearly, the original time record has been modified and the effect in the frequency domain must be considered. The shape of the Hanning window in the frequency domain is the Fourier transform of the window function.

The frequency domain response of the window function determines the passband shape of the individual filters that the FFT produces mathematically. Figure 4-19a shows the overlapped response of several frequency bins using a Hanning window. Notice that the filter shape is rounded off and that the net response of the analyzer drops off somewhat between bins. Therefore, a spectral line falling where the two filters meet will be measured with an error determined by the shape of the filter. The Hanning window introduces a maximum amplitude error of 1.5 dB (16%), which may be a significant error in some applications. The shape of a window is always a compromise between amplitude accuracy (which depends on the flatness of the filter passband) and frequency resolution (which depends on the width of the filter). The Hanning window, compared to other common windows, provides good frequency resolution at the expense of somewhat less amplitude accuracy. Figure 4-20a shows the spectrum of a sine wave measured using the Hanning window.

4.11 FLATTOP WINDOW

A window with a flatter passband which reduces the size of the amplitude dips between bins minimizes the amplitude error. A spectral line that falls halfway between the centers of two bins will be attenuated by a much smaller amount. The *flattop* window has such a characteristic and is shown in Figure 4-19b. Since the response of each bin overlaps considerably more than with the Hanning window, the disadvantage of the flattop window is reduced frequency resolution. The spectral line will appear wider on the spectrum analyzer display, limiting the ability to resolve two closely spaced spectral lines.

The flattop window is considered a high-amplitude accuracy window, having a maximum amplitude error of 0.1 dB (1%). Figure 4-20b shows the spectrum of a sine wave as measured using the flattop window.

4.12 UNIFORM WINDOW

The uniform window is really no window at all; all the samples are left unchanged. Although the uniform window has the potential for severe leakage problems, in some cases the waveform in the time record has the same value at both ends of the record, thereby eliminating the transient introduced by the FFT. Such waveforms are called *self-windowing*. Waveforms such as *pseudorandom noise*[7] (*PRN*), sine bursts, impulses, and decaying sinusoids can all be self-windowing.

The uniform window is appropriate for making network measurements when the internal noise source of the analyzer is used. The noise source is often a PRN generator that produces a noise waveform which is periodic within the time record of the instrument. Since the noise source and the time record are synchronized, no transients occur at the ends of the time record and leakage in the frequency domain is avoided.

4.13 EXPONENTIAL WINDOW

One of the advantages of an FFT analyzer is that it can be used to measure the frequency content of a fast transient. (This is not usually possible in the more conventional swept analyzer since it will miss some of the transient as it is sweeping through its frequency span.) Such a transient might be the step or impulse response of an electrical network. Also, FFT analyzers are often used for mechanical measurements.[8]

A typical transient response is shown in Figure 4-21a. As shown the waveform is self-windowing because it dies out within the length of the time record, reducing the leakage problem. If the waveform does not dissipate within the time record (as shown in Figure 4-21b), then some form of window should be used. If a window such as the Hanning window were applied to the waveform, the beginning portion of the time record would be forced to zero. This is precisely where most of the transient's energy is, so such a window would be inappropriate.

A window with a decaying exponential response is useful in such a situation. The beginning portion of the waveform is not disturbed, but the end of the time record is forced to zero. There still may be a transient at the beginning of the time record, but this transient is not introduced by the FFT, it is, in fact, the transient being measured. Figure 4-21c shows the exponential window and Figure 4-21d shows the resulting time domain function when the exponential window is applied to Figure 4-21b.

The exponential window function is given by

$$w_n = e^{-n/((N-1)k)} \qquad (4\text{–}14)$$

[7] Pseudorandom noise is noise which is not truly random, but instead repeats at some interval.

[8] Mechanical measurements, including vibration and structural analysis, represent an important use of FFT-based spectrum analyzers and are covered extensively in *Modal Testing: Theory and Practice*, by D. J. Ewins, Research Studies Press Ltd., Letchworth, Hertfordshire, England, 1984. (Published in the United States by John Wiley & Sons, Inc.)

where

$$n = \text{bin number}$$
$$N = \text{number of bins}$$
$$k = \text{exponential time constant}$$

The time constant, k, is selected by the user to provide the appropriate amount of exponential decay to prevent leakage.

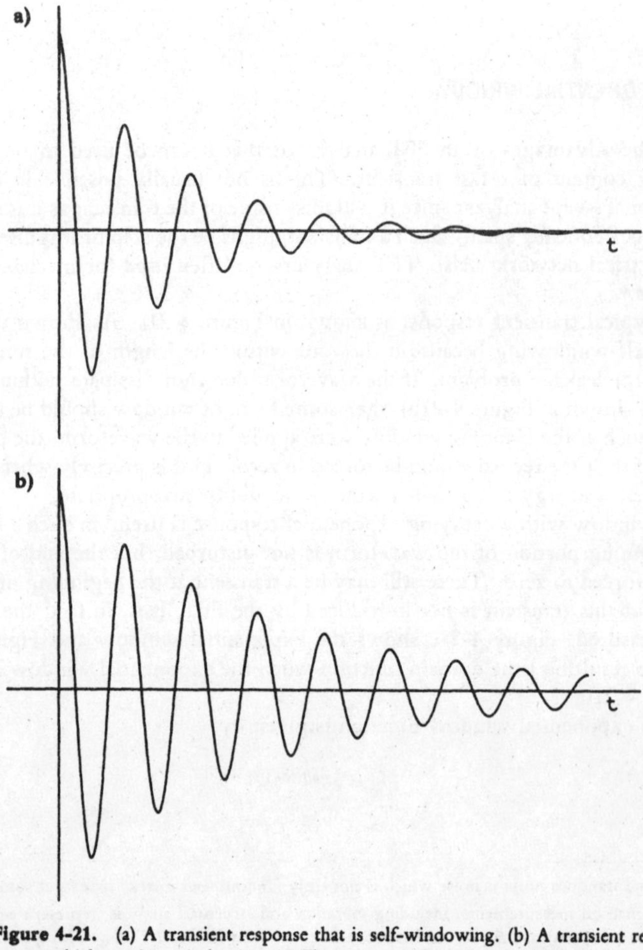

Figure 4-21. (a) A transient response that is self-windowing. (b) A transient response which requires windowing. (c) The exponential window. (d) The windowed transient response.

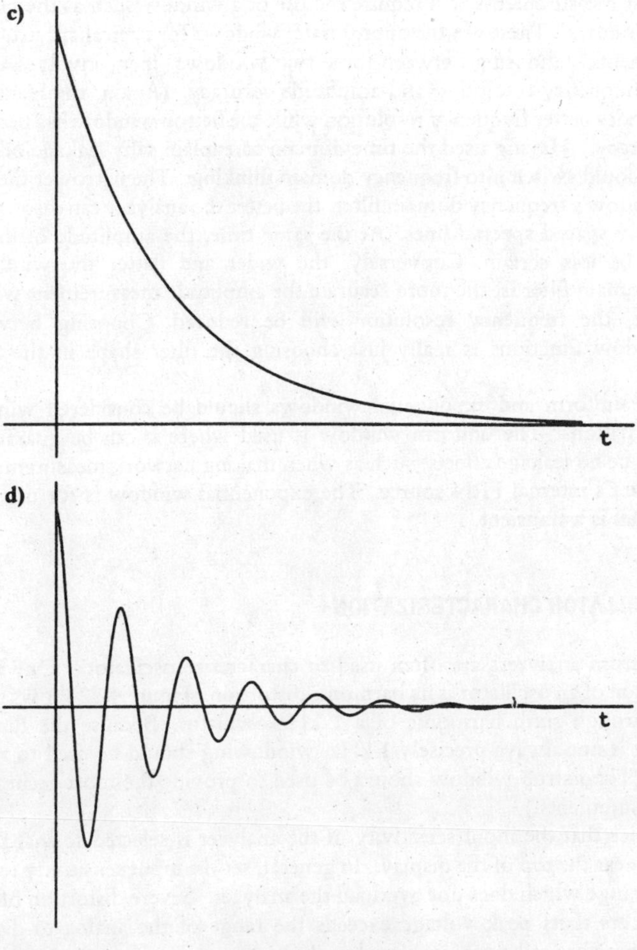

Figure 4-21. *(continued)*

The exponential window is inappropriate for measuring anything but transient waveforms.

4.14 SELECTING A WINDOW FUNCTION

Selecting the appropriate window function for some users (especially those more familiar with swept analyzers) seems like a bewildering task. It need not be such a problem though.

Most measurements will require the use of a window such as the Hanning or flattop windows. These are the appropriate windows for typical spectrum analysis measurements. Choosing between these two windows, then, involves a trade-off between frequency resolution and amplitude accuracy. (Again, the Hanning window provides better frequency resolution while the flattop window has better amplitude accuracy.) Having used the time domain to explain why leakage occurs, here the user should switch into frequency domain thinking. The narrower the passband of the window's frequency domain filter, the better the analyzer can discern between two closely spaced spectral lines. At the same time, the amplitude of the spectral line will be less certain. Conversely, the wider and flatter the window's frequency domain filter is, the more accurate the amplitude measurement will be and, of course, the frequency resolution will be reduced. Choosing between two such window functions is really just choosing the filter shape in the frequency domain.

The uniform and exponential windows should be considered windows for special situations. The uniform window is used where it can be guaranteed that there will be no leakage effects, such as when making network measurements using the analyzer's internal PRN source. The exponential window is for use when the input signal is a transient.

4.15 OSCILLATOR CHARACTERIZATION

FFT spectrum analyzers are often used to characterize oscillators. One important specification of an oscillator is its harmonic distortion. Figure 4-22 shows the fundamental through sixth harmonic of a 1 kHz oscillator. Because the fundamental frequency is not always precisely 1 kHz, windowing should be used to reduce the leakage. (The flattop window should be used to provide the most accurate amplitude measurements.)

Notice that the input sensitivity of the analyzer is selected so that the fundamental is near the top of the display. In general, set the input sensitivity to the most sensitive range which does not overload the analyzer. Severe distortion of the input signal occurs if its peak voltage exceeds the range of the analog-to-digital converter. Therefore, all dynamic signal analyzers warn the user of this condition by some kind of overload indicator.

It is also important to make sure the analyzer is not underloaded. If the signal going into the analog-to-digital converter is too small, much of the useful information of the spectrum may be below the noise level of the analyzer. Therefore, setting the input sensitivity to the most sensitive range that does not cause an overload gives the best possible results.

Figure 4-22a is a display of the spectrum amplitude in logarithmic form to ensure that distortion products far below the fundamental can be seen. All signal amplitudes on this display are in dBV (decibels below 1 Volt RMS). However, since most FFT analyzers have very versatile display capabilities, this spectrum

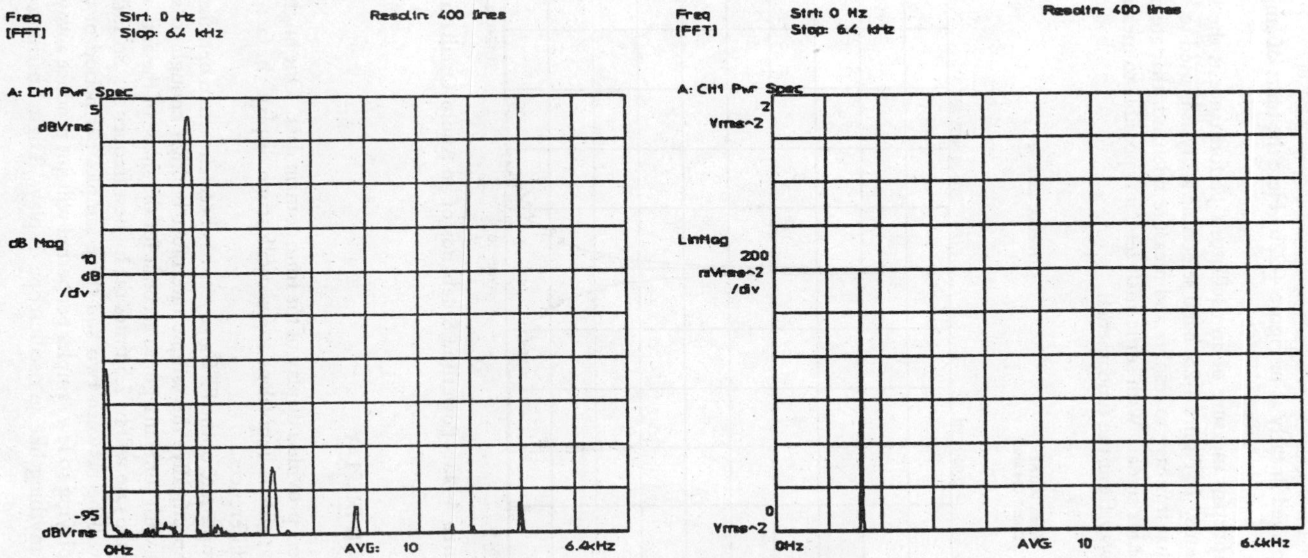

Figure 4-22. Harmonic distortion of an audio oscillator. (a) Logarithmic amplitude scale. (b) Linear amplitude scale.

could also be displayed linearly as in Figure 4-22b. Here the units of amplitude are volts.

Another important measure of an oscillator's performance is the level of its power-line sidebands. In Figure 4-23, band selectable analysis is used to "zoom in" on the signal so that it is easy to resolve and measure sidebands that are only 60 Hz away from our 1 kHz signal. With some analyzers, it is possible to measure signals only millihertz away from the fundamental.

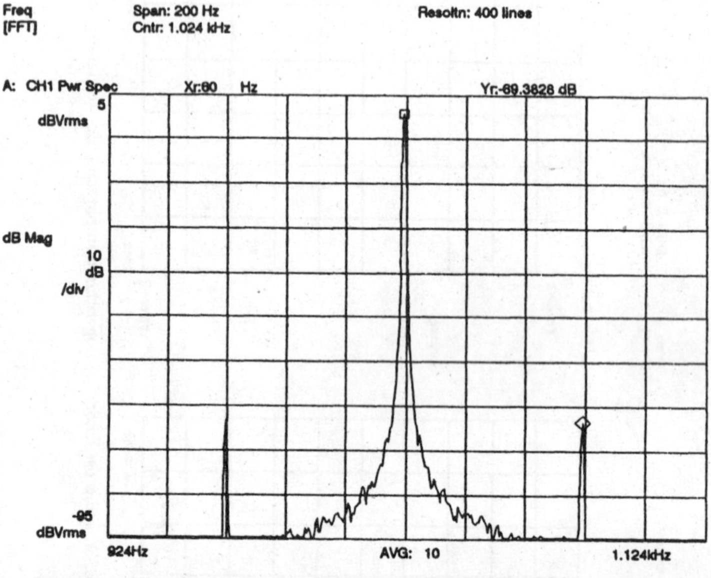

Figure 4–23. Powerline Sidebands of an Audio Oscillator.

4.16 TIME DOMAIN DISPLAY

Many FFT analyzers provide a display of the time domain data. One might jump to the conclusion that this is equivalent to an oscilloscope display. In fact, there are some significant differences.

First, the sample rate of the FFT analyzer has been chosen to optimize for FFT analysis results. Specifically, the sample rate must be high enough to satisfy the sampling theorem with some margin to account for the shape of the anti-alias filter. Typically, the sample rate will be 2.5 times the highest frequency. So at the highest frequency, there will be between two and three samples per period of the waveform. Simply displaying so few samples per period will not produce a waveform on screen that looks anything like an oscilloscope display. (Digital oscilloscopes normally use more samples per period and may provide additional digital signal processing to reconstruct the waveform.)

The anti-alias filter is a steep high-order filter designed to approximate an ideal low-pass filter. It abruptly limits the frequency response of the analyzer and may introduce ringing in the time domain.

In band selectable analysis, the time waveform may be displayed after it has been mixed with the complex sinusoid. The resulting waveform is complex (has a real and an imaginary part) and is often difficult to interpret.

Despite these shortcomings, the time domain display is useful for many applications. The user can monitor the input waveform that is to be FFT'd and the analyzer can be used as a waveform recorder to a limited extent. Some analyzers provide long time buffers for capturing large amounts of time domain data. After capture, portions of the time record can be analyzed for frequency content.

4.17 NETWORK MEASUREMENTS

Traditionally, network measurements are made by supplying a sinusoid to the *device under test* (DUT) and measuring its output, then repeating this at each frequency of interest. This technique can be used with an FFT analyzer, but there is a more efficient and faster technique. An FFT spectrum analyzer can be used to make network measurements using its internal source. The source is connected to the input of the DUT, and the output of the DUT is connected to the input of the analyzer (Figure 4-24).

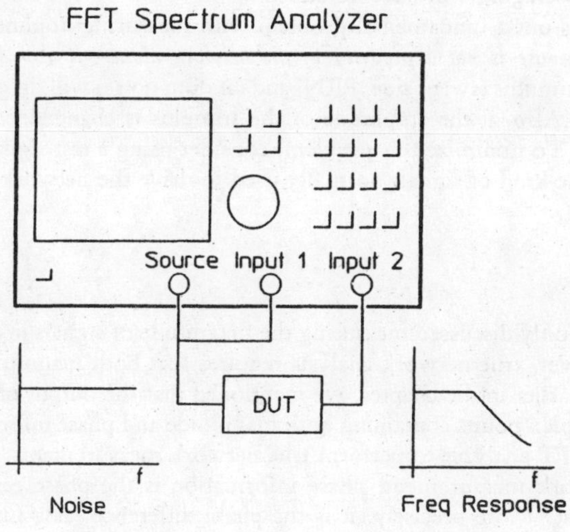

Figure 4-24. An FFT analyzer can perform a network measurement using a noise source.

Recall that the FFT analyzer behaves the same as a bank-of-filters analyzer. To make a network measurement using a sinusoid, we would iteratively set the sinusoid's frequency to be in the center of each of the filters, recording the readings as we went. This requires as many measurements as there are bins. On the other hand, if a signal is supplied which would simultaneously produce energy in each of the FFT bins, then one measurement would suffice. Chirp sine and random noise meet such a requirement.

The chirp sine signal is a swept sine burst designed to fill the time record of the FFT analyzer. This provides energy across the frequency band of interest. The chief advantage of the chirp sine is its relatively high average power—much more than random noise (for the same peak voltage). This produces a better signal-to-noise ratio in the measurement, when compared to the case where random noise is used to excite the network.

Broadband random noise has equal energy in all of the FFT bins and provides a stimulus to the DUT such that the output frequency response will be the frequency response of the network. A PRN (pseudo random noise) signal is often used because it is synchronized with the time record of the analyzer so that it does not produce leakage. Thus, the uniform window is used when making network measurements with a PRN source.

A truly random (not pseudo random) noise source is useful with nonlinear networks. Nonlinear networks produce considerable distortion which corrupts the results of a network measurement. With a random noise source, these distortion effects can be averaged out since they are different for each measurement.[9] With PRN, the noise waveform and the distortion products are the same for every measurement, and averaging will have no effect.

This points out a fundamental problem with measuring nonlinear networks: *the frequency response is not a property of the network alone—it also depends on the stimulus.* Each stimulus (swept sine, PRN, and random noise) will, in general, give a different result. Also, if the amplitude of the stimulus is changed, you will get a different result. To minimize this problem, consider using a test signal that closely approximates the kind of signals normally used to drive the network's inputs.

4.18 PHASE

So far, we have only discussed measuring the magnitude of signals in the frequency domain. However, true network analysis requires that both magnitude and phase be measured. Earlier in the chapter, we mentioned that the output of the FFT was an array of complex points containing both magnitude and phase information. This fact allows an FFT analyzer to perform true network measurements.

In a network measurement, phase information is the phase response of the device under test. More precisely, it is the phase difference (as a function of frequency) between the input stimulus and the measured response. Many FFT analyz-

[9] See Chapter 10, Averaging and Filtering.

ers have two input channels that can be used to simultaneously measure both input to and output of the DUT. In this case, the phase response of the DUT is the phase difference between the two channels.

In a spectrum measurement, the usefulness of the phase information is less obvious. Since phase is a relative concept, one is tempted to ask "phase with respect to what?" Phase displayed on an FFT analyzer depends on the relative position of the waveform in the time record. Shift the waveform 90 degrees in the time record, and the answer on the display will change 90 degrees. Many analyzers provide oscilloscope-like triggering capability to allow some control over the start of the time record. If this feature is used, then the phase of a particular signal can be stabilized. This (for example) allows the phase of harmonics to be compared to the fundamental. If no triggering is used, the analyzer will acquire a time record when ready, which will be uncorrelated to the input signal. In this case, the phase of the signal will vary randomly from measurement to measurement. Use may be made of the relative phases between multiple signals present in a single time record, as with a modulated input signal.

4.19 SPECTRAL MAPS

One feature that has been unique to the FFT analyzer but is finding its way into other analyzers is the spectral map (also known as a waterfall display). This feature displays multiple spectrums as a function of time, giving an almost three-dimensional display (Figure 4-25). For a transient event, this frequency spectrum versus time

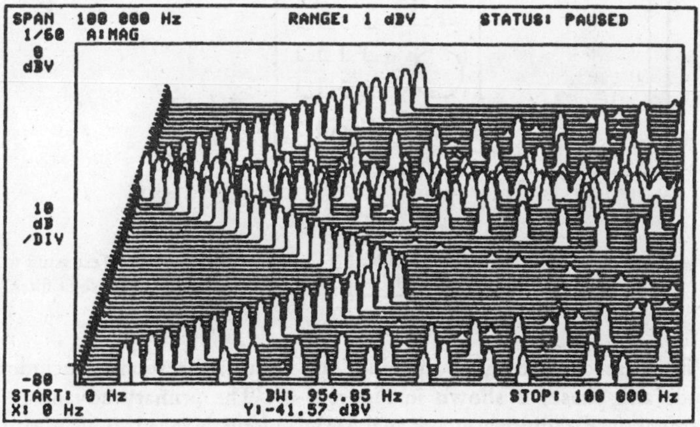

Figure 4-25. A spectral map of a swept sine wave oscillator. The largest responses are caused by the fundamental frequency of the oscillator moving up and down in frequency. The other responses are caused by harmonic distortion and other imperfections in the oscillator.

display characterizes the signal as a function of time. (Traditional spectrum analysis has always assumed a stationary signal.) This feature is particularly appropriate for an FFT analyzer, since it has the ability to produce successive spectra without missing any data. Swept analyzers can easily miss portions of the waveform while sweeping.

Spectral maps are often used to monitor vibrations in the structure of a rotating machine as its speed is steadily increased over time (called run up) or steadily decreased over time (called run down).

4.20 ELECTRONIC FILTER CHARACTERIZATION

Another typical use for an FFT network analyzer is to characterize a low-frequency electronic filter. One possible test setup appears in Figure 4-26. Because the filter is linear, it is possible to use pseudo random noise as the stimulus and get very fast test times. The uniform window is used because the pseudo random noise is periodic in the time record. No averaging is needed since the signal is periodic and reasonably large. Be very careful, as in the single-channel case, to set the input sensitivity for both channels to the most sensitive position that does not overload the analog-to-digital converters.

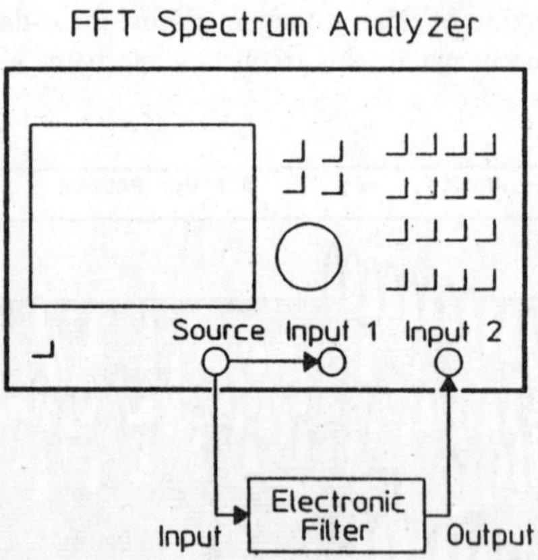

Figure 4-26. Test setup to measure frequency response of filter.

With these considerations in mind, the frequency response, including both magnitude and phase, is shown in Figure 4-27. The primary advantage of this measurement over traditional swept analysis techniques is speed. This measurement can be made in 1/8 second with an FFT analyzer, but would take over 30 seconds with a swept network analyzer. This speed improvement is particularly important when the filter-under-test is being adjusted, or when large volumes are tested on a production line.

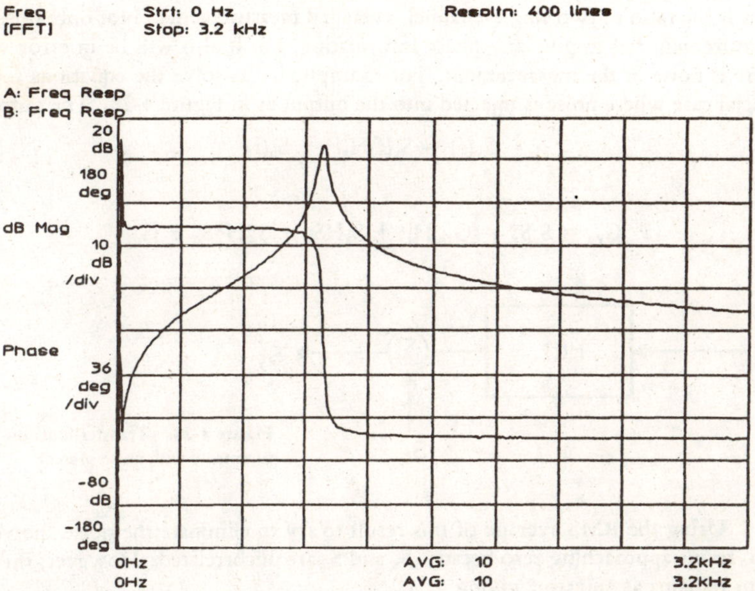

Figure 4-27. Frequency response of electronic filter using PRN and uniform window.

4.21 CROSS POWER SPECTRUM

The cross power spectrum is not often used as a separate measurement, but is used internally by FFT analyzers to compute transfer functions and coherence. The cross power spectrum, G_{xy}, is defined as taking the FFT of two signals separately and multiplying the result together

$$G_{xy}(f) = S_x(f)S_y^*(f)$$

where * indicates the complex conjugate of the function.

With this function, we can define the transfer function, H(f), using the cross power spectrum and the spectrum of the input channel

$$H(f) = \frac{\overline{G_{yx}(f)}}{\overline{G_{xx}(f)}}$$

where ***overbar*** denotes the average of the function.

At first glance it may seem more appropriate to compute the transfer function using

$$|H(f)|^2 = \frac{\overline{G_{yy}}}{\overline{G_{xx}}}$$

This is the ratio of two single-channel, averaged measurements. Not only does this measurement fail to give any phase information, but it also will be in error when there is noise in the measurement.. For example, let us solve the equations for the special case where noise is injected into the output as in Figure 4-28. The output is

$$S_y(f) = S_x(f)H(f) + S_n(f)$$

So

$$G_{yy} = S_yS_y^\star = G_{xx}|H|^2 + S_xHS_n + S_x^\star H^\star S_n + |S_n|^2$$

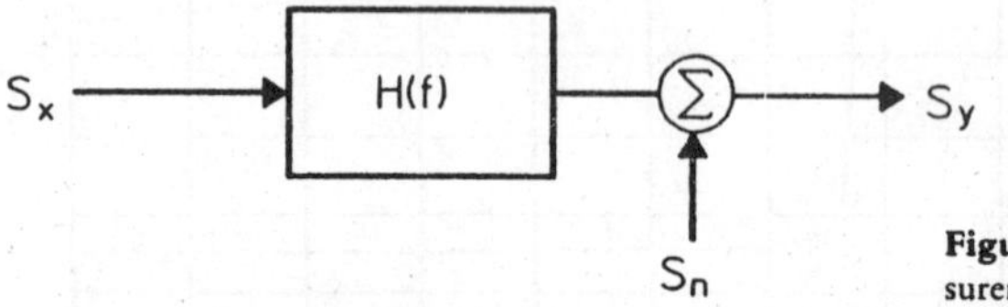

Figure 4-28. Transfer function measurements with noise present.

Using the RMS average of this result to try to eliminate the noise shows the S_xS_n terms approaching zero because S_x and S_n are uncorrelated. However, the $|S_n|^2$ term remains as an error, giving

$$\frac{\overline{G_{yy}}}{\overline{G_{xx}}} = |H|^2 + \frac{\overline{|S_n|^2}}{\overline{G_{xx}}}$$

Therefore, measuring $|H|^2$ by this single-channel technique gives a value which will be high by the value of the noise-to-signal ratio. Averaging the cross power spectrum eliminates this noise error. Using the example in Figure 4-28,

$$\overline{G_{yx}} = \overline{S_yS_x^\star} = \overline{(S_xH + S_n)S_x^\star} = \overline{G_{xx}H} + \overline{S_nS_x^\star}$$

so

$$\frac{\overline{G_{yx}}}{\overline{G_{xx}}} = H(f) + \overline{S_nS_x^\star}$$

Because S_n and S_x are uncorrelated, the second term will average to zero, making this function a much better estimate of the transfer function.

4.22 COHERENCE

Sometimes the components to be tested cannot be isolated from other disturbances. One example is trying to measure the frequency response of a switching power supply that contains a high concentration of power at the switching frequency. Another would be trying to measure the frequency response of a part on one machine in the presence of strong vibration from a nearby machine.

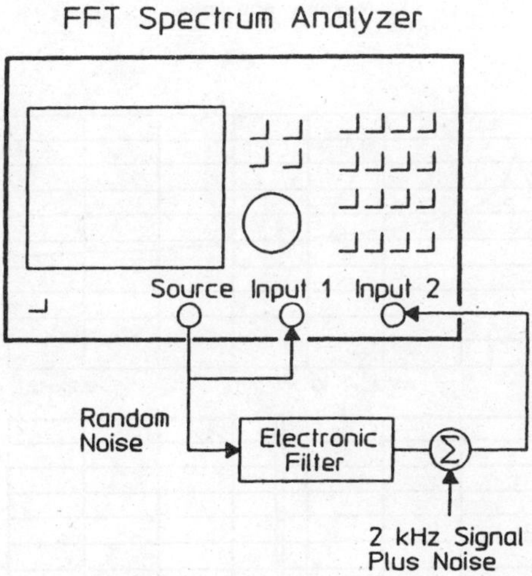

Figure 4-29. Adding noise and a 2 kHz signal to the output of a filter.

Figure 4-29 shows one way to simulate this type of situation by adding noise and a 2 kHz signal to the output of an electronic filter. The measured frequency response is shown in the upper trace in Figure 4-30. RMS averaging has reduced the noise contribution but has not completely eliminated the 2 kHz interference. Further averaging would reduce this interference. If we did not already know the source of this interference, we would think that the filter has an additional resonance at 2 kHz. By using a coherence measurement we can eliminate the unrelated 2 kHz component.

FFT analyzers can often make coherence measurements. This type of measurement is not available with traditional network analyzers. It measures the power in the response channel that is caused by power in the reference channel. Coherence is a unitless value that indicates how much of the output power is coherent with the input power. A coherence value of 1 means that all the power in the output is caused by the input. A coherence value of 0 means that none of the power in the output is caused by the input. (Care must be exercised when using coherence. It does not always imply a causal relationship. For example, if two signals are caused by a third signal, they will be coherent with each other even though one is not caused by the other.)

The lower trace in Figure 4-30 shows the coherence. The coherence goes from 1 (all the output power at that frequency is caused by the input) to 0 (none of the output power at that frequency is caused by the input). The coherence function shows that the response at 2 kHz is not caused by the input but by interference. However, the filter's response near 1 kHz has excellent coherence and so the measurement here is good.

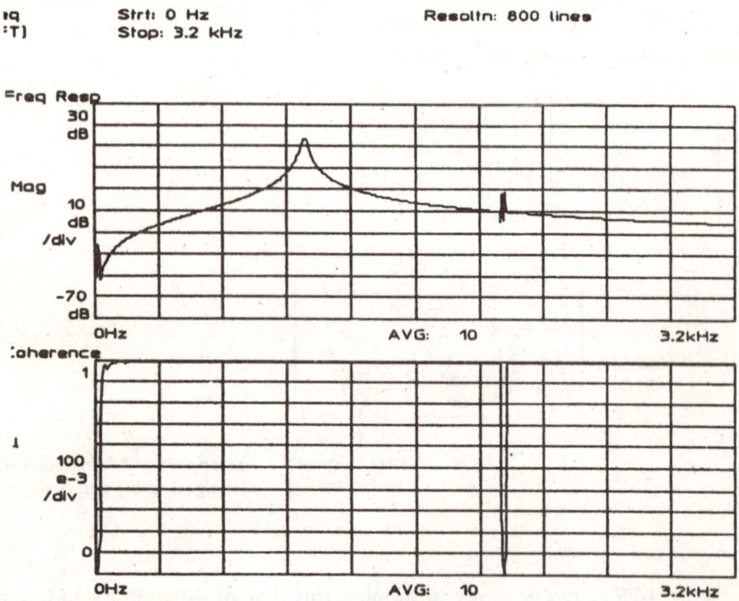

Figure 4-30. Frequency response and coherence with added noise and 2 kHz signal.

The coherence function, $\gamma^2(f)$, is derived from the cross power spectrum by

$$\gamma^2(f) = \frac{\overline{G_{yx}(f)}\,\overline{G^{\star}_{xy}(f)}}{\overline{G_{xx}(f)}\,\overline{G_{yy}(f)}}$$

As stated previously, the coherence function is a measure of the power in the output signal caused by the input. If the coherence is 1, then all the output power is caused by the input. If the coherence is 0, then none of the output is caused by the input.

To explore the mathematics of the coherence function, we will use the example in Figure 4-28. As has been shown before,

$$\overline{G_{yy}} = \overline{G_{xx}}|H|^2 + \overline{S_x H S^{\star}_n} + \overline{S^{\star}_x H^{\star} S_n} + \overline{|S_n|^2}$$

$$G_{yx} = \overline{G_{xx}H} + \overline{S_n S^{\star}_x}$$

As the measurement is averaged, the cross terms S_nS_x approach zero, assuming that the signal and the noise are not related. So the coherence becomes

$$\gamma^2 = \frac{(H\overline{G_{xx}})^2}{\overline{G_{xx}}(|H|^2\overline{G_{xx}} + \overline{|S_n|^2})}$$

$$\gamma^2 = \frac{|H|^2\overline{G_{xx}}}{|H|^2\overline{G_{xx}} + \overline{S_n}^2}$$

This shows that if there is no noise, the coherence function is unity. If there is noise, the coherence will be reduced. Note also that coherence is a function of frequency. Coherence can be unity at frequencies where there is no interference, and low at frequencies where the noise is high.

4.23 CORRELATION MEASUREMENTS

Correlation is a measure of the similarity between two quantities. To understand the correlation between two waveforms, we start by multiplying the waveforms together at each instant in time and adding up all the products. If, as in Figure 4-31a, the waveforms are identical, every product is positive and the resulting sum is large. If however, as in Figure 4-31b, the two records are dissimilar, then some of the products would be positive and some would be negative. There would be a tendency for the products to cancel, so the final sum would be smaller.

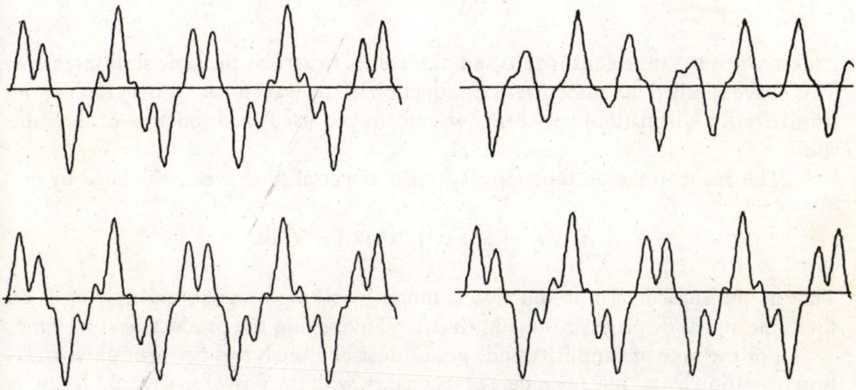

Figure 4-31. (a) Correlation of two identical signals. (b) Correlation of two different signals.

Now consider the waveform in Figure 4-32a, and the same waveform shifted in time, Figure 4-32b. If the time shift were zero, then conditions would be the same as before, that is, the waveforms would be in phase and the final sum of the products would be large. If the time shift between the two waveforms is made large, however, the waveforms appear dissimilar and the final sum is small.

4.24 AUTOCORRELATION

Going one step further, we can find the average product for each time shift by dividing each final sum by the number of products contributing to it. By plotting the average product as a function of time shift, the resulting curve is shown to be

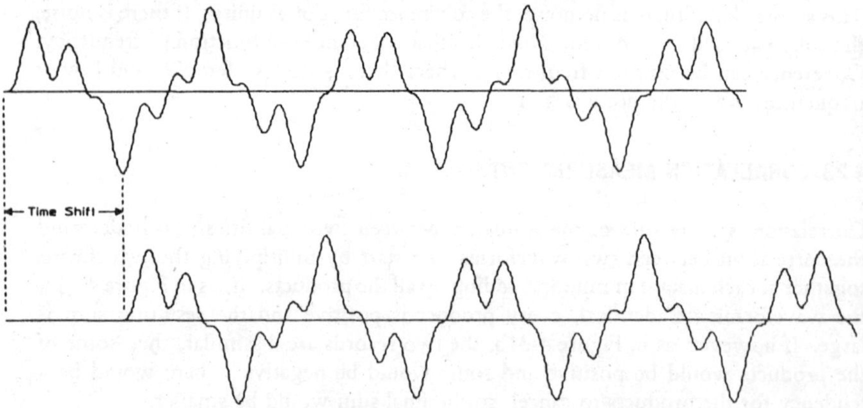

Figure 4-32. Correlation of time displaced signals.

largest when the time shift is zero and diminishes to zero as the time shift increases. This curve is called the autocorrelation function of the waveform. It is a graph of the similarity (or correlation) between a waveform and itself, as a function of the time shift.

The autocorrelation function, $R_{xx}(\tau)$, is a special time average defined by

$$R_{xx}(\tau) = \lim_{T\to\infty} \frac{1}{T} \int_T x(t)x(t + \tau)\, dt$$

That is, the autocorrelation function is found by taking a signal, multiplying it by the same signal displaced τ units in time, and averaging the product over all time.

For the sake of simplicity and speed, most FFT analyzers perform the correlation operation by taking advantage of its duality with the power spectrum. It can be shown that

$$R_{xx}(\tau) = F^{-1}[S_x(f)S_x^\star(f)]$$

where F^{-1} is the inverse Fourier transform and S_x is the Fourier transform of x(t).

The autocorrelation function always has a maximum at $\tau = 0$ equal to the mean square value of x(t). If the signal x(t) is periodic, the correlation function is also periodic with the same period. Random noise, on the other hand, only correlates at $\tau = 0$.

The autocorrelation function can be used to improve the signal-to-noise ratio of periodic signals. The random noise component concentrates near $\tau = 0$ while the periodic component repeats itself with the same periodicity as the signal. Another thing to remember is that impulsive noises such as pulse trains, bearing ping, or gear chatter show up more distinctly in correlation or time record averaging than in a frequency domain analysis.

The autocorrelation function is more easily understood by looking at a few examples. The random noise shown in Figure 4-33 is not similar to itself with any amount of time shift (since it is random) so its autocorrelation has only a single spike at the point of zero time shift.

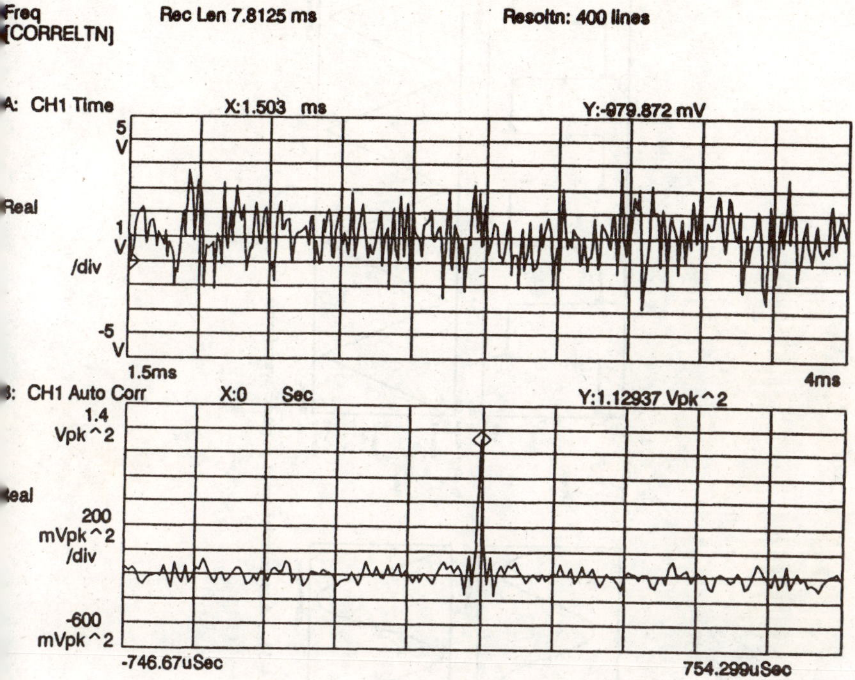

Figure 4-33. Autocorrelation of random noise.

Figure 4-34 shows the autocorrelation of a sine wave and a square wave. These are both special cases of a more general statement; the autocorrelation of any periodic waveform is periodic and has the same period as the waveform itself.

This can be useful when trying to extract a signal hidden by noise. Figure 4-35a shows what looks like random noise, but there is actually a low-level sine wave buried in it. We can see this in Figure 4-35b where we have taken 100 averages of the autocorrelation of this signal. The noise has become the spike around a time shift of zero whereas the autocorrelation of the sine wave is clearly visible, repeating itself with the period of the sine wave.

If a trigger signal is available that is synchronous with the sine wave, it is possible to extract the signal from the noise by linear averaging as in the last exam-

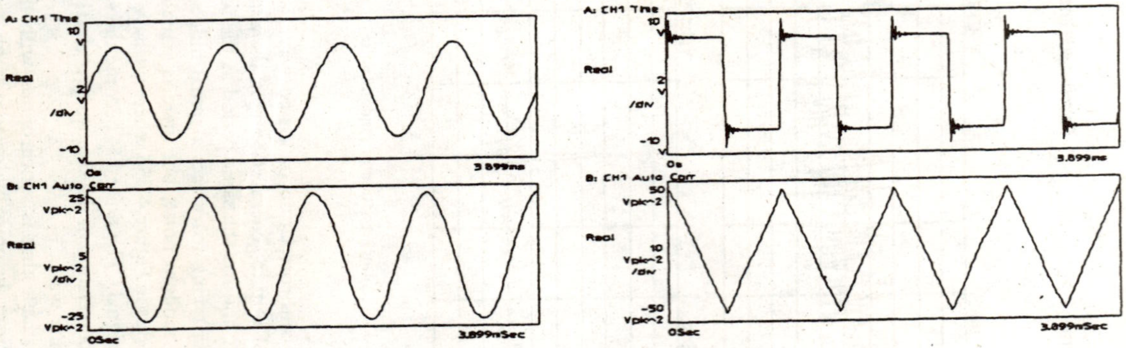

Figure 4-34. Autocorrelation of periodic waveforms.

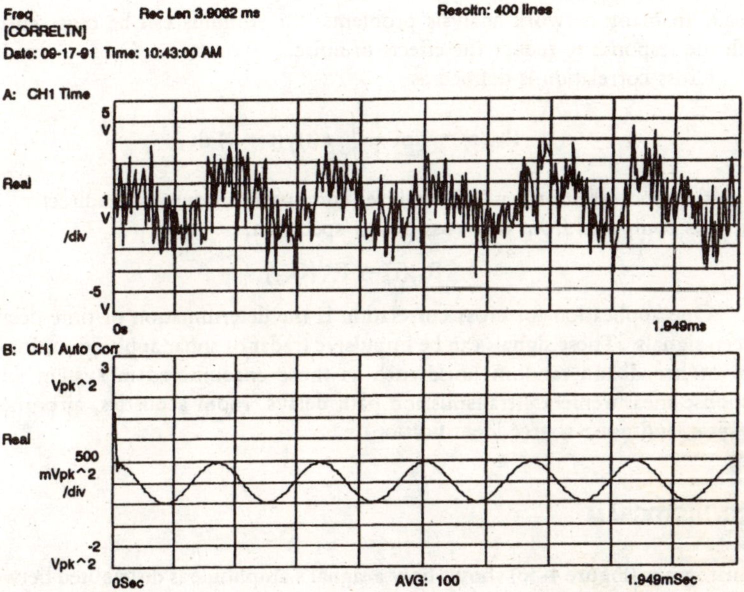

Figure 4-35. Autocorrelation of a sine wave buried by noise.

ple. But the important point about the autocorrelation function is that no synchronizing trigger is needed. In signal identification problems like radio astronomy and passive sonar, a synchronizing signal is not available and so autocorrelation is an important tool. The disadvantage of autocorrelation is that the input waveform is not preserved as it is in linear averaging.

Since any time domain waveform can be transformed into the frequency domain, the reader may wonder what is the frequency transform of the autocorrelation function? It is the magnitude of the input spectrum squared. Thus, there is really no new information in the autocorrelation function; the same information existed in the spectrum of the signal. But, as always, a change in perspective between these two domains often clarifies problems. In general, impulsive-type signals like pulse trains, bearing ping, or gear chatter show up better in correlation measurements. Signals containing several sine waves of different frequencies, like structural vibrations and rotating machinery, are clearer in the frequency domain.

4.25 CROSS CORRELATION

Cross correlation is a measure of the similarity between two signals as a function of the time shift between them. If the same signal is present in both waveforms, it is reinforced in the cross correlation function, while any uncorrelated noise is reduced. In many network analysis problems, the stimulus can be cross correlated with the response to reduce the effects of noise.

Cross correlation is defined as

$$R_{xy}(\tau) = \lim_{T\to\infty} \frac{1}{T} \int_T x(t)y(t + \tau)\, dt$$

As with autocorrelation, an FFT analyzer computes this quantity indirectly. In this case it is computed from the cross power spectrum.

$$R_{xy}(\tau) = F^{-1}[G_{xy}]$$

One application for cross correlation is the determination of time delays between signals. These signals can be impulsive (radar or sonar application, for example) or broadband random noise such as those encountered in system stimulus response measurements (transmission path delays, room acoustics, airborne noise analysis, and noise source identification).

4.26 HISTOGRAM

A histogram (Figure 4-36) shows how a signal's amplitude is distributed between its minimum and maximum values. A histogram displays number of samples versus

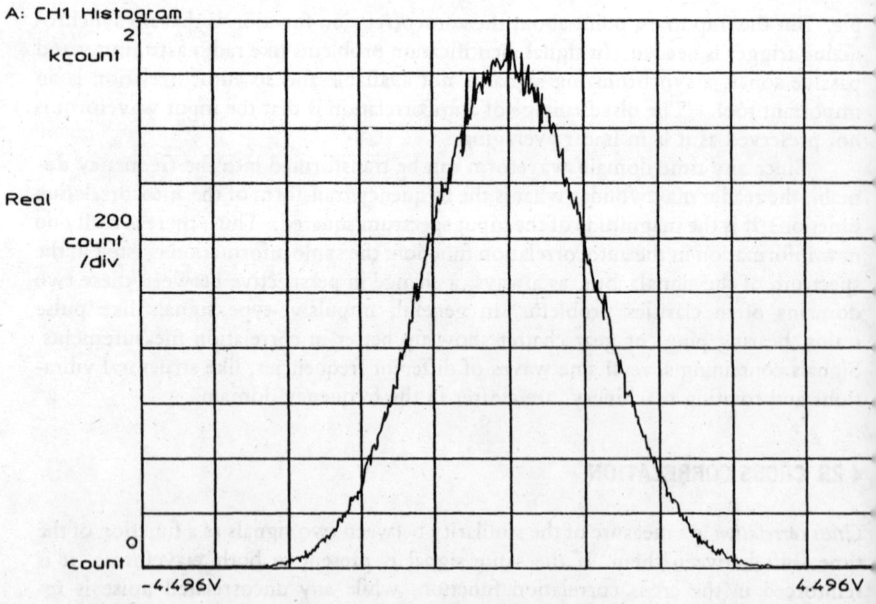

Figure 4-36. Histogram of random noise.

amplitude. This measurement is useful for determinng the statistical properties of noise and monitoring the performance of electromechanical positioning systems. Other measurement data derived from a histogram measurement are probability density function and cumulative density function.

The probability density function, or PDF (Figure 4-37) is the histogram data normalized to unit area. It is a statistical measurement of the probability that a specific level occurred. PDF is calculated using the following formula: histogram/ (number of records * blocksize * ΔV) (where blocksize is the number of points per record). The probability of an input signal falling between two points is equal to the integral of the curve between those points. For more information see Chapter 8, section 8.1.

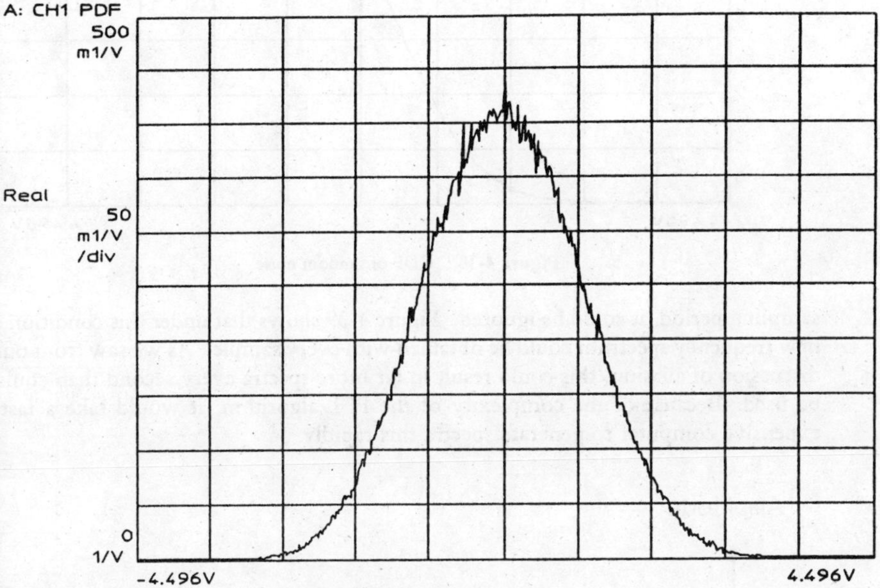

Figure 4-37. PDF of random noise.

The cumulative density function, or CDF (Figure 4-38) is a measure of the probability that a level equal to (or less than) a specific level occurred. It is calculated by integrating the PDF.

4.27 REAL-TIME BANDWIDTH

Until now we have ignored the fact that it takes a finite time to compute the FFT of a time record. In fact. if the transform could be computed in less time than our

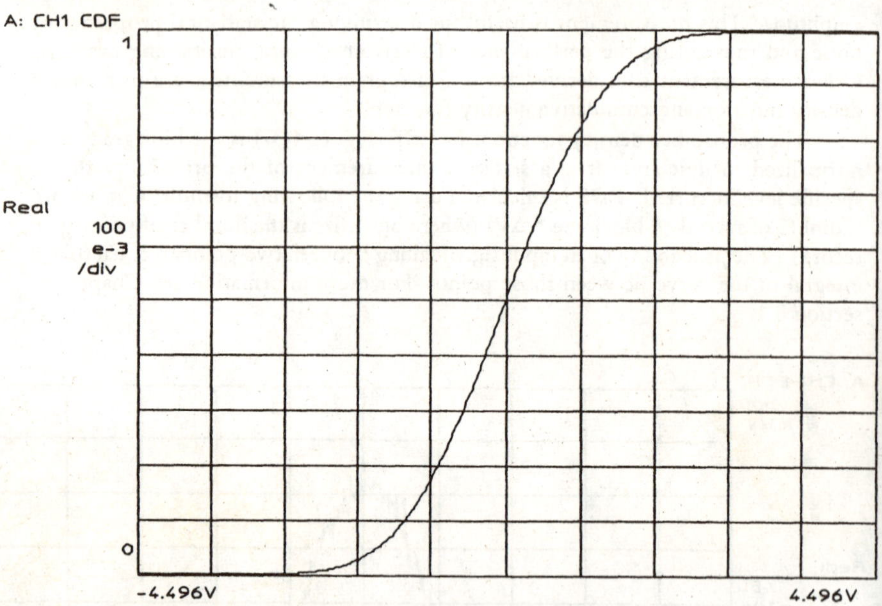

Figure 4-38. CDF of random noise.

sampling period, it could be ignored. Figure 4-39 shows that under this condition a new frequency spectrum could be obtained with every sample. As we saw from our discussion of aliasing, this could result in far more spectra every second than could be used. Because of the complexity of the FFT algorithm, it would take a fast, expensive computer to generate spectra this rapidly.

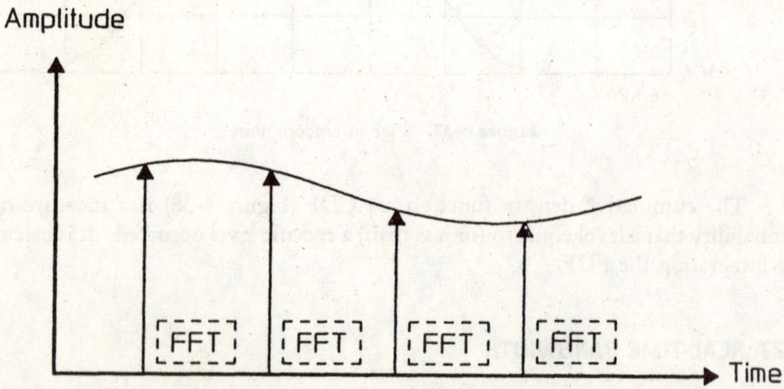

Figure 4-39. A new transform every sample.

A reasonable alternative is to add a time-record buffer (Figure 4-40) before the FFT. Figure 4-41 shows how this arrangement allows the analyzer to compute the frequency spectrum of the previous time record while gathering the current time record. If the transform can be computed before the time-record buffer fills, then the analyzer is said to be operating in real time.

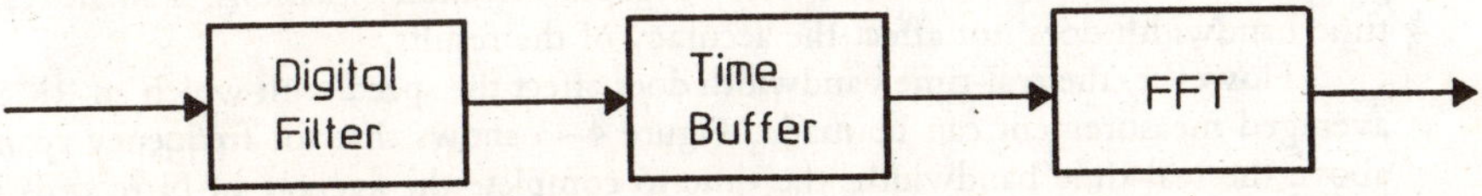

Figure 4-40. Time buffer added to block diagram.

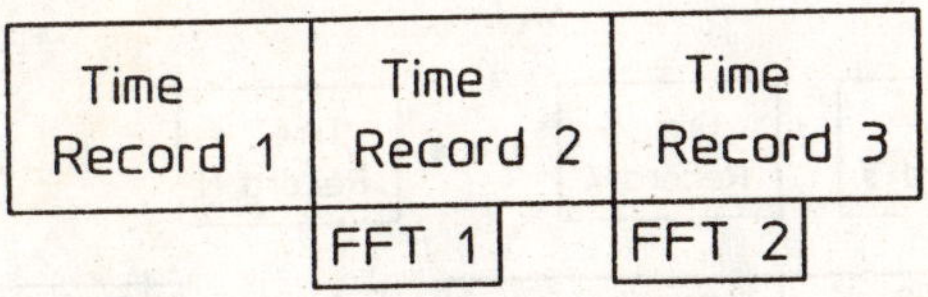

Figure 4-41. Real-time operation.

To see what this means, consider the case where the FFT computation takes longer than the time required to fill the time record. This is shown in Figure 4-42. Although the buffer is full, the last transform has not been completed, so data collection must stop. When the transform is finished, the time record can be transferred to the FFT and collection of another time record begun. Because some input data was missed, the analyzer is said to be no longer operating in real time.

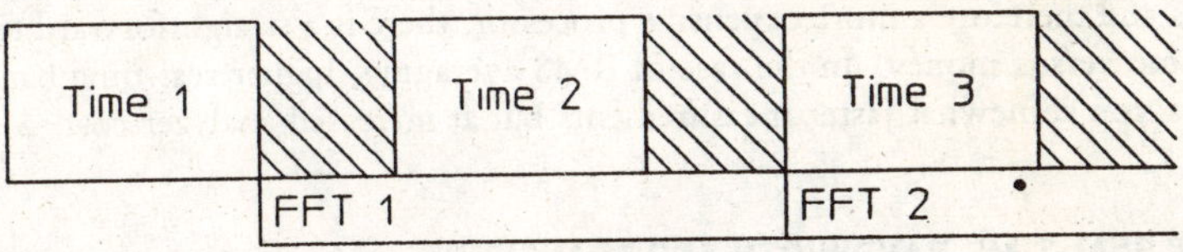

Figure 4-42. Non-real-time operation.

Recall that the time record is not constant but deliberately varied to change the frequency span of the analyzer. For wide frequency spans, the time record is shorter. Therefore, as the frequency span of the analyzer is increased, a span is reached where the time record is equal to the FFT computation time. This frequency span is called the real-time bandwidth, or RTBW. For frequency spans at or below the real-time bandwidth, the analyzer does not miss any data.

4.28 REAL-TIME BANDWIDTH AND RMS AVERAGING

There are situations when a measurement requires RMS averaging. It is important to know how the real-time specifications of the FFT analyzer affect the measurement. We might be interested in the spectral distribution of noise, or in reducing the variation of a signal contaminated by noise. There is no requirement in averaging that records must be consecutive with no gaps.[10] In these situations, a small real-time bandwidth does not affect the accuracy of the results.

However, the real-time bandwidth does affect the speed with which an RMS-averaged measurement can be made. Figure 4-43 shows that for frequency spans above the real-time bandwidth, the time to complete the average of N records is dependent only on the time to compute the N transforms. Rather than continually reducing the time to compute the RMS average as we increase our span, we reach a fixed time to compute N averages.

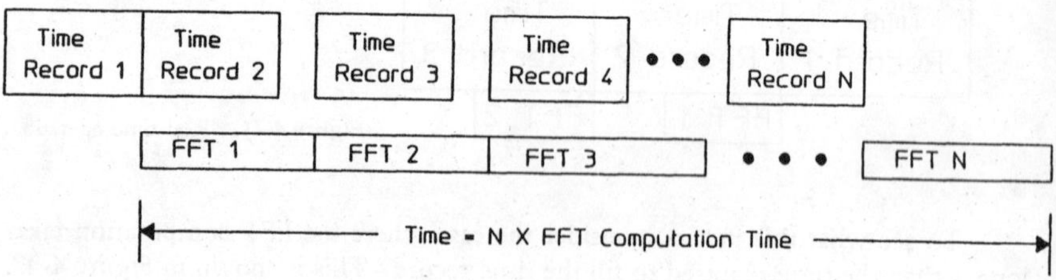

Figure 4-43. RMS averaging time.

Therefore, a small real-time bandwidth is only a problem when RMS averaging large spans using a large number of averages. Under these conditions it takes longer to get the answer. Since wider real-time bandwidths require faster computations and therefore a more expensive processor, there is a straightforward trade-off of time versus money. In the case of RMS averaging, higher real-time bandwidth gives you somewhat faster measurements but at increased analyzer cost.

4.29 REAL-TIME BANDWIDTH AND TRANSIENTS

Real-time bandwidth is an important consideration when analyzing transient events. If the entire transient fits within the time record (Figure 4-44), the FFT computation time is of little interest. The analyzer can trigger on the transient and store the event in the time-record buffer. The time to compute its spectrum is not important.

---

[10] This is because averaging is useful only if the signal is periodic and the noise stationary.

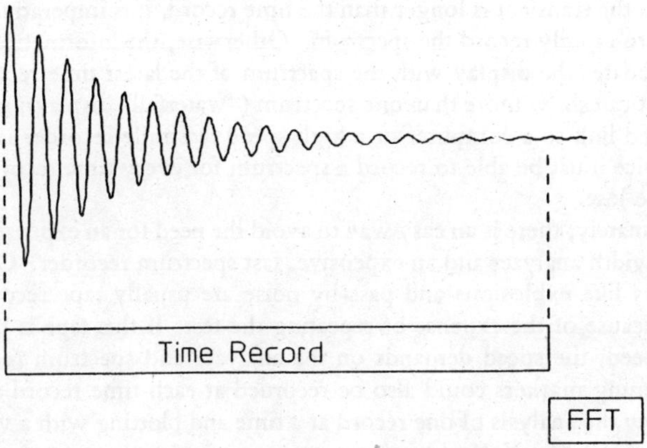

Figure 4-44. Transient fits in time record.

However, if a transient event contains high-frequency energy and lasts longer than the time record necessary to measure the high-frequency energy, the processing speed of the analyzer is critical. As shown in Figure 4-45, not all of the transient will be analyzed if the computation time exceeds the time-record length.

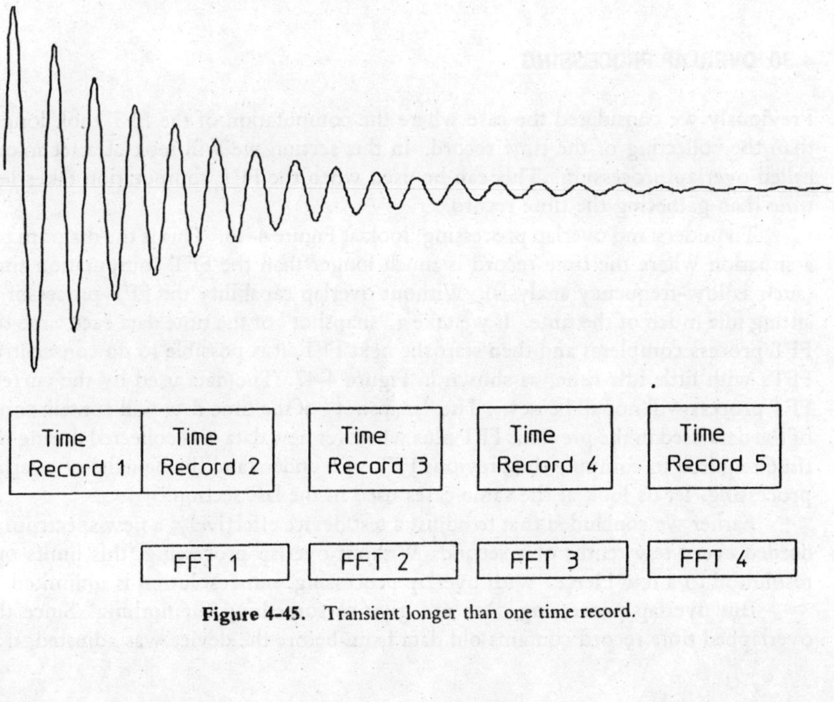

Figure 4-45. Transient longer than one time record.

When the transient is longer than the time record, it is imperative that there be some way to rapidly record the spectrum. Otherwise, the information is lost as the analyzer updates the display with the spectrum of the latest time record. A special display that can show more than one spectrum ("waterfall" display), mass memory, a high-speed link to a computer or a high-speed facsimile recorder is needed. The output device must be able to record a spectrum for every time record or information will be lost.

Fortunately, there is an easy way to avoid the need for an expensive, wide real-time bandwidth analyzer and an expensive, fast spectrum recorder. One-time transient events like explosions and pass-by noise are usually tape recorded for later analysis because of the expense of repeating the test. If this tape is played back at reduced speed, the speed demands on the analyzer and spectrum recorder are reduced. Timing markers could also be recorded at each time record interval. This would allow the analysis of one record at a time and plotting with a very slow (and commonly available) X-Y plotter.

It is clear that there is no single answer as to what real-time bandwidth is necessary in an FFT analyzer. Except in analyzing long transient events, the added expense of a wide real-time bandwidth gives little advantage. It is possible to analyze long transient events with a narrow real-time bandwidth analyzer, but it does require the recording of the input signal. This method is slow and requires operator care, but it removes the need to purchase an expensive analyzer and fast spectrum recorder.

4.30 OVERLAP PROCESSING

Previously we considered the case where the computation of the FFT took longer than the collecting of the time record. In this section we will look at a technique called overlap processing. This can be used when the FFT computation takes less time than gathering the time record.

To understand overlap processing, look at Figure 4-46. This is the diagram for a situation where the time record is much longer than the FFT computation time (such as low-frequency analysis). Without overlap capability the FFT processor is sitting idle much of the time. If we take a "snapshot" of the time data each time the FFT process completes and then start the next FFT, it is possible to do consecutive FFTs with little idle time, as shown in Figure 4-47. The data used by the current FFT process will not all be new. The "snapshot" of the time data will contain some of the data used in the previous FFT plus whatever new data was collected during the time required to compute the previous FFT. To understand the benefits of overlap processing, let us look at the same cases used in the last section.

Earlier we concluded that to adjust a test device effectively, a new spectrum is needed every few tenths of a second. Without overlap processing, this limits our resolution to a few Hertz. With overlap processing, our resolution is unlimited.

But overlap processing does not give us something for nothing. Since the overlapped time record contains old data from before the device was adjusted, it is

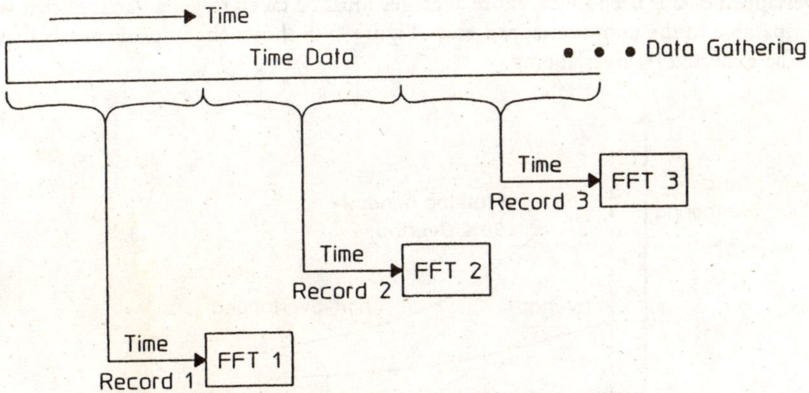

Figure 4-46. Non-overlapped processing is performed only on completely new data (time records).

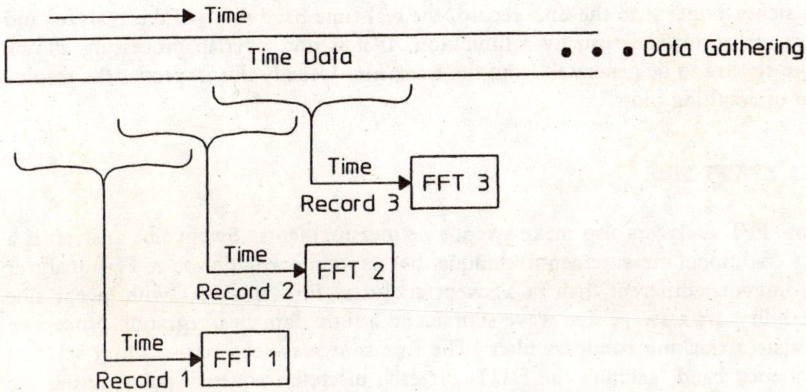

Figure 4-47. Overlapped processing is performed on data that combines old and new. The time between FFTs represents display processing.

not completely correct. It does indicate the direction and the amount of change, but we must wait one full time record after the change for the new spectrum to be accurately displayed. Nonetheless, by indicating the direction and magnitude of the changes every few tenths of a second, overlap processing greatly increases the speed with which these devices can be adjusted.

Overlap processing can dramatically reduce the time needed to compute RMS averages with a given variance. Recall that window functions reduce the effects of leakage by weighting the ends of the time record to zero. Overlapping eliminates most (if not all) of the time that would be wasted taking this data. Since some overlapped data is used twice, more averages must be taken to get a variance that is comparable to the non-overlapped case. Figure 4-48 shows the improvements that can be expected by overlapping.

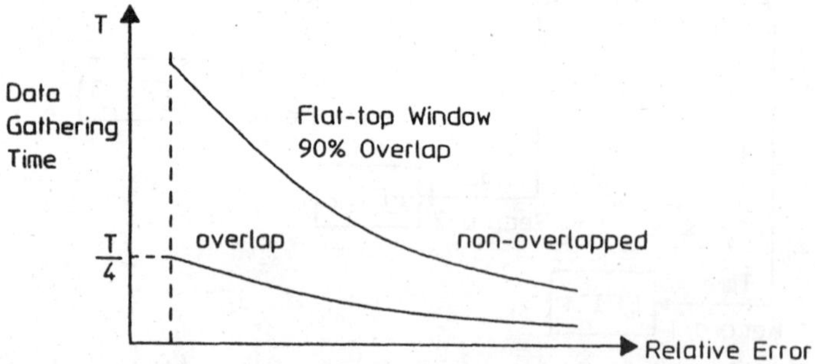

Figure 4-48. RMS averaging is faster with overlap processing.

For transients shorter than the time record, overlap processing is useless. For transients longer than the time record, the real-time bandwidth of the analyzer and spectrum recorder is usually a limitation. If it is not, overlap processing allows more spectra to be generated from the transient. Usually this improves the resolution of resulting plots.

4.31 SWEPT SINE

Some FFT analyzers also make swept sine measurements. Swept sine analysis is a very traditional measurement technique, but its implementation in an FFT analyzer is somewhat different than in a swept analyzer. FFT analyzers with swept sine capability use a swept sine wave source and a time domain integration process to emulate a tracking bandpass filter. The sine source sweeps across a user-selected frequency band, exciting the DUT. At each discrete frequency point during the sweep, the analyzer measures and displays the relative magnitudes and phases of the DUT's sinusoidal response.

Swept sine measurements provide a very good signal-to-noise ratio and can characterize nonlinear systems. At each point in the sweep, the DUT exhibits a transient response and a steady-state response. The analyzer waits for the transient response to settle out, then measures the steady-state response. The swept sine measurement allows the analyzer to characterize nonlinearities and better excite the DUT because the energy is concentrated in a narrow frequency band. To reduce the effect of noise, the analyzer integrates the input signal over several cycles.

Most FFT analyzers that provide swept sine measurements include some automatic adjustments to optimize the results. These include automatic source level adjust, automatic input range adjust, and automatic resolution adjust.

For nonlinear devices, the transfer function varies depending on the input level. With autoleveling, the analyzer adjusts the signal source level to keep the DUT output level within a specified range.

With autoranging, the analyzer adjusts the input range up or down when the DUT output level goes above or below the optimum for the current range. This can greatly increase the dynamic range.

With autoresolution, the analyzer adjusts the spacing between adjacent measurement points, taking finer or coarser steps where necessary. Coarser steps minimize measurement time, but autoresolution can narrow the steps where there are rapid changes in the response. This minimizes the sweep time while still catching fast changes in amplitude or phase.

4.32 OCTAVE MEASUREMENTS

Some FFT analyzers also make octave measurements. An octave measurement computes power in bands using banks of filters covering several octaves. Each higher filter has a wider bandwidth than the previous filter. The filter spacing is logarithmic rather than linear. The most common spacing is 1/3 octave, but some analyzers also provide full octave and 1/12 octave spacing. Octave measurements are provided because many acoustic noise regulations are specified in terms of third-octave measurements.

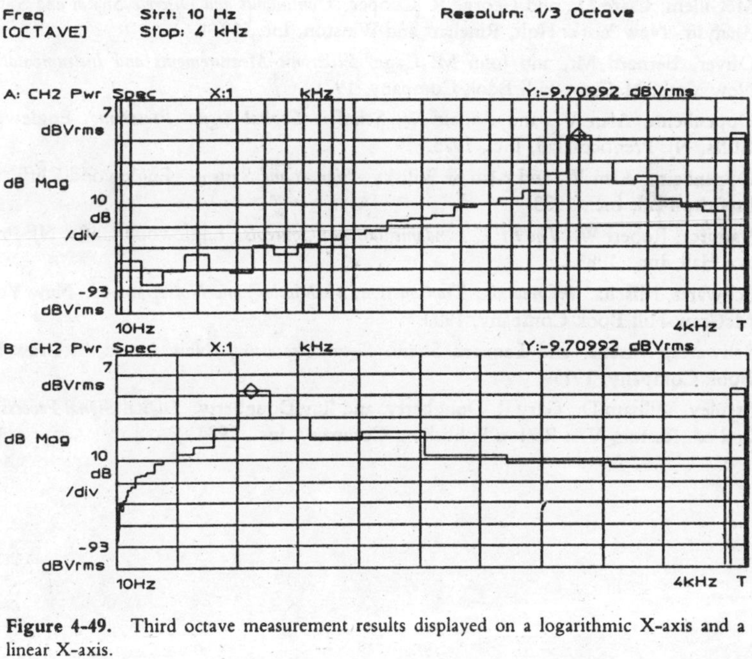

Figure 4-49. Third octave measurement results displayed on a logarithmic X-axis and a linear X-axis.

Octave measurements are displayed on a logarithmic X-axis, so each band appears to be the same width. In Figure 4-49, the top trace shows a third octave with a log X-axis. The bottom trace shows the same data on a linear X-axis. Notice that the higher frequency bands are much wider than the lower frequency bands.

FFT analyzers that make octave measurements usually include an A-weight filter to simulate the frequency response of the human ear. Third-octave measurements show how the human ear perceives the frequency content of a signal, but the frequency resolution does not reveal the exact spectral component of the signal. To diagnose the specific cause of a noise problem, the analyzer's FFT measurement is more useful.

REFERENCES

1. Brigham, E. Oran. *The Fast Fourier Transform and Its Applications.* Englewood Cliffs, NJ: Prentice Hall, Inc., 1988.
2. Hewlett-Packard Company. "Understanding the HP3582A Spectrum Analyzer," Application Note, Publication Number 5952-8773, Palo Alto, CA, 1978.
3. Hewlett-Packard Company. "Fundamentals of Signal Analysis," Application Note 243, Publication Number 5952-8898, Palo Alto, CA, 1981.
4. McGillem, Clare D., and George R. Cooper. *Continuous and Discrete Signal and System Analysis.* New York: Holt, Rinehart and Winston, Inc., 1974.
5. Oliver, Bernard M., and John M. Cage. *Electronic Measurements and Instrumentation.* New York: McGraw-Hill Book Company, 1971.
6. Oppenheim, Alan V., and Ronald W. Schafer. *Digital Signal Processing.* Englewood Cliffs, NJ: Prentice-Hall, Inc., 1975.
7. Oppenheim, Alan V., and Alan S. Willsky. *Signals and Systems.* Englewood Cliffs, NJ: Prentice-Hall, Inc., 1983.
8. Ramirez, Robert W. *The FFT, Fundamentals and Concepts.* Englewood Cliffs, NJ: Prentice Hall, Inc., 1985.
9. Schwartz, Mischa. *Information, Transmission, Modulation, and Noise,* 3rd ed. New York: McGraw-Hill Book Company, 1980.
10. Schwartz, Mischa, and Leonard Shaw. *Signal Processing.* New York: McGraw-Hill Book Company, 1975.
11. Stanley, William D., Gary R. Dougherty, and Ray Dougherty. *Digital Signal Processing,* 2nd ed. Reston, VA: Reston Publishing Company, Inc., 1984.

5

Swept Spectrum Analyzers

he traditional method for implementing a spectrum analyzer is the swept hetero-yne block diagram. Similar to a radio receiver, the spectrum analyzer is automati-lly tuned (swept) over the band of interest. This type of analyzer has been gradu-ly replaced by the FFT analyzer at low frequencies, but the swept analyzer remains e dominant technology in the radio frequency range and above.

1 THE WAVE ANALYZER

he bank-of-filters analyzer, which was examined in the previous chapter, uses a rge number of fixed filters to implement a spectrum analyzer. Another approach to use one filter, but to make it tunable over the frequency range of interest (Figure 1). Since this technique allows only one frequency to be measured at a time, it is ot a true spectrum analyzer, but is called a *wave analyzer* or *wave meter*.

The user tunes the wave analyzer to the frequency of interest and reads the gnal level present at that frequency. The bandwidth of the tunable filter deter-ines the resolution bandwidth, BW_{RES}, of the wave analyzer. It is desirable for the ter to be as flattopped as possible, with steep skirts so that equal amplitude signals ithin the passband of the filter produce the same meter reading.

This type of instrument has been used extensively for making simple "tuned ltmeter" measurements and still exists today in the form of a *selective level meter*. lective level meters have very flattopped passbands, resulting in excellent ampli-de accuracy.

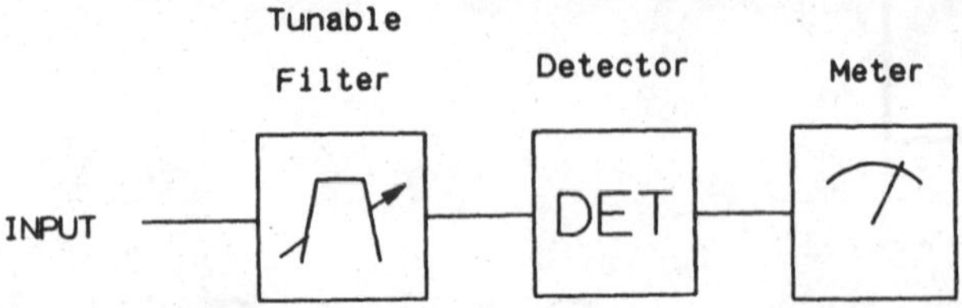

Figure 5-1. A conceptual block diagram of the wave analyzer.

5.2 HETERODYNE BLOCK DIAGRAM

Practical tunable bandpass filters have severe restrictions on the tuning range of filter's center frequency. Thus, the wave analyzer is rarely implemented using actual tunable filter. Instead of moving the filter in frequency, the input signa translated in frequency and the filter's frequency remains fixed. The bandpass ter's center frequency is called the *intermediate frequency (IF)* and the filter is called *IF filter*.

The simplified block diagram of a practical wave analyzer is shown in Fig 5-2. The key component of this block diagram is the mixer. The mixer is a th port device which is driven by the input signal of the analyzer (usually called the *signal*) and the *local oscillator (LO)* signal. The output of the mixer is at the inter diate frequency.

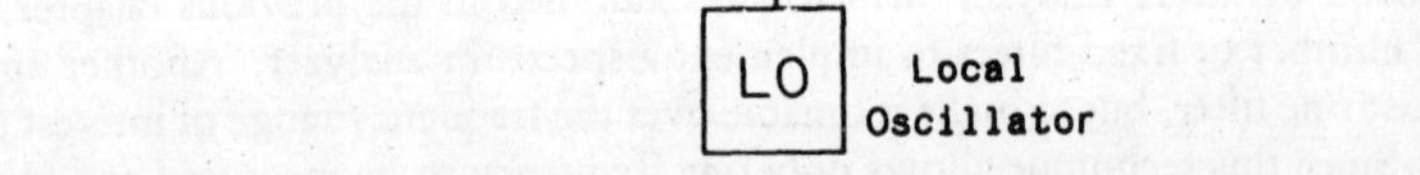

Figure 5-2. A more practical block diagram of the wave analyzer.

Ideally, the mixer functions as a multiplier. Suppose the input is a cosine

$$v_{RF}(t) = A\cos(2\pi f_{RF}t)$$

and

$$v_{LO}(t) = \cos(2\pi f_{LO}t)$$

The output of the mixer is

$$v_{IF}(t) = A\cos(2\pi f_{RF}t)\cos(2\pi f_{LO}t) \tag{5-3}$$

$$v_{IF}(t) = \frac{A}{2}[\cos(2\pi f_{LO}t + 2\pi f_{RF}t) + \cos(2\pi f_{LO}t - 2\pi f_{RF}t)] \tag{5-4}$$

Therefore, the mixer's output is the sum and difference frequencies of the local oscillator and RF signals.[1]

This characteristic is used to implement the superheterodyne block diagram. The IF filter always remains tuned to the same center frequency, and the mixer is used to shift the input signal in frequency so that it falls on the center of the IF filter. This makes the IF filter easier to build since it does not require a tunable center frequency. Of course, the LO must be made tunable, but this is usually an easier task than building a filter which tunes over a wide range.

The mixer produces both the sum and difference frequencies of the input and LO. Only the sum or the difference frequency is used since, by design, it will fall directly on the IF. The other frequency will be rejected by the IF filter. This requires some careful choices in defining the LO and IF frequencies.

A numerical example should help explain the operation of the superheterodyne block diagram. Suppose a wave analyzer is required to measure signals from 0 to 10 MHz. The IF frequency chosen is 20 MHz and the local oscillator operates between 20 MHz and 30 MHz. Now, suppose that the input frequency happens to be 5 MHz. To measure this frequency, the LO is tuned to 25 MHz, producing the sum and difference frequencies of 20 MHz and 30 MHz. The 20 MHz signal is exactly the IF frequency (by design) and is passed through the IF filter and detected and displayed on the meter. The 30 MHz signal falls outside of the IF filter and is rejected.

If the input frequency were changed to 1 MHz, the LO would have to be tuned to 21 MHz, producing sum and difference frequencies of 20 MHz and 22 MHz. Again, the 20 MHz signal is the IF frequency and is measured while the 22 MHz signal falls outside the IF filter and is ignored.

The low-pass filter at the input of the block diagram is known as the *image filter*. If this filter were not included, undesirable frequencies could enter the mixer and be translated down to the IF frequency, corrupting the measurement. Suppose the wave analyzer is still tuned to 5 MHz. If a 45 MHz signal made its way into the mixer, it would mix with the LO frequency (25 MHz) and produce sum and difference frequencies of 20 MHz and 70 MHz. The 70 MHz signal would be ignored, but the 20 MHz signal would fall directly on the IF filter and would be included in the measurement. Without an image filter the wave analyzer could not distinguish between the desired 5 MHz signal and the 45 MHz image frequency. (As a side note, the image filter serves exactly the same function as the anti-alias filter used in FFT analyzers, discussed in Chapter 4.)

The image frequency (for this block diagram) causes the difference frequency

---

[1] Practical mixers will usually have other higher-order products which are ignored here.

(when mixed with the LO) to fall on the IF frequency.

$$f_{IF} = f_{IMAGE} - f_{LO} \tag{5-5}$$

$$f_{IMAGE} = f_{IF} + f_{LO} \tag{5-6}$$

The LO frequency is the input frequency plus the IF frequency or

$$f_{LO} = f_{RF} + f_{IF} \tag{5-7}$$

Thus,

$$f_{IMAGE} = f_{RF} + 2\,f_{IF} \tag{5-8}$$

The image frequency is twice the IF frequency away from the desired input frequency. This holds for the case shown where the IF frequency is higher than the input frequency.

5.3 THE SWEPT SPECTRUM ANALYZER

The wave analyzer can measure only one frequency at a time. An obvious enhancement is to have the analyzer automatically sweep through the frequency range of interest. In a spectrum analyzer this is accomplished by sweeping the local oscillator. Figure 5-3 shows how the wave analyzer block diagram could be converted

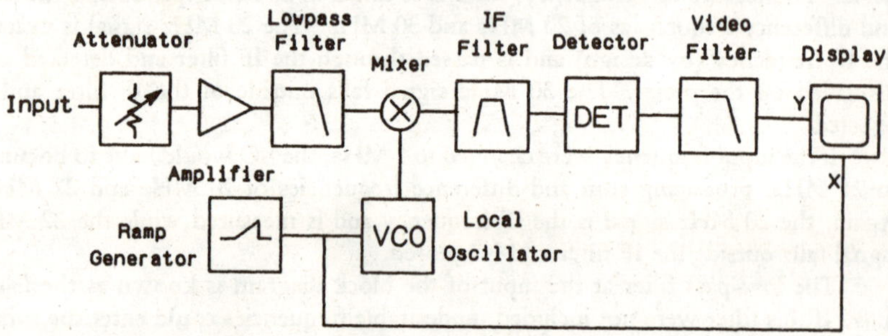

Figure 5-3. A simplified block diagram of a swept spectrum analyzer.

into a spectrum analyzer by using a voltage-controlled oscillator (VCO) as the local oscillator. A ramp generator is used to produce a linearly increasing voltage, which drives the tuning port of the VCO. The same ramp voltage is applied to the horizontal (X) axis of the display, while the detector output is low-pass filtered and connected to the vertical (Y) axis. As the LO is swept in frequency, the spectrum of the input signal is automatically plotted on the display. The low-pass filter at the output of the detector is called the *video filter* and is a postdetection filter (as discussed in Chapter 10) which serves to smooth out the response as the analyzer sweeps.

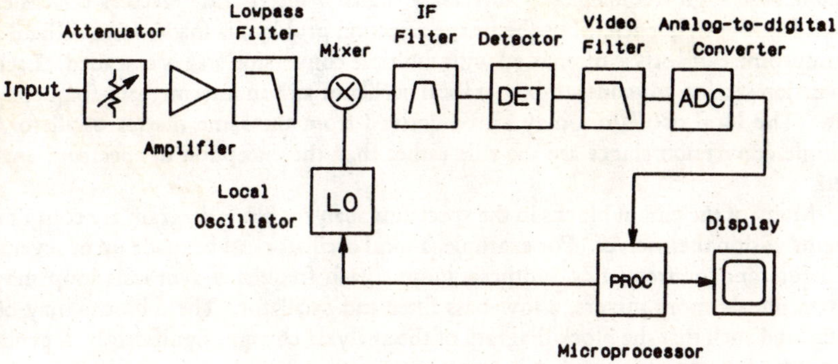

Figure 5-4. A simplified block diagram of a spectrum analyzer using microprocessor control.

As shown, the block diagram is implemented in a totally analog fashion. Although this is a practical technique, the advent of the microprocessor and the digital display has caused the block diagram to take on a digital flavor (Figure 5-4). For example, the LO is often implemented using digital synthesis techniques which lend themselves to microprocessor control. (The LO may be stepped or swept in frequency under microprocessor control.) The output of the IF filter (or the detector) may be sampled and converted to digits by an analog-to-digital converter, which is read by the microprocessor. The display in a modern spectrum analyzer is almost always a digital graphics display. That is, the display graphics information is written to a designated area of memory and the display is refreshed from this memory. This eliminates any problems with display refresh rate being too slow due to a slow sweep rate, since the display can be refreshed much faster than the sweep rate.

Although the analog block diagram of Figure 5-3 has been largely replaced by the digital equivalent, it still represents a good conceptual basis for understanding the operation of the spectrum analyzer.

5.4 PRACTICAL CONSIDERATIONS

The block diagram previously discussed uses a single mixer/IF stage and is therefore called a *single conversion receiver*. This simple block diagram can be used to implement a spectrum analyzer, but its performance is limited. Modern analyzers use much more complicated block diagrams to achieve state-of-the-art performance.

Some factors in the block diagram design call for a high-IF frequency, while others require a low-IF frequency. A high-IF frequency makes the rejection of image frequencies easier, but narrow-IF filters and detectors are more difficult to implement at high frequencies. Conversely, narrow filters and detectors are easier to build at low frequencies, but the image rejection problem is made more difficult. A compromise of sorts is often used with multiple conversion stages cascaded. Each conversion section contains a mixer, a local oscillator and an intermediate-frequency filter. (The local oscillators may all be derived from the same master oscillator.) Multiple conversion stages are the rule rather than the exception in spectrum analyzers.

Many of the circuit blocks in the spectrum analyzer block diagram are complex systems within themselves. For example, a local oscillator can be made up of several oscillators and/or frequency synthesis loops. Each frequency synthesis loop may contain one or more mixers, a low-pass filter and oscillator. These blocks may be configured such that the block diagram of the analyzer changes significantly depending on the frequency range that is being measured.

Regardless of the complexity of the actual spectrum analyzer block diagram, conceptually it simply implements a sweeping tuned filter.

5.5 INPUT SECTION

The input to the spectrum analyzer block diagram has a variable attenuator, often followed by an amplifier. The purpose of this input section is to control the signal level applied to the rest of the instrument. If the signal level is too large, the analyzer circuits will distort the signal, causing distortion products to appear along with the desired signal. If the signal level is too small, the signal may be masked by noise present in the analyzer. Either problem tends to reduce the dynamic range of the measurement.

Some instruments provide an autorange feature which automatically selects an appropriate input attenuation. Other instruments require the user to select the appropriate input attenuation. The input circuitry of a typical analyzer is very sensitive and will not withstand much abuse. Careful attention should be paid to the allowable signal level at the input, particularly for microwave analyzers. Some instruments tolerate DC voltages at their inputs, but others require that no DC be applied, or be restricted to small values.

5.6 RESOLUTION BANDWIDTH

The bandwidth of the last IF filter usually determines the resolution bandwidth, BW_{RES}, of the instrument. If multiple IF filters are used, the composite response of the IF chain determines the resolution bandwidth. Usually, one of the IF filters will be significantly narrower than the others and alone will determine the resolution bandwidth.

Multiple resolution bandwidths are supplied by simply switching in different filters. Wider-bandwidth filters settle faster, providing faster measurements. Narow-bandwidth filters take longer to settle but produce better frequency resolution nd better signal-to-noise ratio (see Chapter 10).

.7 SWEEP LIMITATIONS

'he swept spectrum analyzer generally provides a significant increase in measurenent speed over the wave analyzer since the entire frequency range of interest can be isplayed at once. This is not meant to imply that the spectrum analyzer can be wept arbitrarily fast. The IF filter (resolution bandwidth filter) must have time to espond to the changing signal level that it experiences at its input.

Consider the case where the spectrum analyzer sweeps past a sinusoidal signal Figure 5-5). In this case, we will analyze the situation by considering the IF filter to e fixed and the signal to be moving. The signal starts well outside of the passband f the filter (Figure 5-5a). Then, the signal starts up the skirt of the filter with the lter's output level increasing accordingly (Figure 5-5b). Finally, the signal enters he passband of the filter and starts down the other side (Figure 5-5c).

If the signal is swept slowly enough, the shape of the IF filter is traced out on he spectrum analyzer display. (Normally, the IF filter bandwidth is small comared to the frequency span being swept, so the IF filter shape will appear as a pectral line on the display.) If the signal is swept too fast, the filter does not have ime to respond and two types of display errors occur (Figure 5-6). The amplitude f the spectral line is smaller than the slowly swept case and the spectral line will shift o the right slightly, causing a frequency error. Additionally, there may be filter ringing" down the back edge of the filter shape.

How fast is too fast of a sweep rate? Ideally, the filter should be swept infiitely slow since the filter response time will always degrade the measurement. In ractice, the filter can be swept at some finite rate as long as some small error can be olerated. If this error is small compared to other errors in the analyzer, then there is o penalty for sweeping. A typical error limit due to sweep induced errors is 0.1 B. The maximum sweep rate (with such an error limit) is proportional to the quare of the resolution bandwidth.

$$\text{sweep rate (max)} = BW_{\text{RES}}^2/k \tag{5-9}$$

vhere k is a factor depending on the resolution bandwidth filter characteristics. A ypical value for k is 2 (for Gaussian filters), and the sweep rate has units of Hz/sec. f a steep-walled filter is used, the response time of the filter increases, causing k to e larger.

In a wave analyzer, the shape of the IF filter is usually designed to be steep valled and as flattopped as possible. However, this is inappropriate for an analyzer ntended to sweep, due to the increased sweep time required. In swept analyzers a nore rounded filter such as a Gaussian filter is used to minimize the sweep time Figure 5-7).

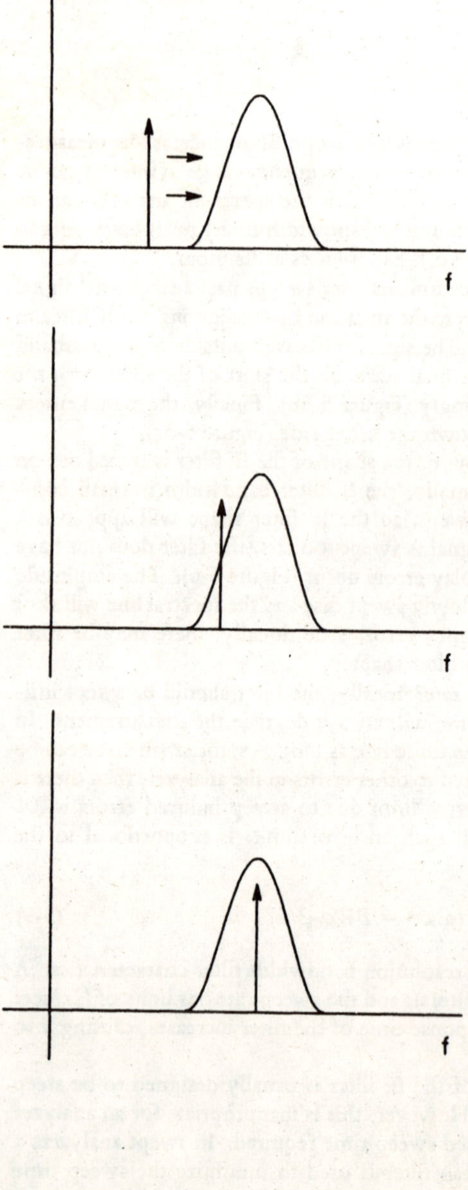

Figure 5-5. A sine wave passes through the IF filter of a spectrum analyzer.

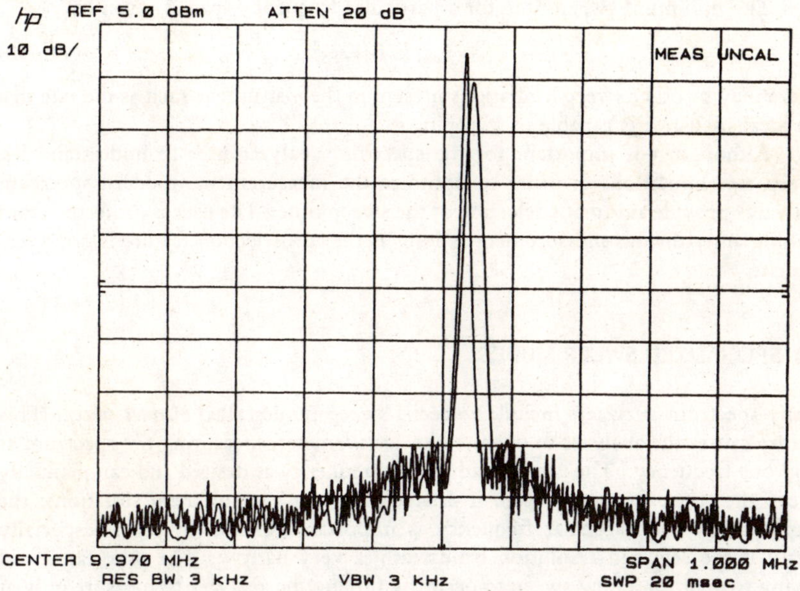

Figure 5-6. These two measurements show the effects of sweeping too fast. The leftmost spectral line was swept correctly. Sweeping too fast causes the spectral line to be smaller in amplitude and shifted to the right.

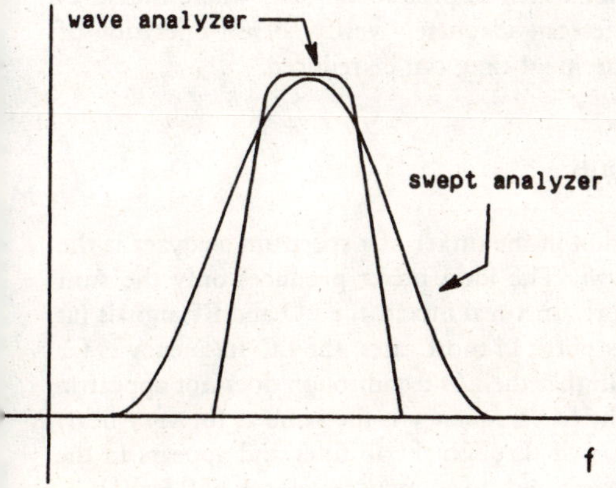

Figure 5-7. The IF (resolution bandwidth) response of a spectrum analyzer is more rounded and less selective than a wave analyzer.

The minimum sweep time for a particular frequency span is given by

$$T_s = f_{span}\, k/BW_{RES}^2 \qquad (5\text{-}10)$$

There may be other sweep limitations present in the instrument such as the rate that the local oscillator is capable of sweeping.

Although it is important for the spectrum analyzer user to understand the sweep rate limitations in order to optimize the measurement, modern spectrum analyzers provide automatic selection of the sweep time. The user is protected from making an erroneous measurement as long as the autoselection feature is not overridden.

5.8 SPECIALIZED SWEEP MODES

Many spectrum analyzers include a special sweep mode called *manual sweep*. This feature causes the analyzer to operate like a wavemeter, measuring the spectrum at only one frequency. The user can adjust the frequency as desired and can manually sweep the entire frequency span, if desired. In some measurement situations, the amplitude at one particular frequency is important. Manual sweep is especially useful if the required resolution bandwidth is very narrow. The user can avoid having to wait for a long sweep to occur by forcing the analyzer to measure only at the frequency of interest.

Some spectrum analyzers provide another sweep mode, called *discrete sweep* or *program sweep*, which lets the user specify a list of frequencies to test. The analyzer automatically hops from frequency to frequency, measuring the spectral content at each one. For measurement applications such as production test, where measurements at a small number of frequencies can adequately verify correct operation of the device under test, the total measurement time can be reduced.

5.9 LOCAL OSCILLATOR FEEDTHROUGH

One particularly noticeable imperfection in the mixer of a spectrum analyzer is the phenomenon known as *LO feedthrough*. The ideal mixer produces only the sum and difference frequencies at the IF port. In a real mixer, the LO and RF signals (at reduced amplitude) also appear at this port. In most cases, the LO frequency is far enough away from the center of the IF that the LO feedthrough does not appear in the measurement. However, when the LO frequency is the same as (or very near) the IF frequency, the LO signal is passed through the IF filter and appears in the measurement. This LO frequency corresponds to an input frequency of 0 Hz (DC).

LO feedthrough is also known as the *DC response*, since that is where it appears on the spectrum analyzer display. In nonsynthesized spectrum analyzers (with limited frequency accuracy in the LO), this is used as a method of locating 0 Hz on the display.

If the IF filter were infinitely narrow, the LO feedthrough would appear only at exactly 0 Hz. With a finite-width IF filter, the LO feedthrough extends from 0 Hz to approximately $BW_{RES}/2$, following the shape of the IF filter (Figure 5-8). It may be necessary to reduce the resolution bandwidth in order to prevent the LO feedthrough from interfering with the measurement.

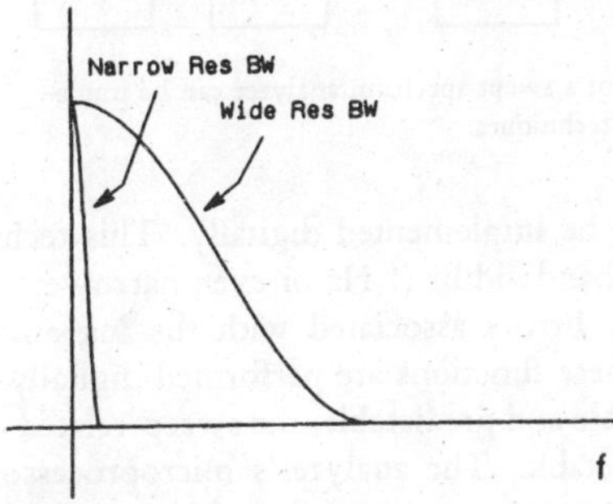

Figure 5-8. Local oscillator feedthrough appears as a response at DC whose width depends on the resolution bandwidth.

5.10 DETECTOR

The output of the IF filter is connected to the detector. The detector produces a DC level proportional to the AC level of the signal in the IF section. The detector could be of several different types, but a peak detector is commonly used.

The detector usually has a log amp in front of it, which compresses the signal level according to a logarithmic function. (For an input voltage amplitude ν, the output voltage amplitude is $\log(\nu)$.) This greatly reduces variation in signal level seen by the detector and simultaneously provides the user with a logarithmic vertical scale which is calibrated to read in decibels. The logarithmic scale is desirable in a spectrum analyzer due to the large variation in signal levels.

When the output of the detector is sampled and converted to digital form, it is important that the output be sampled often enough so that spectral components are not missed. Usually, the detector is sampled many times between displayed points, with the maximum value encountered shown on the display. Without this technique, a response might appear at the detector output between display samples and disappear as the analyzer sweeps past it.

5.11 THE DIGITAL IF SECTION

In recent instruments, the resolution bandwidth filters and the detector have been implemented using digital signal processing (Figure 5-9). The signal is digitized while it is still at the last IF frequency. A digital filter algorithm is then used to provide the resolution bandwidth function and the filtered signal is detected digi-

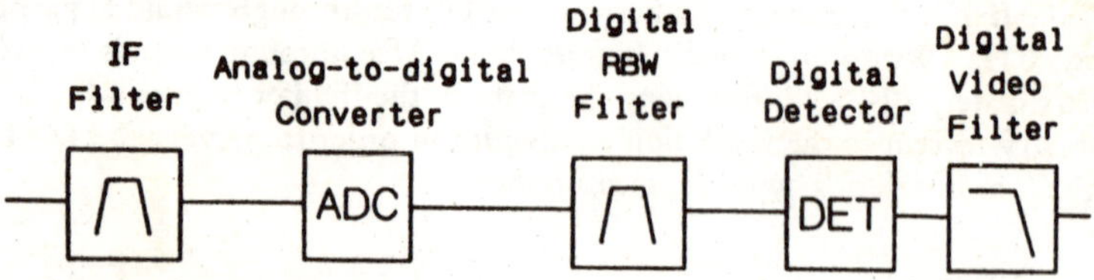

Figure 5-9. The IF/detector section of a swept spectrum analyzer can be implemented using digital signal processing techniques.

tally. If desired, a video filter can also be implemented digitally. This technique provides very stable narrow resolution bandwidths (1 Hz or even narrower), since digital filters do not exhibit any drift. Errors associated with the log amp and detector are eliminated, since both of these functions are performed digitally.

Since the filter response is very stable and predictable, the sweep-related errors discussed in Section 5.7 are also predictable. The analyzer's microprocessor can remove these errors providing a more accurate measurement. More importantly, the analyzer can be "overswept" to increase measurement speed.

5.12 THE TRACKING GENERATOR

A *tracking generator* is a very useful addition to the basic spectrum analyzer block diagram. A tracking generator, as the name implies, provides a sinusoidal output whose frequency is the same as the analyzer's input frequency. A tracking generator allows a spectrum analyzer to perform basic network measurements. The output of the tracking generator is connected to the input of the device under test and the response is measured with the analyzer's receiver. As the analyzer sweeps, the tracking generator is always operating at the receiver's frequency and the transfer characteristics of the device can be measured.

5.13 FFT VERSUS SWEPT

Besides the inherently simple block diagram in the FFT approach, the FFT analyzer provides a speed improvement over the swept analyzer. As previously discussed, the swept analyzer measurement speed is limited by its resolution bandwidth, with the measurement time being inversely proportional to BW_{RES}^2. At low frequencies, very narrow resolution bandwidths are required to separate closely spaced spectral lines. Narrow resolution bandwidths require a longer sweep time so the total measurement time can get unacceptably long. On the other hand, the FFT analyzer's speed is limited by the time it takes to acquire the data and the time it takes to compute the FFT. For equivalent-frequency resolution, the FFT analyzer is much faster than the swept analyzer

The FFT analyzer is limited in frequency range due to the need for a high-resolution analog-to-digital converter (ADC) to sample somewhat above the Nyquist rate. With current ADC technology, FFT analyzers cover up to several hundred kilohertz, while swept analyzers can measure frequencies up to hundreds of gigahertz.

As mentioned in Chapter 3, any practical measurement is limited to a finite time. For a signal that is changing, it may be desirable to measure the spectrum instantaneously, so that its frequency content at an instant in time can be determined. Unfortunately, this is often not possible. Swept analyzers, in particular, may take several seconds or even minutes to perform one swept measurement. During this time, the signal may change. Since the swept analyzer is measuring only one frequency at a time as it sweeps, it may miss portions of the signal.

An FFT analyzer acquires a time record which contains the entire spectral content of a signal for that particular slice of time. The FFT computation transforms this time domain data into its spectrum. As long as the FFT is performed at least as fast as new time domain data are acquired, the analyzer can continue to capture and display the spectral content of the signal without ever missing any portion of the signal. Thus, an FFT analyzer is more effective at measuring changing signals.

5.14 THE HYBRID ANALYZER

Some spectrum analyzers have combined swept analyzer technology with FFT digital signal processing in order to provide the benefits of both techniques. The analyzer uses the digital IF structure of Figure 5-9, with the rest of the analyzer block diagram unchanged. The IF signal of the analyzer is digitized by an ADC, and an FFT is performed by the microprocessor or dedicated digital logic. Using the excellent frequency resolution of the FFT and the expanded frequency range of a swept analyzer, this scheme allows very narrowband spectral analysis at very high frequencies.

High-frequency narrowband analysis is particularly useful in measuring close-in sidebands on carriers, such as very-low-frequency modulation sidebands and close-in phase noise.

REFERENCES

1. Clarke, Kenneth K., and Donald T. Hess. *Communication Circuits: Analysis and Design*. Reading, MA: Addison-Wesley Publishing Company, 1971.
2. Hayward, W. H. *Introduction to Radio Frequency Design*. Englewood Cliffs, NJ: Prentice Hall, Inc., 1982.

3. Hewlett-Packard Company. "Spectrum Analysis Basics," Application Note 150, Publication Number 5952-0292, Palo Alto, CA. November 1989.

4. Engelson, Morris. *Modern Spectrum Analyzer Theory and Applications*. Dedham, MA. Artech House, 1984.

5. Oliver, Bernard M., and John M. Cage. *Electronic Measurements and Instrumentation*. New York: McGraw-Hill Book Company, 1971.

6 Modulation Measurements

Ever since the early days of radio, modulation techniques have played an important part in electronic communications. A low-frequency voice or data signal is used to modulate some characteristic (usually the amplitude, phase, or frequency) of a carrier signal. This type of system represents an intentional use of modulation. In addition, unintentional modulation may be present, such as power line sidebands on an oscillator output or residual frequency modulation on an amplitude-modulated signal. Whether the modulation is intentional or not, a spectrum analyzer can be used to characterize and measure it.

6.1 THE CARRIER

Analog modulation techniques start with a carrier which is a pure sinusoid.

$$v(t) = A \cos(2\pi f_c t + \theta) \tag{6-1}$$

where

A = carrier amplitude (zero-to-peak)

f_c = carrier frequency (hertz)

θ = the carrier phase

The carrier may be modulated in a variety of ways, but the various techniques fall into two main categories: *amplitude modulation* and *angle modulation*. Amplitude modulation implies that the amplitude is no longer simply a constant but is a function of time. Similarly, angle modulation occurs when the angle of the cosine term is varied. Angle modulation may take the form of *frequency modulation* or of *phase modulation*, depending on the particular modulation technique.

All the modulation techniques result in increasing the occupied bandwidth of the carrier by spreading out sidebands in the frequency domain. Originally, the carrier is presumed to be a single spectral line, infinitely thin, occupying only one exact frequency. When modulated, the signal bandwidth increases depending on the type of modulation and the modulating signal. Sidebands appear beside the carrier, either in the form of discrete frequencies or, for nonperiodic modulation such as voice or music, more complex spectral shapes.

6.2 AMPLITUDE MODULATION

Amplitude modulation (AM) is generally considered the simplest modulation system. Although usually lumped under the general label of AM, there are several distinct variations.

An AM signal with carrier[1] is represented by the equation

$$v(t) = A_c[1 + am(t)] \cos(2\pi f_c t) \qquad (6\text{-}2)$$

where

A_c = constant which determines the overall signal amplitude

a = modulation index ($0 \leq a \leq 1$)

$m(t)$ = normalized modulating signal

f_c = carrier frequency (hertz)

Note that the modulating signal is normalized, meaning that it is always within the range of -1 and $+1$. $A_C[1 + am(t)]$ defines the amplitude of the carrier envelope. With the stated restrictions on a and $m(t)$, the zero-to-peak amplitude of the carrier is always in the range of 0 to 2 A_C, inclusive. Thus, the amplitude of the carrier can be driven to zero, but it cannot go negative and change the sign of the envelope.[2] Figure 6-1 shows a modulating signal and the resulting modulated carrier.

[1] This is the most common type of AM as it is used for standard AM radio broadcasting and in other AM voice communications systems.

[2] In most communication systems, if this happens, the signal is overmodulated and will not be recovered properly at the receiver.

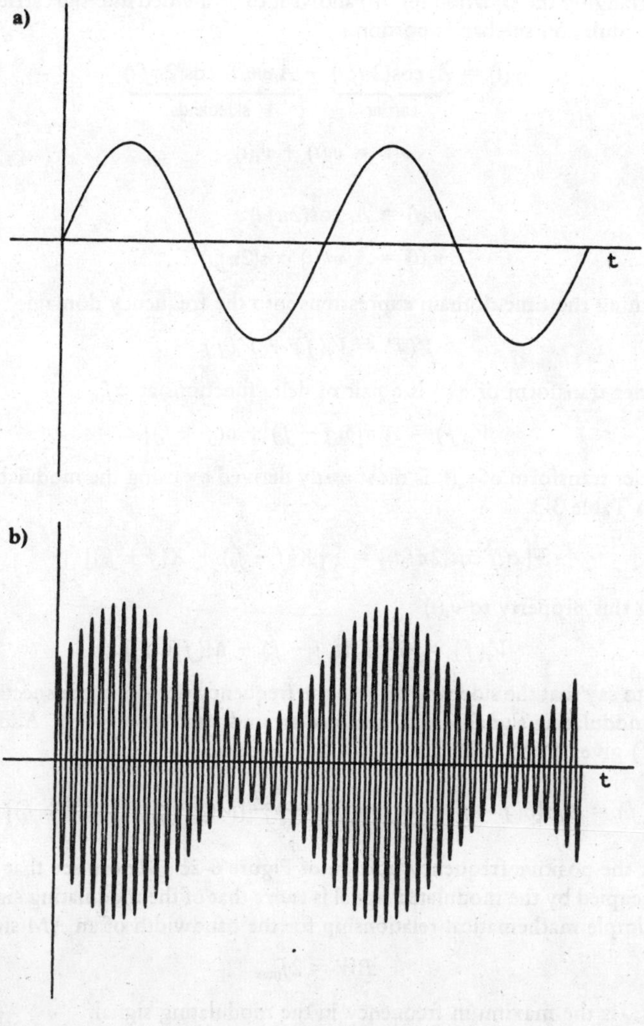

Figure 6-1. Amplitude modulation causes the amplitude of the carrier to be determined by the modulating signal. (a) The modulating signal. (b) The amplitude-modulated carrier.

Rearranging the equation for $v(t)$ allows it to be divided into the carrier portion and the modulation sidebands portion.

$$v(t) = \underbrace{A_c \cos(2\pi f_c t)}_{\text{carrier}} + \underbrace{A_c a m(t) \cos(2\pi f_c t)}_{\text{sidebands}} \tag{6-3}$$

$$v(t) = v_c(t) + v_s(t) \tag{6-4}$$

where

$$v_c(t) = A_c \cos(2\pi f_c t)$$

$$v_s(t) = A_c a m(t) \cos(2\pi f_c t)$$

Transforming the time domain expressions into the frequency domain,

$$V(f) = V_c(f) + V_s(f) \tag{6-5}$$

The Fourier transform of $v_c(t)$ is a pair of delta functions at $\pm f_c$.

$$V_c(f) = A_c\pi[\delta(f - f_c) + \delta(f + f_c)] \tag{6-6}$$

The Fourier transform of $v_s(t)$ is most easily derived by using the modulation property from Table 3-3.

$$\mathcal{F}[x(t)\cos(2\pi f_0 t)] = \frac{1}{2}[X(f - f_0) + X(f + f_0)] \tag{6-7}$$

Applying this property to $v_s(t)$

$$V_s(f) = \frac{A_c a}{2} 2[M(f - f_c) + M(f + f_c)] \tag{6-8}$$

which is to say that the sideband term in the frequency domain is the spectrum of the original modulating signal, $M(f)$, centered around $\pm f_c$ (Figure 6-2b). Adding $V_c(f)$ and $V_s(f)$ gives $V(f)$, which is shown in Figure 6-2c.

$$V(f) = A_c\pi[\delta(f - f_c) + \delta(f + f_c)] + \frac{A_c a}{2}[M(f - f_c) + M(f + f_c)] \tag{6-9}$$

Consider the positive frequency portion of Figure 6-2c. We can see that the bandwidth occupied by the modulated signal is twice that of the modulating signal. This gives a simple mathematical relationship for the bandwidth of an AM signal.

$$BW = 2f_{max}$$

where f_{max} is the maximum frequency in the modulating signal.

Sinusoidal Modulation

The case where the modulating signal is a sinusoid is an important and common occurrence in electronic systems. This case can be analyzed using Fourier transforms, but it can also be easily explained using trigonometry. Since the trig approach is instructive and gives a result which is inherently one-sided in the frequency domain, we will use it here.

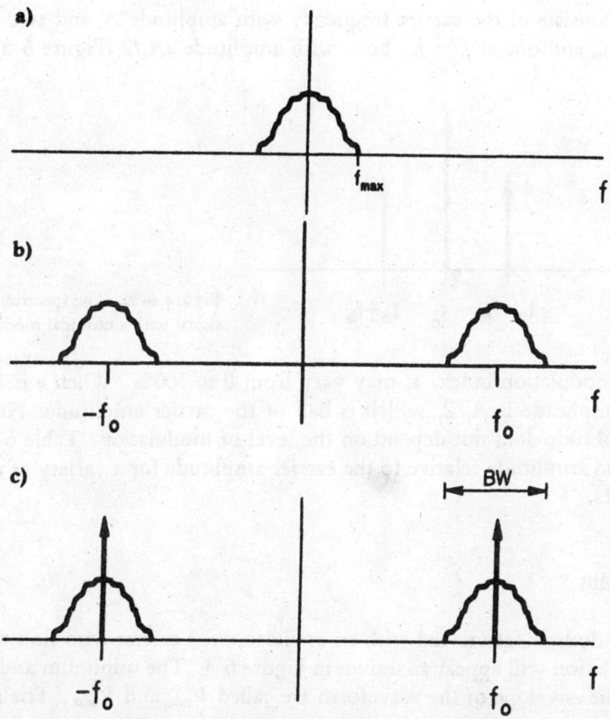

Figure 6-2. (a) The spectrum of the modulating signal. (b) The spectrum of the modulating signal centered around f_0. (c) The spectrum of the AM signal with carrier.

$$m(t) = \cos(2\pi f_m t) \tag{6-10}$$

Recall that

$$v(t) = A_c \cos(2\pi f_c t) + A_c a m(t) \cos(2\pi f_c t) \tag{6-11}$$

$$v(t) = A_c \cos(2\pi f_c t) + A_c a \cos(2\pi f_m t) \cos(2\pi f_c t) \tag{6-12}$$

Using the trig identity,

$$\cos A \cos B = 1/2[\cos(A + B) + \cos(A - B)] \tag{6-13}$$

$$v(t) = A_c \cos(2\pi f_c t) + \frac{aA_c}{2} [\cos 2\pi(f_m + f_c)t + \cos 2\pi(f_m - f_c)t] \tag{6-14}$$

Since $\cos(A - B) = \cos(B - A)$,

$$v(t) = A_c \cos(2\pi f_c t) + \frac{aA_c}{2} [\cos 2\pi(f_c + f_m)t + \cos 2\pi(f_c - f_m)t] \tag{6-15}$$

Thus, $v(t)$ consists of the carrier frequency with amplitude A_c and two sidebands, one at $f_c + f_m$ and one at $f_c - f_m$, both with amplitude $aA_c/2$ (Figure 6-3).

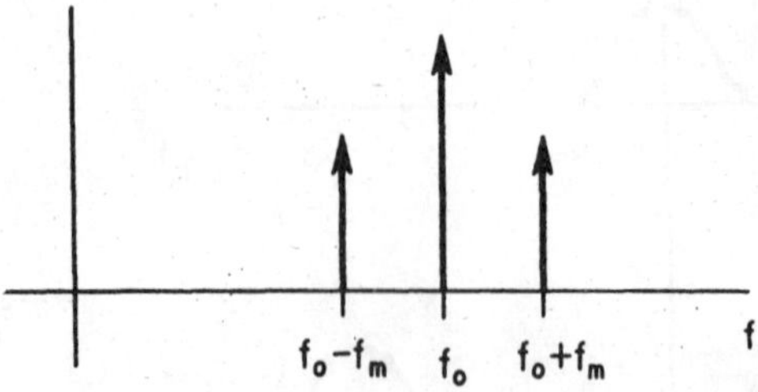

Figure 6-3. The spectrum of an AM signal with sinusoidal modulation.

The modulation index, a, may vary from 0 to 100%. When a is 100%, each sideband amplitude is $A_c/2$, which is half of the carrier amplitude. Note that the carrier amplitude does not depend on the level of modulation. Table 6-1 tabulates the sideband amplitude relative to the carrier amplitude for a variety of modulation index values.

Time Domain

In the time domain, as viewed with an oscilloscope, a carrier with sinusoidal amplitude modulation will appear as shown in Figure 6-4. The minimum and maximum values of the envelope of the waveform are called V_{min} and V_{max}. The modulation index can be computed from these two parameters.

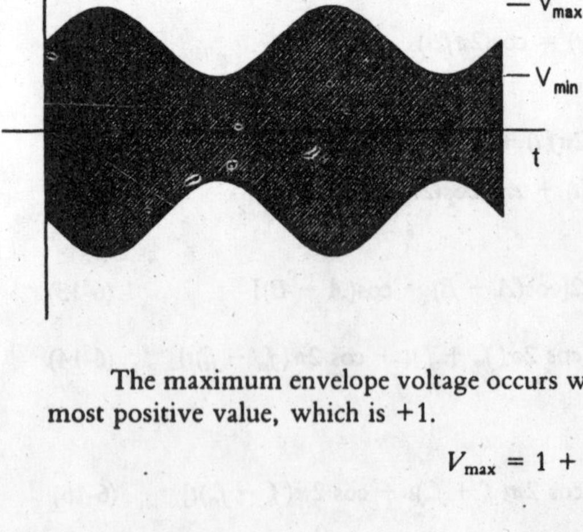

Figure 6-4. The envelope of an AM signal in the time domain can be used to determine the modulation index.

The maximum envelope voltage occurs when the modulating sinusoid is at its most positive value, which is +1.

$$V_{max} = 1 + a \tag{6-16}$$

TABLE 6-1 MODULATION INDEX AND RELATIVE SIDEBAND AMPLITUDE

Modulation index (%)	Sideband amplitude relative to carrier (%)	dB
100	50.0	−6.02
95	47.5	−6.47
90	45.0	−6.94
85	42.5	−7.43
80	40.0	−7.96
75	37.5	−8.52
70	35.0	−9.12
65	32.5	−9.76
60	30.0	−10.46
55	27.5	−11.21
50	25.0	−12.04
45	22.5	−12.96
40	20.0	−13.98
35	17.5	−15.14
30	15.0	−16.48
25	12.5	−18.06
20	10.0	−20.00
15	7.5	−22.50
10	5.0	−26.02
9	4.5	−26.94
8	4.0	−27.96
7	3.5	−29.12
6	3.0	−30.46
5	2.5	−32.04
4	2.0	−33.98
3	1.5	−36.48
2	1.0	−40.00
1	0.5	−46.02

The minimum envelope voltage occurs when the modulating sinusoid reaches its most negative value, which is −1.

$$V_{min} = 1 - a \tag{6-17}$$

Solving for a,

$$a = \frac{V_{max} - V_{min}}{V_{max} + V_{min}} \tag{6-18}$$

6.3 AM MEASUREMENTS

The spectrum analyzer can be used to characterize an amplitude-modulated signal in the frequency domain. The parameters which can be measured are the carrier amplitude and frequency, the modulating frequency, and the modulation index.

The carrier amplitude and frequency are measured just like any other spectral component, either by reading the values using the graticule or with the help of a marker or cursor readout. The modulating frequency is the difference between the carrier frequency and one of the sidebands. (The sidebands are symmetrical around the carrier.) Measuring the difference between the carrier and the sideband is made easier by the use of a marker that has offset (delta) capability. The modulation index is determined by measuring the sideband amplitude relative to the carrier amplitude. Usually this is expressed in dB. Table 6-1 or the following equation allows the user to convert relative sideband amplitude back to modulation index.

$$a = 2 \times 10^{(A_{dB}/20)} \tag{6-19}$$

where A_{dB} is the sideband amplitude relative to the carrier, expressed in decibels.

Example 6.1

A spectrum analyzer measurement of an amplitude-modulated signal is shown in Figure 6-5. Determine the modulating frequency and modulation index of the signal.

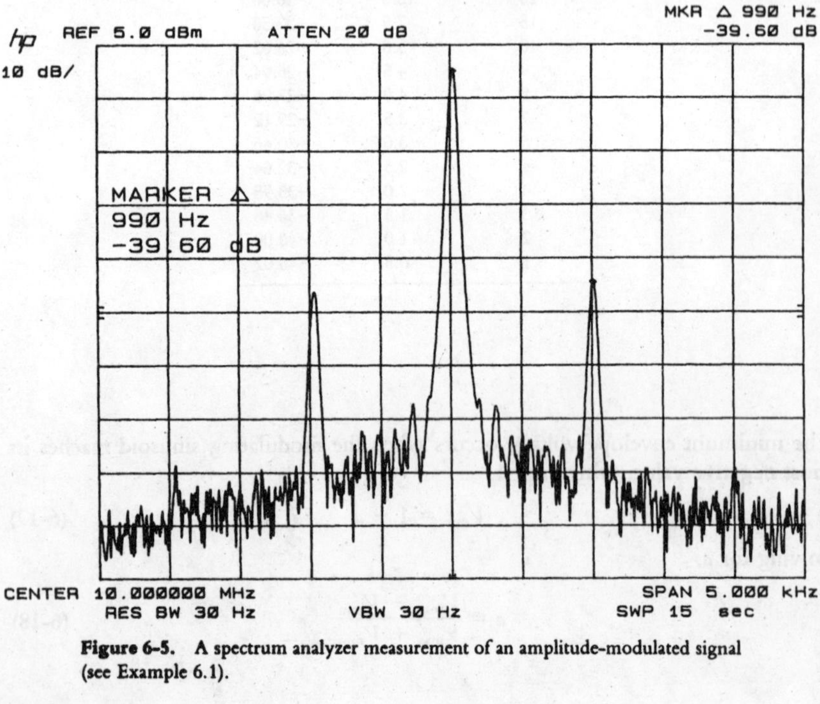

Figure 6-5. A spectrum analyzer measurement of an amplitude-modulated signal (see Example 6.1).

The delta marker feature is used to measure the amplitude and frequency differences between the carrier and the sidebands in Figure 6-5. The sidebands are −39.6 dB relative to the carrier and are offset by 990 Hz. Therefore, the modulating frequency is 990 Hz. The modulation index can be found by

$$a = 2 \times 10^{(A_{dB}/20)} = 2 \times 10^{(-39.6/20)} = 2.1\%$$

6.4 ZERO SPAN OPERATION

Most swept spectrum analyzers provide a simple, but powerful feature for observing slow amplitude variations in signals. The spectrum analyzer is set for a frequency span of zero (*zero span*),[3] with some nonzero sweep time. The center frequency is set to the carrier frequency and the resolution bandwidth must be set large enough to allow the sidebands to be included in the measurement. The analyzer will plot the amplitude of the signal versus time, within the limitations of its detector and video and resolution bandwidths. Since the minimum sweep time is about 25 msec on the fastest analyzers and may be as slow as 300 msec, this feature cannot be used

[3] This is also known as *synchrotune* operation.

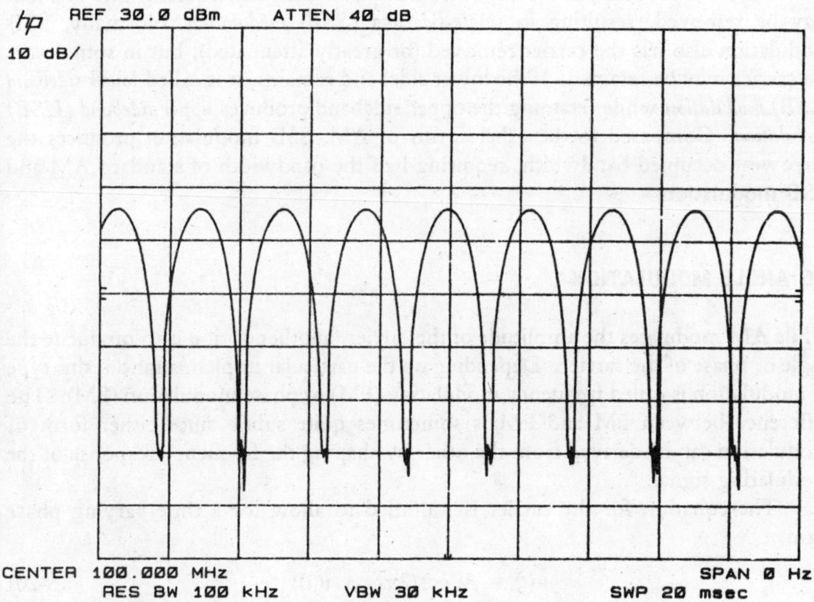

Figure 6-6. The zero span mode of a spectrum analyzer is used here to view the amplitude variation in an AM signal.

for quickly varying signals. The highest modulating frequency that can be observed is approximately 1/(sweep time), which will put only one cycle of the modulation on screen.

A spectrum analyzer measurement using zero span is shown in Figure 6-6. One can see the variation due to amplitude modulation on the signal as it is plotted across the display. The sweep time is 20 msec, so the horizontal axis is 2 msec/div. One cycle of the modulating signal occurs in 2.5 msec, indicating a modulating frequency of 400 Hz.

Zero span operation is not limited to modulation measurements. It can be used to characterize any signal that is slowly varying in amplitude.

6.5 OTHER FORMS OF AMPLITUDE MODULATION

We will briefly mention some of the other varieties of amplitude modulated signals. Note that "standard" AM includes a carrier and two sidebands which are symmetrical about the carrier. A significant amount of power is used to generate the carrier, which contains no information content since it does not vary with the modulating signal. All the information is in the sidebands. Sometimes the carrier is removed from the modulating signal, which produces *double-sideband (DSB) modulation* (also called *AM suppressed carrier* or *double-sideband suppressed carrier modulation*).

Since the modulation sidebands contain redundant information, one of them may be removed, resulting in *single-sideband (SSB) modulation*. Normally, SSB modulation also has the carrier removed (or greatly attenuated), but in some cases the carrier may be retained. If the lower sideband remains, it is called *lower sideband (LSB) modulation* while retaining the upper sideband produces *upper sideband (USB) modulation*. Compared to the other forms of AM, SSB modulation produces the narrowest occupied bandwidth, requiring half the bandwidth of standard AM and DSB modulation.

6.6 ANGLE MODULATION

While AM modulates the amplitude of the carrier, another option is to modulate the angle or phase of the carrier. Depending on the particular implementation, this type of modulation is called frequency modulation (FM) or phase modulation (PM). The difference between FM and PM is sometimes quite subtle since either form of modulation can be derived from the other by shaping the frequency response of the modulating signal.

The equation for the carrier is modified to allow for a time-varying phase term:

$$v(t) = A_c \cos(2\pi f_c t + \theta(t)) \qquad (6\text{-}20)$$

where $\theta(t)$ is the time-varying phase containing the modulation information.

For phase modulation, the phase term is directly proportional to the modulating signal:

$$\theta(t) = k_p m(t) \tag{6-21}$$

where

$$k_p = \text{deviation constant}$$
$$m(t) = \text{modulating signal}$$

The phase modulated carrier is

$$v(t) = A_c \cos(2\pi f_c t + k_p m(t)) \tag{6-22}$$

For frequency modulation, the frequency must be proportional to the modulating signal. Since the frequency is the time derivative of phase,

$$\frac{d\theta}{dt} = k_f m(t) \tag{6-23}$$

Solving for phase,

$$\theta(t) = k_f \int_{t_0}^{t} m(x)\, dx + \theta_0 \tag{6-24}$$

where

$$k_f = \text{frequency deviation constant}$$
$$\theta_0 = \text{initial phase at } t = 0$$

Setting the initial phase to zero, the frequency-modulated carrier is

$$v(t) = A_c \cos\left(2\pi f_c t + k_f \int^{t} m(x)\, dx\right) \tag{6-25}$$

Figure 6-7 shows a modulating signal, the resulting phase-modulated carrier, and the resulting frequency-modulated carrier. Notice the difference between changing the carrier phase and changing the carrier frequency.

The previous mathematical discussion centered on converting the modulating signal into a phase term in order to produce a frequency-modulated carrier. Thus, integrating the modulating signal in a phase-modulated system is equivalent to frequency modulating the carrier. This technique may be used in actual circuit implementations. The converse is also true. If a frequency modulator circuit was available, it could be used to produce a phase-modulated signal by taking the derivative of the modulating signal before applying it to the modulator. Since the integrator and differentiator operations can be approximated with high-pass and low-pass filters, respectively, the difference between FM and PM is often just the frequency shaping of the modulator circuits.

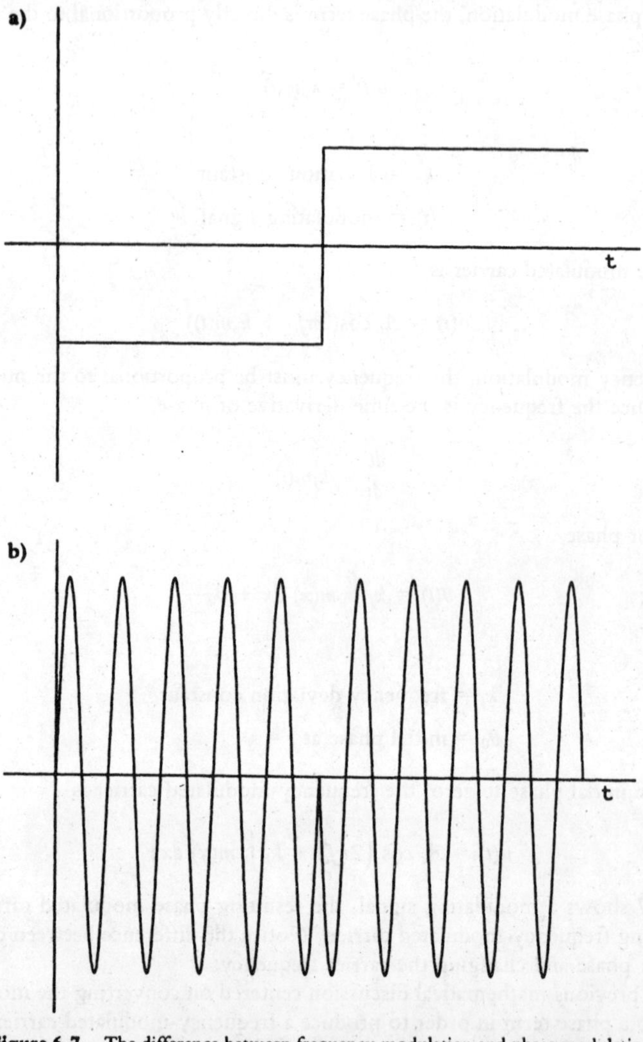

Figure 6-7. The difference between frequency modulation and phase modulation is easily shown by considering the case where the modulating signal is a step. (a) The modulating signal. (b) The phase-modulated carrier stays at the same frequency, but changes phase with the modulating signal. (c) The frequency-modulated carrier changes frequency when the step in the modulation occurs.

c)

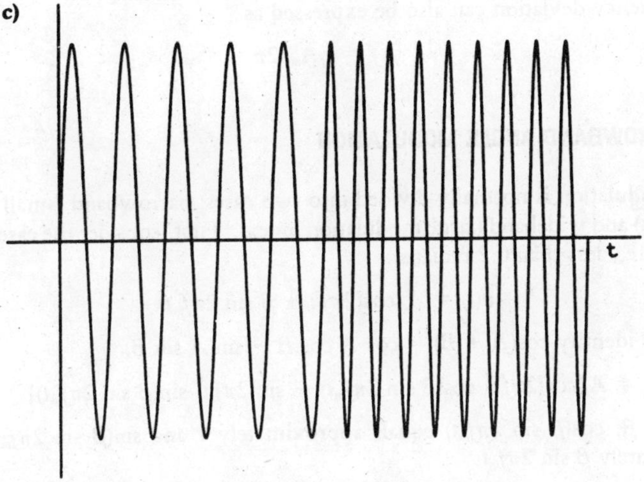

Figure 6-7. *(continued)*

Sinusoidal Modulation

Consider the important case where the modulating signal is a sinusoid, in an FM system:

$$m(t) = A_m \cos(2\pi f_m t) \tag{6-26}$$

The frequency-modulated carrier is

$$v(t) = A_c \cos\left(2\pi f_c t + k_f \int^t A_m \int^t \cos 2\pi f_m\, x\, dx\right) \tag{6-27}$$

Taking the integral of the modulating signal,[4]

$$v(t) = A_c \cos\left(2\pi f_c t + \frac{k_f A_m}{2\pi f_m} \sin 2\pi f_m t\right) \tag{6-28}$$

Introducing the *modulation index*, $\beta = k_f A_m/(2\pi f_m)$

$$v(t) = A_c \cos(2\pi f_c t + \beta \sin 2\pi f_m t) \tag{6-29}$$

The modulation index is defined as

$$\beta = \frac{\Delta f}{f_m} \tag{6-30}$$

where Δf is the frequency deviation (hertz).

---

[4] Note that this same modulated signal would result in a phase-modulated system with

$$m(t) = \frac{A_m}{2\pi f_m} \sin 2\pi f_m t$$

The frequency deviation can also be expressed as

$$\Delta f = k_f A_m / 2\pi \tag{6-31}$$

6.7 NARROWBAND ANGLE MODULATION

Angle modulation is normally divided into two cases: narrowband (small modulation index) and wideband (large modulation index). First, consider the case where β is small (i.e., less than 0.2 radians).

$$v(t) = A_c \cos(2\pi f_c t + \beta \sin 2\pi f_m t) \tag{6-32}$$

Using the identity $\cos(A + B) = \cos A \cos B - \sin A \sin B$,

$$v(t) = A_c[\cos(2\pi f_c t) \cos(\beta \sin 2\pi f_m t) - \sin(2\pi f_c t) \sin(\beta \sin 2\pi f_m t)] \tag{6-33}$$

For small β, $\cos(\beta \sin 2\pi f_m t)$ equals approximately 1 and $\sin(\beta \sin 2\pi f_m t)$ equals approximately $\beta \sin 2\pi f_m t$:

$$v(t) = A_c[\cos(2\pi f_c t) - \beta \sin(2\pi f_c t) \sin(2\pi f_m t)] \tag{6-34}$$

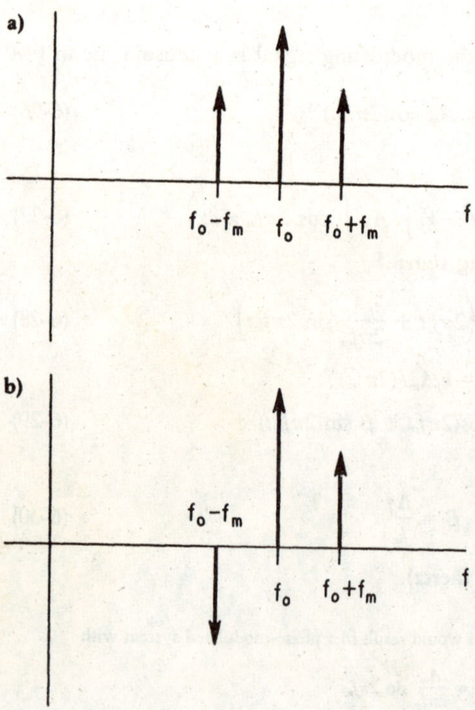

Figure 6-8. Narrowband FM is very similar to AM, except that one sideband is out of phase. (a) Spectrum of an AM signal. (b) Spectrum of a narrowband FM signal.

This can be broken down further into its spectral components by the use of sin A sin $B = 1/2\,[\cos(A - B) - \cos(A + B)]$

$$v(t) = A_c\left\{\cos(2\pi f_c t) - \frac{\beta}{2}\,[\cos 2\pi(f_c - f_m)t - \cos 2\pi(f_m + f_c)t]\right\} \tag{6-35}$$

$$v(t) = A_c\cos(2\pi f_c t) + \frac{A_c\beta}{2}\,[\cos 2\pi(f_c + f_m)t - \cos 2\pi(f_c - f_m)t] \tag{6-36}$$

This result should be reminiscent of the AM formula. Just like the AM case, the narrowband FM signal has frequency components at the carrier and at $\pm f_m$ away from the carrier. The subtle difference between AM and narrowband FM is that the phase of the lower sideband $(f_c - f_m)$ is changed by 180 degrees as indicated by the minus sign in front of the term (Figure 6-8). If no phase information is available (as with most spectrum analyzers) the two types of signals are indistinguishable in the frequency domain.

6.8 WIDEBAND ANGLE MODULATION

For the case where the modulation index is large, wideband angle modulation will result. As the name implies, in the frequency domain the signal will occupy a much larger bandwidth than the narrowband case.

As previously noted, an angle modulated carrier with sinusoidal modulation is

$$v(t) = A_c\cos[2\pi f_c t + \beta\sin(2\pi f_m t)] \tag{6-37}$$

The mathematics to expand this equation into individual frequency components is tedious and cannot be solved in closed form. However, it can be expanded into a series of sinusoids with Bessel functions (of the first kind) as coefficients.

$$\begin{aligned} v(t) = {} & J_0(\beta)\cos 2\pi f_c t \\ & - J_1(\beta)[\cos 2\pi(f_c - f_m)t - \cos 2\pi(f_c + f_m)t] \\ & + J_2(\beta)[\cos 2\pi(f_c - 2f_m)t + \cos 2\pi(f_c + 2f_m)t] \\ & - J_3(\beta)[\cos 2\pi(f_c - 3f_m)t - \cos 2\pi(f_c + 3f_m)t] + \cdots \end{aligned} \tag{6-38}$$

where $J_n(\beta)$ is the Bessel function of order n, evaluated at β.

Although the Bessel functions cannot be solved in closed form, they have been tabulated, and a small set of them is listed in Table 6-2. Examining the previous equation, note that the spectral components fall at the original carrier frequency and multiples of the modulating frequency away from the carrier. One is tempted to think of an FM signal as a carrier moving back and forth across some frequency range according to the modulating signal. But the actual effect (for sinusoidal modulation) is to set up discrete sidebands spaced every f_m. Ideally, the number of sidebands extends out infinitely, but in practice they will gradually decrease in

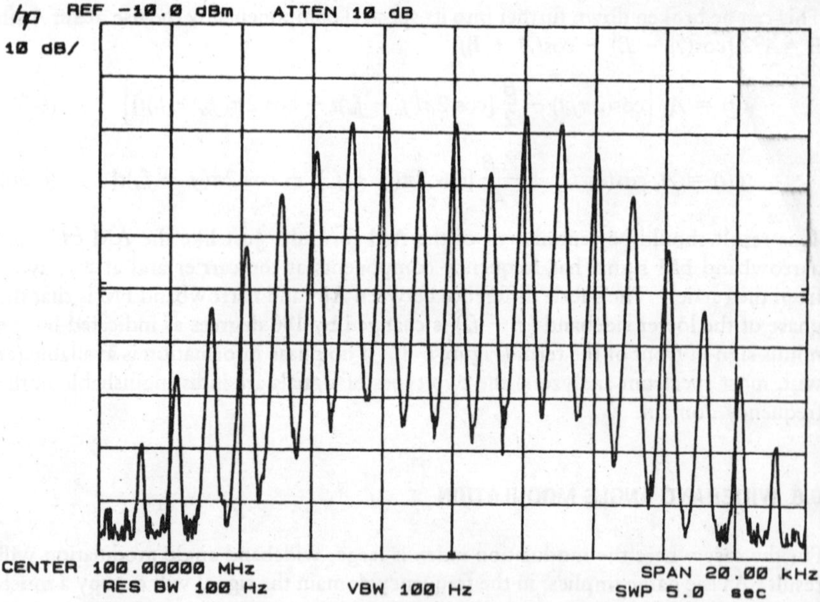

Figure 6-9. A spectrum measurement of a sine wave modulated wideband FM signal, with numerous sidebands spaced at multiples of the modulating frequency.

TABLE 6-2 BESSEL FUNCTIONS OF THE FIRST KIND $J_n(\beta)$

	β						
n	0.1	0.2	0.5	1	2	5	10
0	0.997	0.990	0.938	0.765	0.224	−0.178	−0.246
1	0.050	0.100	0.242	0.440	0.577	−0.328	0.043
2	0.001	0.005	0.031	0.115	0.353	0.047	0.255
3	0.000	0.000	0.003	0.020	0.129	0.365	0.058
4	0.000	0.000	0.000	0.002	0.034	0.391	−0.220
5	0.000	0.000	0.000	0.000	0.007	0.261	−0.234
6	0.000	0.000	0.000	0.000	0.001	0.131	−0.014
7	0.000	0.000	0.000	0.000	0.000	0.053	0.217
8	0.000	0.000	0.000	0.000	0.000	0.018	0.318
9	0.000	0.000	0.000	0.000	0.000	0.006	0.292
10	0.000	0.000	0.000	0.000	0.000	0.001	0.207
11	0.000	0.000	0.000	0.000	0.000	0.000	0.123
12	0.000	0.000	0.000	0.000	0.000	0.000	0.063

amplitude with distance from the carrier and at some point will be small enough to be ignored. The spectrum of a typical wideband FM signal is shown in Figure 6-9.

Carson's Rule

With a single modulating frequency, the table of Bessel functions can be used to obtain the exact spectrum of the modulated signal. The signal bandwidth can then be inferred from the spectrum. With multiple modulating frequencies (such as voice modulation), the analysis quickly gets unmanageable. However, *Carson's rule* can be used to estimate the bandwidth of a frequency-modulated signal.

$$BW = 2(\Delta f + f_m) \tag{6-39}$$

Recall that Δf is the peak frequency deviation and f_m is the modulating frequency. For the single tone case, f_m retains this definition, but for multitone modulation, the highest modulating frequency is substituted for f_m.

Example 6.2

Determine the spectrum of a 10 MHz carrier frequency modulated by a 5 kHz signal, with a frequency deviation of 10 kHz.

The modulation index is given by $\beta = \Delta f/f_m = 10\text{ kHz}/5\text{ kHz} = 2$. Since the modulating signal has a frequency of 5 kHz, the sidebands will be spaced at multiples of 5 kHz relative to the carrier frequency. Table 6-2 shows the following coefficients for the sidebands (modulation index = 2):

n	Table coefficient	Frequencies (MHz)	Amplitude relative to unmodulated carrier
0	0.224	10.000	−13.0 dB
1	0.577	9.995, 10.005	−4.78 dB
2	0.353	9.990, 10.010	−9.04 dB
3	0.129	9.985, 10.015	−17.8 dB
4	0.034	9.980, 10.020	−29.4 dB
5	0.007	9.975, 10.025	−43.1 dB
6	0.001	9.970, 10.030	−60.0 dB

6.9 FM MEASUREMENTS

The individual spectral components of an FM signal can be measured directly using a spectrum analyzer. Both the frequency and amplitude of the spectral lines can be determined, either absolutely or relative to the carrier. Determining the frequency deviation is more difficult.

Carrier Null Method

For certain values of β, the carrier frequency of an FM signal (with sinusoidal modulation) will disappear. These carrier null points are listed in Table 6-3.

TABLE 6-3. TABLE OF FM CARRIER NULLS

Null	Modulation index
First	2.405
Second	5.520
Third	8.654
Fourth	11.792
Fifth	14.931
Sixth	18.071

A radio transmitter or signal generator's frequency deviation can be set by using the *carrier null method*. A modulating frequency is chosen such that the desired deviation level causes a null on the carrier frequency. The output is monitored with a spectrum analyzer or other instrument to detect the null. Since carrier nulls occur at many different values of modulation index, it is important to use the correct carrier null. Normally, the deviation level is set to zero and then gradually increased while the carrier nulls are noted.

Example 6.3

A signal generator is to be adjusted such that its FM deviation is 5 kHz. What frequency should the modulating signal be to cause the first carrier null to occur at this frequency deviation?

The first carrier null occurs at $\beta = 2.405$. $\beta = \Delta f/f_m = \Delta f/\beta = 5000/2.405 =$ 2079 Hz.

6.10 COMBINED AM AND FM

In many high-frequency circuits, signals may be inadvertently amplitude modulated and/or frequency (or phase) modulated. When the modulation is purely amplitude or purely angle modulation, the previous sections of this chapter can be used to measure and understand it. However, when different forms of modulation appear simultaneously, the measurements may be very confusing.

Recall that the AM signal and the narrowband angle-modulated signal are identical except for the phase of the lower sideband. A carrier which has simultaneous AM and narrowband FM can be described by combining the AM and narrowband FM equations, Eq. 6-15 and 6-36. (We will assume that these two signals combine with negligible interaction that would produce new frequency components.)

$$
\begin{aligned}
v(t) = {} & A_c \cos(2\pi f_c t) \\
& + \frac{aA_c}{2}[\cos 2\pi(f_c + f_m)t + \cos 2\pi(f_c - f_m)t] \\
& + \frac{A_c\beta}{2}[\cos 2\pi(f_c + f_m)t - \cos 2\pi(f_c - f_m)t]
\end{aligned} \tag{6-40}
$$

$$
\begin{aligned}
v(t) = {} & A_c \cos(2\pi f_c t) + \frac{A_c(a+\beta)}{2}[\cos 2\pi(f_c + f_m)t] \\
& + \frac{A_c(a-\beta)}{2}[\cos 2\pi(f_c - f_m)t]
\end{aligned} \tag{6-41}
$$

If $a = \beta$, then cancellation of the lower sideband may occur:

$$
v(t) = A_c \cos(2\pi f_c t) + \frac{A_c(a+\beta)}{2} \cos 2\pi(f_c + f_m)t \tag{6-42}
$$

Several assumptions were made in this analysis. The modulation sources were assumed to be the same and there was no phase shift between the two modulation mechanisms. The two modulation indexes must also match exactly. In practice, these conditions will not usually be met, and cancellation will not be complete. However, it is common to find some partial cancellation (Figure 6-10), causing modulation sidebands which are not symmetrical.

The individual amounts of AM and FM can be estimated by assuming that the larger sideband is due to the AM and FM sidebands adding and the smaller sideband is due to the subtraction of the AM and FM sidebands.

Example 6.4

The carrier level of a signal with both AM and FM is 0.1 volt. The upper sideband amplitude is 0.05 volts and the lower sideband amplitude is 0.02 volts. Estimate the AM and FM modulation indexes.

The upper sideband is larger so it represents the addition of the AM and FM sidebands.

$$A_c(a + \beta) = 0.05, \quad a + \beta = 0.05/0.1 = 0.5$$

The lower sideband is smaller and represents the subtraction of the AM and FM sidebands.

$$A_c(a - \beta) = 0.02, \quad a - \beta = 0.02/0.1 = 0.2$$

Solving simultaneously, this implies that $a = 0.35$ and $\beta = 0.15$.

Devices which limit the amplitude of a signal (such as mixers and overdriven amplifiers) are notorious for converting amplitude modulation to phase modula-

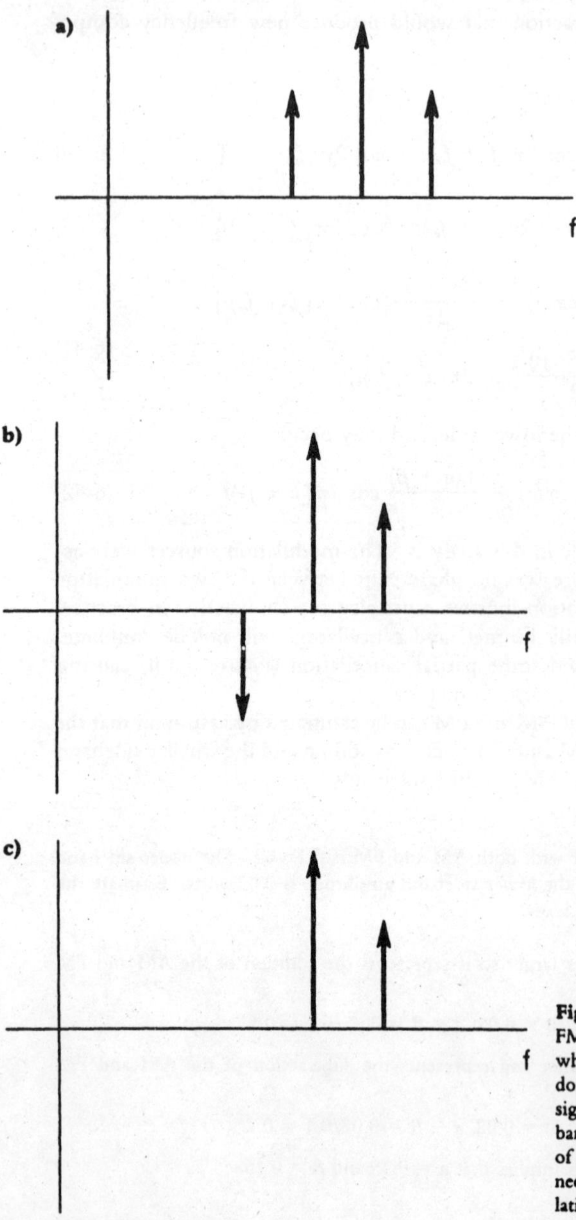

Figure 6-10. Simultaneous AM and FM can produce modulated signals which are asymmetrical in the frequency domain. (a) The spectrum of an AM signal. (b) The spectrum of a narrowband FM signal. (c) Perfect cancellation of the lower sideband due to simultaneous AM and FM. (d) Partial cancellation of the lower sideband.

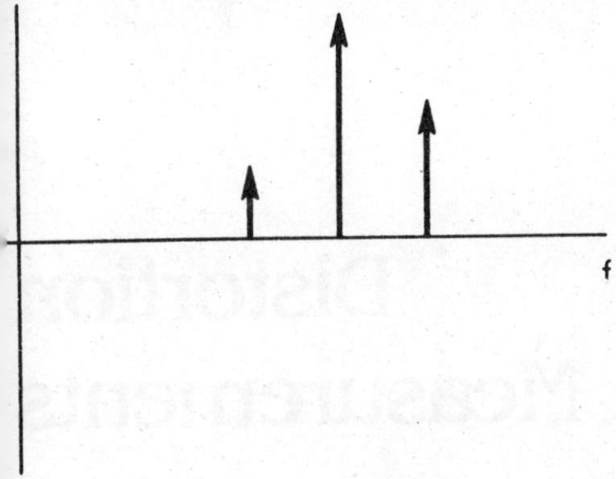

Figure 6-10. *(continued)*

tion. Often only a portion of the AM is converted to PM, causing a combined AM/PM signal at the device's output. In the frequency domain, this may cause a single sideband spectrum or, more likely, a spectrum with asymmetrical sidebands.

REFERENCES

1. Adam, Stephen F. *Microwave Theory and Applications.* Englewood Cliffs, NJ: Prentice Hall, Inc., 1969.
2. Engelson, Morris. *Modern Spectrum Analyzer Theory and Applications.* Dedham, MA: Artech House, 1984.
3. Hewlett-Packard Company, "Spectrum Analysis . . . Amplitude and Frequency Modulation," Application Note 150-1, Publication Number 5954-9130, January 1989.
4. Kinley, R. Harold. *Standard Radio Communications Manual.* Englewood Cliffs, NJ: Prentice Hall, Inc., 1985.
5. Malvino, Albert Paul. *Electronic Principles,* 3rd ed. New York: McGraw-Hill Book Company, 1984.
6. Schwartz, Mischa. *Information, Transmission, Modulation, and Noise,* 3rd ed. New York: McGraw-Hill Book Company, 1980.
7. Ziemer, R. E., and W. H. Tranter. *Principles of Communications.* Boston: Houghton Mifflin Company, 1976.

7

Distortion Measurements

Many of the circuits that are used in electronic systems are considered to be linear. This means that for a sinusoidal input, the output will also be sinusoidal with perhaps a different amplitude and phase.[1] In the time domain, the user expects to see an output waveform that has the exact same shape as the input waveform. In the frequency domain, we expect to see at the output the same frequency that was at the input (and only that frequency). Any other frequencies that are generated due to the input signal are considered distortion.[2]

7.1 THE DISTORTION MODEL

Most of the distortion mechanisms measured with spectrum analyzers are low level. That is, the devices producing the distortion are mostly linear and have only a slight nonlinear behavior. Such a weakly nonlinear system can be modeled with a power series.

$$V_{out} = k_0 + k_1 V_{in} + k_2 V_{in}^2 + k_3 V_{in}^3 + k_4 V_{in}^4 + \cdots \tag{7-1}$$

[1] One definition of a distortionless system is that the transfer function of the system is constant in amplitude and linear in phase (as a function of frequency).

[2] This does not include frequency components which are generated in the circuit independent of the input signal, such as spurious responses.

The first coefficient, k_0, represents the DC offset in the system. The second coefficient, k_1, is the gain of the circuit associated with linear circuit theory. The remaining coefficients, k_2 and above, represent the nonlinear behavior of the circuit. If the circuit were completely linear, all of the coefficients except k_1 would be zero.

The model can be simplified by ignoring the terms that come after the k_3 term. For gradual nonlinearities, the size of k_n decreases rapidly as n gets larger. For many applications the reduced model is sufficient, since the second-order and third-order effects dominate. (Expansion of the model to higher order is discussed later.)

$$V_{out} = k_0 + k_1 V_{in} + k_2 V_{in}^2 + k_3 V_{in}^3 \tag{7-2}$$

7.2 SINGLE-TONE INPUT

The simplest distortion test of a system is to input a pure sinusoid and measure the frequency content of the output signal:

$$V_{in} = A \cos \omega t \tag{7-3}$$

The angular frequency, $\omega = 2\pi f$, where f is the frequency in hertz.

Inserting this into the distortion model gives

$$V_{out} = k_0 + k_1 A \cos \omega t + k_2 A^2 \cos^2 \omega t + k_3 A^3 \cos^3 \omega t \tag{7-4}$$

$$\begin{aligned} V_{out} = {} & k_0 + k_1 A \cos \omega t + (k_2 A^2/2)(1 + \cos 2\omega t) \\ & + k_3 A^3 (3/4 \cos \omega t + 1/4 \cos 3\omega t) \end{aligned} \tag{7-5}$$

Collecting terms,

$$\begin{aligned} V_{out} = {} & k_0 + k_2 A^2/2 + (k_1 A + 3k_3 A^3/4) \cos \omega t \\ & + (k_2 A^2/2) \cos 2\omega t + (k_3 A^3/4) \cos 3\omega t \end{aligned} \tag{7-6}$$

This leaves us with an output voltage containing a DC component, the original (fundamental) frequency, and its second and third harmonics. Had we used a higher-order model, the analysis would have shown even higher-order harmonics present at the output. Note that the fundamental amplitude is affected by the nonlinear third-order coefficient of the model, k_3. Similarly, the DC component of the equation is affected by the second-order coefficient. The fundamental is mostly proportional to A, the second harmonic is proportional to A^2, and the third harmonic is proportional to A^3.

The model is somewhat limited since we do not usually know the values of k_0, k_1, k_2, and k_3 for a particular device. However, we can infer some useful information from the model anyway. Consider what happens when the signal level, A, is reduced. The fundamental will be reduced almost in direct proportion to the signal amplitude. We might say that the fundamental is reduced 1 dB per dB of change in the signal level. The second harmonic will go down as the square of A, or converting to dB

$$20 \log (A^2) = 2(20 \log A) = 2\ A_{dB} \quad (7\text{-}7)$$

This means that the second harmonic will be changed 2 dB per dB of signal level change. Similarly, the third harmonic term has an amplitude proportional to A^3. Converting to dB,

$$20 \log (A^3) = 3(20 \log A) = 3\ A_{dB} \quad (7\text{-}8)$$

which means that the third harmonic will be reduced by 3 dB per dB of signal level reduction.

Figure 7-1 shows the spectrum of a typical signal having harmonic distortion. (Ideally, a pure sine wave would have no harmonics.) Note that the odd harmonics, particularly the third harmonic, is larger than the even harmonics. Distortion that maintains the 50% duty cycle of the ideal waveform will create only odd harmonics. (Recall the case of the square wave from Chapter 3.) Distortion mechanisms that upset the symmetry of the signal produce even harmonics.

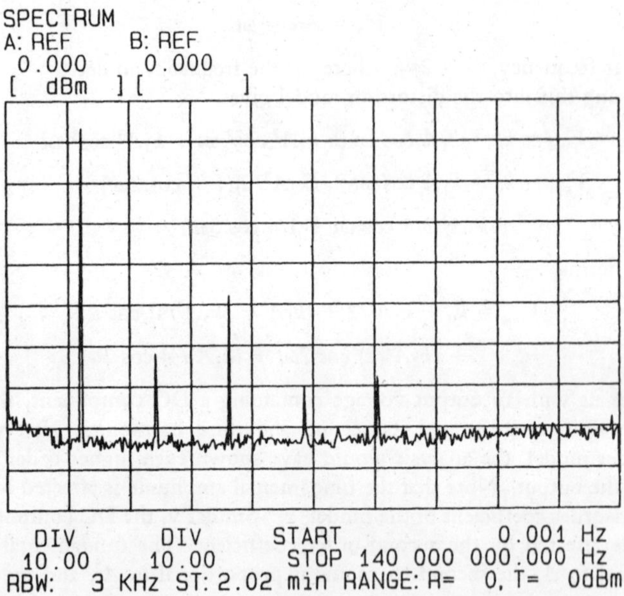

Figure 7-1. Measurement of harmonic distortion in a signal.

When using good-quality spectrum analyzers, one must get accustomed to the fact that there are very few pure sine waves. For example, a good signal or function generator may have a third harmonic that is 30 or 40 dB lower than the fundamental. When viewed on an oscilloscope, this signal will appear to be a pure sine wave since the distortion is not discernable. When measured with even a moderate performance spectrum analyzer, the harmonics will be easily visible. This illustrates the advantage of a narrowband receiver (the spectrum analyzer) versus a wideband receiver (the oscilloscope).

7.3 TWO-TONE INPUT

Another input signal commonly used for distortion tests is the two-tone signal.

$$V_{in} = A_1 \cos \omega_1 t + A_2 \cos \omega_2 t \qquad (7\text{-}10)$$

Using our distortion model,

$$V_{out} = k_0 + k_1 V_{in} + k_2 V_{in}^2 + k_3 V_{in}^3 \qquad (7\text{-}11)$$

The result is in the form

$$\begin{aligned} V_{out} = {} & c_0 + c_1 \cos \omega_1 t + c_2 \cos \omega_2 t + c_3 \cos 2\omega_1 t \\ & + c_4 \cos 2\omega_2 t + c_5 \cos 3\omega_1 t + c_6 \cos 3\omega_2 t \\ & + c_7 \cos(\omega_1 t + \omega_2 t) + c_8 \cos(\omega_1 t - \omega_2 t) \\ & + c_9 \cos(2\omega_1 t + \omega_2 t) + c_{10} \cos(2\omega_1 t - \omega_2 t) \\ & + c_{11} \cos(2\omega_2 t + \omega_1 t) + c_{12} \cos(2\omega_2 t - \omega_1 t) \end{aligned} \qquad (7\text{-}12)$$

where $c_0, \ldots, c_{12}$ are coefficients determined by $k_0, \ldots, k_3$, A_1, and A_2.

Besides the harmonics of the two tones (as in the single-tone case), there are also sum and difference frequencies. These new frequency components are called *intermodulation distortion (IMD)*, because they result from the two tones modulating together. The frequencies present in the output satisfy the following criteria

$$\omega_{nm} = |n\omega_1 \pm m\omega_2| \qquad (7\text{-}13)$$

where n and m are positive integers such that $n + m \leq 3$ or, in units of hertz.

$$f_{nm} = |nf_1 \pm mf_2| \qquad (7\text{-}14)$$

If the distortion model is expanded from the third-order model to a higher-order model, the limit on the sum of $n + m$ is raised accordingly.

The order of a particular frequency component is the sum of the n and m values used to obtain that frequency (e.g., f_{12} and f_{21} are third-order terms and f_{20} and f_{11} are second-order terms). As in the single-tone case, second-order terms will be reduced 2 dB in amplitude when the input tones are reduced by 1 dB. Equivalently, second-order terms are reduced 2 dB/dB of input signal reduction. Third-order terms are reduced 3 dB/dB of signal reduction and so on for higher-order terms, if present.

Example 7.1

Assuming a third-order distortion model, what frequencies will be present at the output with a two-tone input signal with frequencies of 10.7 MHz and 10.8 MHz?

The output frequencies are given by $f = |n f_1 \pm m f_2|$. For $n = 1$ and $m = 0$,

$$f_{10} = |10.7 \text{ MHz} \pm 0| = 10.7 \text{ MHz}$$

For $n = 2$ and $m = 0$,

$$f_{20} = |2(10.7 \text{ MHz}) \pm 0| = 21.4 \text{ MHz}$$

For $n = 3$ and $m = 0$,

$$f_{30} = |3(10.7 \text{ MHz}) \pm 0| = 32.1 \text{ MHz}$$

For $n = 0$ and $m = 1$,

$$f_{01} = |0 \pm 10.8 \text{ MHz}| = 10.8 \text{ MHz}$$

For $n = 0$ and $m = 2$,

$$f_{02} = |0 \pm 2(10.8 \text{ MHz})| = 21.6 \text{ MHz}$$

For $n = 0$ and $m = 3$,

$$f_{03} = |0 \pm 3(10.8 \text{ MHz})| = 32.4 \text{ MHz}$$

The previous frequencies are simply the first three harmonics of the two input tones. Now the various sum and difference frequencies will be calculated.

For $n = 1$ and $m = 1$,

$$f_{11} = |10.7 \text{ MHz} \pm 10.8 \text{ MHz}| = 0.1 \text{ MHz}, 21.5 \text{ MHz}$$

For $n = 2$ and $m = 1$,

$$f_{21} = |2(10.7 \text{ MHz}) \pm 10.8 \text{ MHz}| = 10.6 \text{ MHz}, 32.2 \text{ MHz}$$

For $n = 1$ and $m = 2$,

$$f_{12} = |10.7 \text{ MHz} \pm 2(10.8 \text{ MHz})| = 10.9 \text{ MHz}, 32.5 \text{ MHz}$$

The spectrum of the output signal is shown in Figure 7-2. The amplitudes of the frequency components will depend on the levels of the input tones and the coefficients of the distortion model. However, the amplitudes that are shown in the figure are typical of a distorted signal.

A few comments are in order now that a numerical example has been given. The two input tones were chosen to be close to each other in frequency, as is usually the case for two-tone testing. An examination of Figure 7-2 will reveal that the

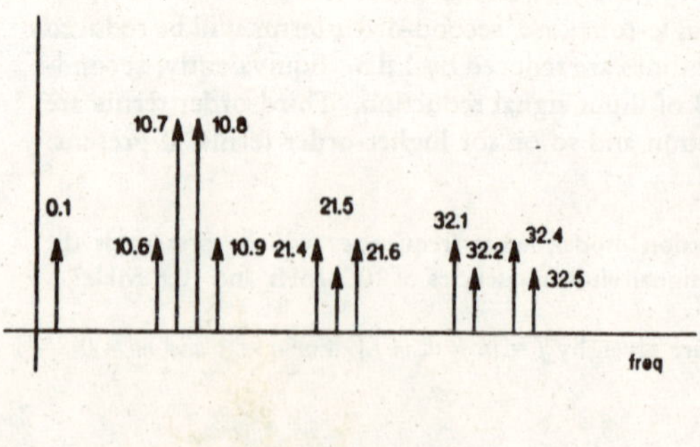

Figure 7-2. The spectrum of a two-tone signal with third-order intermodulation distortion products.

spectral lines tend to fall in four groupings. The $f_1 - f_2$ frequency (0.1 MHz) will fall down near DC. The other frequencies fall in groups near the fundamentals (near 10.7 MHz), the second harmonics (near 21.5 MHz), and the third harmonics (near 32.4 MHz) of the original two tones. Depending on the system involved, some of these distortion components can be neglected since they will be filtered out at some point. For instance, an intermediate amplifier (IF) stage will usually be narrowband, centered on the two input tones. Spectral components out at the second and third harmonics will be easily filtered out. The distortion components close to the original tones (f_{21} and f_{12}) will be more troublesome since they fall near the desired frequencies. In general, odd-order intermodulation products are of the most concern to RF designers, since the distortion products fall "in-band."

7.4 HIGHER-ORDER MODELS

We have chosen to limit the number of terms in the distortion model to produce a third-order behavior. Even with such a simple model, the derivation of the output signal frequency components is lengthy and expanding the model to a higher order only makes the situation worse. Fortunately, for many situations, a third-order model is sufficient.

But what if the third-order model is insufficient? For instance, it is common to have significant energy in the fifth, sixth, or seventh harmonic of a single tone, yet the third-order model does not show this effect. The analytical approach used previously can simply be expanded to include the higher-order terms, with the penalty of the mathematics getting more difficult. Another approach is to simply expand on the concepts demonstrated by the third-order model, even though they have not been proven rigorously. As stated previously, the frequencies generated by the distortion model obey the $n f_1 \pm m f_2$ rule, where the maximum value of $m + n$ is the order of the model. So it is possible to predict the frequency components of higher-order systems without extensive mathematics.

7.5 THE INTERCEPT CONCEPT

Increasing the signal level at the input to a weakly nonlinear device will cause the distortion products to increase at the output. Not only do the distortion products increase in amplitude, they increase in amplitude faster than the input signal. Figure 7-3 shows a plot of the output power versus the input power for the fundamental, second-order frequency components, and third-order frequency components. For increasing fundamental input power, the fundamental output power increases in a linear manner, according to the gain or loss of the device. At some point, gain compression occurs and the fundamental output power no longer increases with input power. The output power of the second-order distortion products also increases with fundamental input power, but at a faster rate. Recall that the distortion

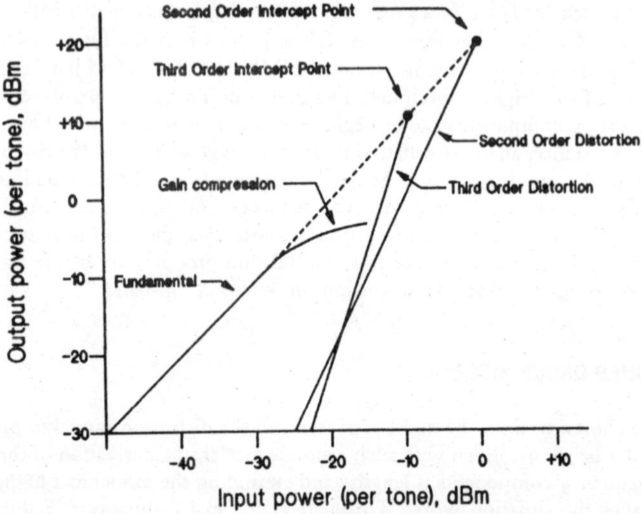

Figure 7-3. Plot of fundamental, second-order distortion product, and third-order distortion product power levels illustrates the concept of second- and third-order intercept points.

model shows that second-order terms change 2 dB per 1 dB of change in the fundamental. Thus, on a decibel plot, the line representing the second-order output power has twice the slope of the fundamental line. Similarly, the third-order distortion products change 3 dB per 1 dB of change in the fundamental, so that line has a slope that is three times the slope of the fundamental line.

If there was no gain compression, the fundamental input power could be increased until the second-order distortion products would eventually catch up with it and the two output power levels would be equal. This point is referred to as the *second-order intercept point*. The third-order distortion products also increase faster than the fundamental, and those two lines will intersect at the *third-order intercept point*. Rarely can either of these two points be measured directly, due to the gain compression of the fundamental. Instead, the intercept points are extrapolated from measurements of the fundamental and distortion products at power levels below where gain compression occurs. The intercept points are usually specified in dBm and may refer either to the output or the input. (It is important to always specify whether the intercept point refers to the output power or the input power. The two points will differ by the gain of the device.)

The utility of the intercept concept is in specifying and predicting the distortion level in a system. One might be tempted to specify the distortion of a circuit or system directly by stating the level of the distortion products in decibels relative to the signal level. This can be done, but is not very meaningful unless the signal level is also specified. One circuit's distortion might be −80 dB relative to the signal while another circuit might achieve only −40 dB. However, these two values are not a fair comparison unless the same signal level is used. The second-order and third-order intercept points are figures of merit which are independent of signal level. Therefore, the distortion performance of two different circuits can be compared quite easily if their intercept points are known.

Most often, an engineer is interested in the level of the distortion products relative to the signal level. The intercept points do not indicate this directly and may seem cumbersome to use, but a few observations will show how the relative distortion level can be easily determined from the intercept point. The difference between the level of the second-order distortion products and the fundamental signal level is the same as the difference between the fundamental signal level and the intercept point. Suppose the second-order intercept point is +15 dBm and the fundamental signal level is −10 dBm (both referred to the output of the device). The difference between these two values is 25 dB. Therefore, the second-order distortion products will be 25 dB below the fundamental, or −35 dBm. So the intercept point allows easy conversion between fundamental signal level and distortion level. Often the distortion level is specified relative to the fundamental power level, and the conversion to absolute power (dBm) is not necessary.

The difference between the level of the third-order distortion products and the fundamental signal level is *twice* the difference between the fundamental signal level and the third-order intercept point. (Note that the second-order intercept point is *not* the same as the third-order intercept point.) Suppose that the third-order intercept point is +5 dBm and the fundamental signal level is −25 dBm, both referred to the output of the device. The difference between the intercept and the fundamental is 30 dB, so the third-order distortion products will be two times 30 dB down from the fundamental. The relative distortion level is −60 dB and the absolute power level of the distortion products is −85 dBm.

Example 7.2

What is the maximum allowable power level of the input signal if the third-order distortion products are to be less than −70 dB relative to the fundamental? The third-order intercept point is +10 dBm, referred to the input.

The third-order distortion products are to be 70 dB below the fundamental, so the fundamental must be 70/2 dB or 35 dB below the intercept point. The intercept point is +10 dBm, so the signal level should be −25 dBm at the input.

7.6 HARMONIC DISTORTION MEASUREMENTS

Harmonic distortion measurements can easily be made with a spectrally pure signal source and a spectrum analyzer. The quality of the measurement is limited by the harmonic distortion of both the signal source and spectrum analyzer. The signal source is most often the limiting factor, with harmonic distortion performance often not much better than 40 db below the fundamental.

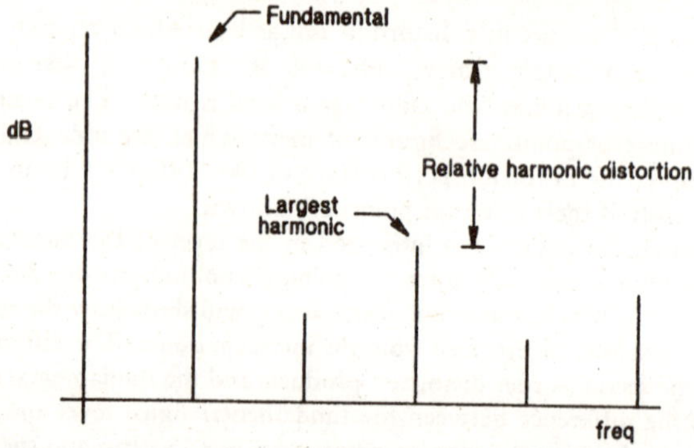

Figure 7-4. The harmonic distortion of a signal is often specified by stating the amplitude of the largest harmonic in dB relative to the fundamental.

The source provides a signal to the device under test and the spectrum analyzer is used to monitor the output. Figure 7-4 shows a typical harmonic distortion measurement. The distortion level may be specified by expressing the largest harmonic in dB relative to the fundamental, as shown in Figure 7-4.

Alternatively, the distortion may be specified as *total harmonic distortion (THD)*, usually as a percent of the fundamental. THD takes into account the power in all the harmonics:

$$\text{THD} = \sqrt{V_2^2 + V_3^2 + \cdots}/V_1 \tag{7-15}$$

where V_1 is the RMS voltage of the fundamental and V_2, V_3, . . . are the RMS voltages of the harmonics.

All harmonics of the fundamental are summed in a root-mean-square manner and are divided by the fundamental RMS voltage. Since an infinite number of harmonics cannot be measured, a finite number will have to suffice. Fortunately, the harmonic amplitudes tend to decrease with higher harmonic numbers. The calculation is somewhat tedious for a large number of harmonics, but some spectrum analyzers include an automatic THD function. If not, the user must determine each harmonic amplitude and compute the THD.

Example 7.3

Determine the total harmonic distortion of a signal with the following spectral components: 1 MHz, 3.5 volts RMS; 2 MHz, 0.1 volts RMS; 3 MHz, 0.2 volts RMS; 4 MHz, 0.05 volts RMS. Express the largest harmonic in decibels relative to the fundamental.

The fundamental frequency is 1 MHz.

$$\text{THD} = \sqrt{(0.1)^2 + (0.2)^2 + (0.05)^2}/3.5 = 0.229/3.5$$

$$= 0.065 \text{ or } 6.5\%$$

The largest harmonic is the third harmonic (3 MHz). In decibels, this harmonic is 20 log (0.2/3.5) = −24.9 dB relative to the fundamental.

7.7 USE OF LOW-PASS FILTER ON SOURCE

The signal source is often the limiting factor in a harmonic distortion measurement, due to its own harmonic distortion. A typical signal generator has harmonics on the order of −40 dB relative to the fundamental,[3] while a typical spectrum analyzer may have a dynamic range of 70 or 80 dB.

A low-pass filter can be used to improve the source's effective harmonic distortion, as shown in Figure 7-5. The cutoff frequency of the low-pass filter is chosen such that the fundamental frequency is passed largely intact, while the harmonics are attenuated significantly. The performance of the source/filter combination can be verified directly by the spectrum analyzer. The passband attenuation of the filter should be kept to a minimum, but the exact value is not critical. If the loss through the filter at the fundamental frequency is significant, it should be accounted for when setting the source output level. The spectrum analyzer can be used to check directly the fundamental level at the output of the filter.

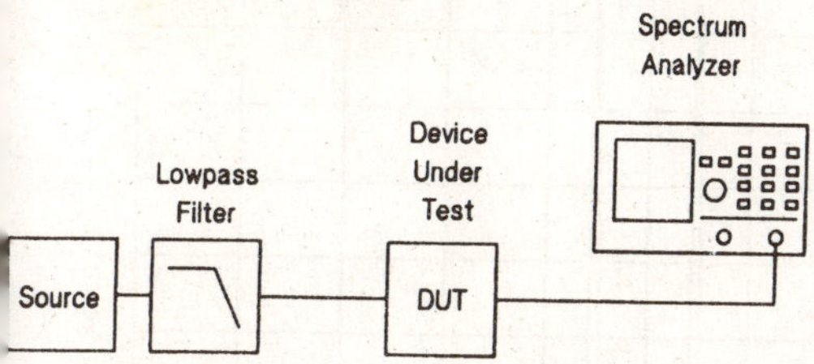

Figure 7-5. The harmonic distortion of a signal source can be improved by installing a low-pass filter at the source's output.

7.8 INTERMODULATION DISTORTION MEASUREMENTS

To test for intermodulation distortion, two stimulus sine waves are required. The test setup shown in Figure 7-6 has two independent signal sources connected with a power splitter (used as a combiner) to drive the device under test. The sources are set at the same output level, but at different frequencies. The 6 dB loss of the combiner should be accounted for when setting the output amplitudes of the sources. A typical spectrum analyzer display of the two-tone distortion test is shown in Figure 7-7. As shown, the third-order products (f_{21} and f_{12}) that fall close to the original two tones are being measured. This is a common measurement since the two distortion products fall close to the original two tones and are difficult to remove by filtering.

In some cases, the two sources may interact and produce intermodulation distortion. This problem can be detected with the spectrum analyzer and can be cured by inserting fixed attenuators at the outputs of the sources. These attenuators

[3] Sources that are designed with distortion measurements in mind may have considerably better harmonic distortion, but are usually restricted to frequencies below 10 MHz with best distortion performance in the audio range.

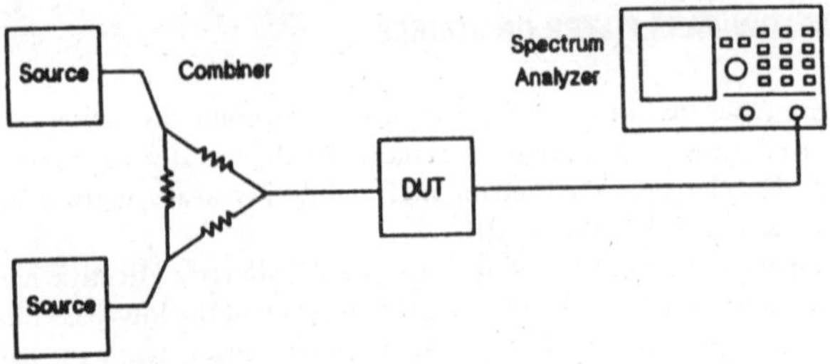

Figure 7-6. The outputs of two signal sources can be combined to a two-tone signal for intermodulation distortion tests.

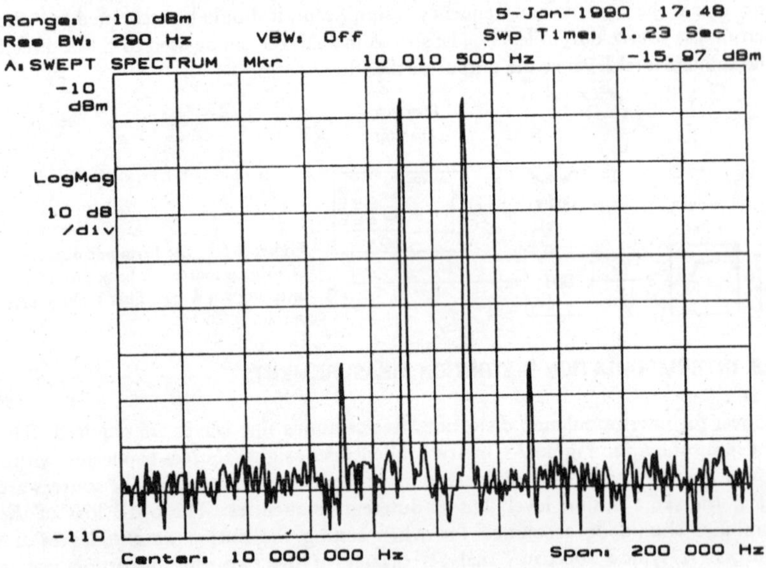

Figure 7-7. A typical two-tone intermodulation distortion measurement which measures the third-order products close to the original two tones.

increase the isolation between the sources and prevent internally generated intermodulation distortion. The output levels of the sources should be increased to compensate for the signal loss in the attenuators.

It should be kept in mind that the two sine waves will combine to create a signal which is 6 dB larger than the individual tones (after accounting for the combiner loss). The device under test is often sensitive to the peak instantaneous voltage applied to it and the user may inadvertently supply twice the desired peak input voltage.

7.9 DISTORTION INTERNAL TO THE ANALYZER

The preceding discussion was oriented toward understanding and measuring distortion in the device under test. However, the internal circuits of the analyzer are imperfect and will also produce distortion products. The distortion performance of the analyzer is specified by the manufacturer, either directly or lumped into a dynamic range specification. The instrument user can stretch the performance of the analyzer by understanding the nature of these distortion products.

As shown in this chapter, distortion products can be reduced in amplitude by reducing the signal level. Not only do the absolute levels of the distortion products decrease, they decrease more than the decrease in signal level. So as the signal level decreases, the relative distortion level also decreases, depending on the order of the distortion product. Higher-order distortion products decrease the fastest. This implies that the distortion products internal to the analyzer can be reduced by reducing the signal level into the analyzer.[4] The internal input attenuators of the analyzer may be used or an external attenuator may be attached, improving the distortion measurement range of the analyzer. The most obvious disadvantage of reduced signal level is reduced signal-to-noise ratio. The user may find that the low-level distortion products are buried in the noise. Reducing the resolution bandwidth of the analyzer will reduce the measured noise, but at the expense of a slower sweep rate.

In some measurement situations, the amount of distortion is not of concern, and the signal level at the input of the analyzer can be increased to provide a better signal-to-noise ratio. For many measurements, the distortion products are known to occur at frequencies which are not of interest. For example, a narrowband measurement around the fundamental frequency of a sine wave will not be degraded by the presence of harmonic distortion, since the harmonics will fall far away from the frequency range of interest. The instrument user must always be careful not to apply too large a signal to the input of an analyzer, so that the damage level is not exceeded.

REFERENCES

1. Bartz, Manfred. "Designing Effective Two-Tone Intermodulation Distortion Test Systems." *RF Design* (November 1987).
2. Hardy, James K. *High Frequency Circuit Design*. Reston, VA: Reston Publishing Company, Inc., 1979.
3. Hayward, W. H. *Introduction to Radio Frequency Design*. Englewood Cliffs, NJ: Prentice-Hall, Inc., 1982.

---

[4] This may not be true of some spectrum analyzers which use digital IF sections (see Chapter 5). The analog-to-digital conversion process can introduce low-level distortion products that do not decrease in amplitude in response to decreasing signal level.

8

Noise and Noise Measurements

In frequency domain measurements, electronic noise is a concern in two distinctly different ways. The first case is when the measurement of a particular parameter is affected by the presence of unwanted noise. Here, the noise is a nuisance. For example, we could be measuring the distortion of a particular amplifier with the amplifier's noise getting in the way. The second case occurs when the noise present in the system is the parameter to be measured. In that same amplifier, we may want to measure how much noise is present at the output. Many of the same principles apply to both cases, but it is important to focus on whether the noise *is* the measurement or whether it degrades the accuracy of the measurement.

The electronic noise present in our measurements may come from the Device Under Test (DUT) that is being measured or may be generated internally by the analyzer. If the noise of the DUT is to be measured, the analyzer's internal noise must be significantly lower than the DUT noise.

8.1 STATISTICAL NATURE OF RANDOM NOISE

Many waveforms that we wish to measure can be easily characterized in the time domain. For instance, a sine wave can be completely described by its amplitude, frequency, and phase. Once we know these values, the instantaneous voltage of the waveform can be predicted for any arbitrary instant in time. Such a waveform is said to be *deterministic*. Noise, on the other hand, is random in nature such that the instantaneous voltage cannot be predicted for arbitrary points in time.[1] Thus, random noise is *nondeterministic*.

Noise cannot be characterized in the time domain by simple parameters such as amplitude and phase since the voltage at any point in time is a random function. Noise can be characterized in the time domain, but with a statistical approach. We can describe the noise by tabulating how often a certain voltage appears. In a continuous form, this results in the *probability density function (PDF)* of a random waveform. Figure 8-1 shows the probability density function of a particular waveform. The PDF shown happens to have a Gaussian shape, which is very common, but other PDF shapes are possible. The PDF does not define the shape of the time domain waveform, but tells us the probability of a certain voltage occurring.

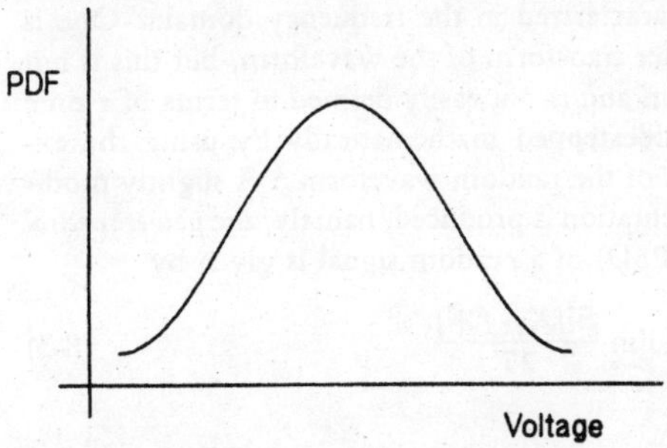

Figure 8-1. The probability density function shows the probability of a particular voltage occurring in a waveform.

8.2 MEAN, VARIANCE, AND STANDARD DEVIATION

The statistical characteristics of a random waveform can be described by a few simple parameters. First, the waveform will have an average or *mean* value given by

$$\bar{x} = E[x] = \int_{-\infty}^{\infty} xp(x)\, dx \tag{8-1}$$

where

$E[x]$ = expected value of x

$p(x)$ = probability density function of $x(t)$

Of course, the mean value has a less mathematical and more intuitive definition of being the average value of the waveform.

With the mean value of the waveform defined, a measure of how much the voltage of the waveform varies is in order. The *variance* of a waveform is given by

[1] The definition of noise is restricted in this chapter to include only truly random noise. Other noise processes may exist which are not completely random.

$$\sigma^2 = E[(x - \bar{x})^2] = \overline{x^2} - (\bar{x})^2 \quad (8\text{-}2)$$

The variance is a measure of how far the instantaneous value of x strays from the mean value of x. If the variance is zero, the waveform is a DC level which never changes from its mean value. Closely related to the variance is the standard deviation, σ. Since the square of the standard deviation is equal to the variance, the two quantities are redundant. The variance describes the power in the random waveform, while the standard deviation is related to the voltage.

8.3 POWER SPECTRAL DENSITY

A random waveform can also be characterized in the frequency domain. One is tempted to simply compute the Fourier transform of the waveform, but this is not possible since the waveform is random and is not easily defined in terms of a time domain function. This problem is sidestepped mathematically by using the expected value of the Fourier transform of the random waveform.[2] A slightly modified form of frequency domain representation is produced, namely, the *power spectral density*. The power spectral density (PSD) of a random signal is given by

$$S_x(f) = \lim_{T \to \infty} \frac{E[|X_T(f)|^2]}{2T} \quad (8\text{-}3)$$

where

$E[\ ]$ = expected value

$X_T(f)$ = Fourier transform of the random waveform, $x(t)$, evaluated over the time interval, $-T < t < T$

A less rigorous but more useful definition is that the PSD gives the density of power in a signal as a function of frequency. The power over a particular frequency range is given by

$$P_{12} = \int_{f_1}^{f_2} S_x(f)\, df \quad (8\text{-}4)$$

The total power in the signal is found by integrating over all frequencies:

$$P_T = \int_{-\infty}^{\infty} S_x(f)\, df = \overline{x^2(t)} \quad (8\text{-}5)$$

The power spectral density is a two-sided function, having values for both positive and negative frequencies. The PSD gives the power in the signal referenced to 1 Ω. That is, since no resistance is specified, $x(t)$ is interpreted as a voltage (or current)

[2] This concept is described in more detail in Chapter 6 of McGillem and George R. Cooper, *Continuous and Discrete Signal and System Analysis* (New York: Holt, Rinehart and Winston, Inc., 1974).

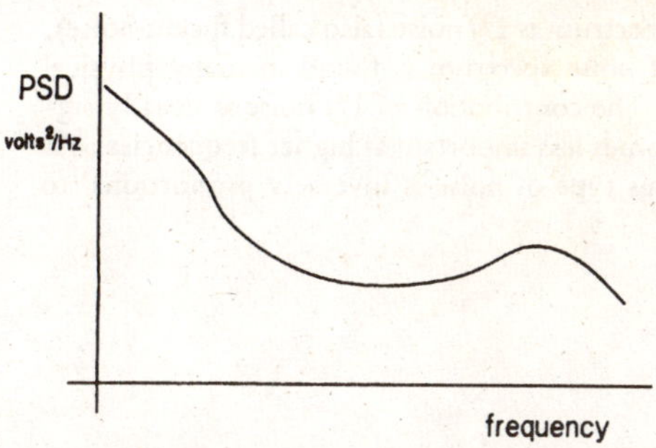

Figure 8-2. The power spectral density function shows the power density of a signal as a function of frequency.

across (through) a one ohm resistor, with the power in the resistor equal to $x^2(t)$. Figure 8-2 shows the power spectral density of a particular random signal.

The spectrum or frequency domain representation of a signal has been discussed previously in this book. Here, the emphasis should be placed on the word "density" that appears in power spectral density. The frequency domain representation of noise does not have discrete spectral lines, but instead is a continuous function of frequency which has a certain density per unit frequency. The basic units of PSD emphasize the significance. PSD has units of volts^2/Hz, so in order to state the voltage or power of a noise waveform, the measurement bandwidth must be specified.

8.4 FREQUENCY DISTRIBUTION OF NOISE

In general, noise may have any arbitrary frequency content, resulting in a variety of possible PSD shapes. Noise that has equal power density at all frequencies is called *white noise* (Figure 8-3). A strict definition of white noise requires that the flat power density characteristic extend out for an infinite bandwidth. A more practical definition requires the noise to have a flat PSD over some frequency range.

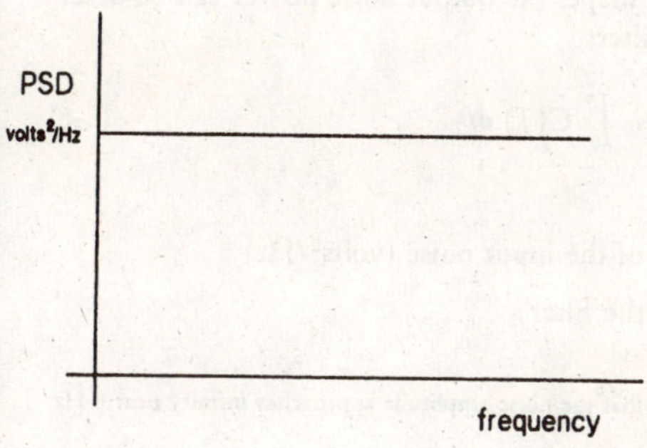

Figure 8-3. White noise has a power spectral density which is constant with frequency.

Another common type of noise spectrum is $1/f$ noise (also called flicker noise), as shown in Figure 8-4. This type of noise spectrum is found in many physical systems, including electronic circuits. The contribution of $1/f$ noise is usually significant only at low frequency and becomes less important at higher frequencies. As the name implies, the amplitude of this type of noise is inversely proportional to frequency.[3]

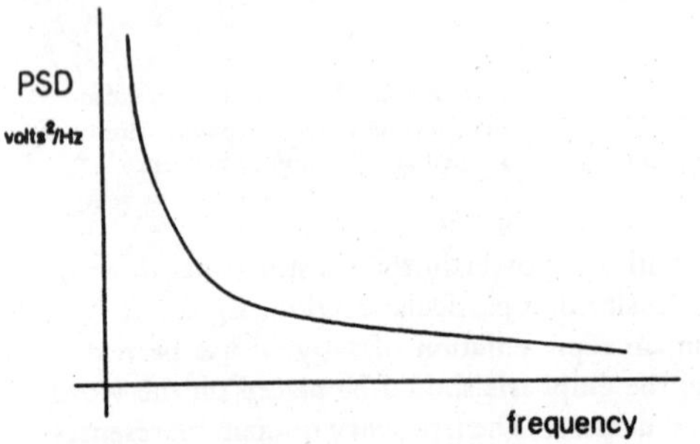

Figure 8-4. Another common type of noise is $1/f$ noise.

Other PSD shapes are possible, since they may result from a combination of electronic noise sources. In addition, the noise PSD will be affected by the components of the systems (such as filters). Usually, we can consider the noise power density to be constant over a small frequency range, which simplifies the mathematical complexity involved.

8.5 NOISE EQUIVALENT BANDWIDTH

One problem which presents itself in designing and understanding spectrum analyzers is how a filter with some arbitrary shape will respond to noise. More specifically, how much noise will be present at the output of a filter with a known power density of noise at its input? Consider the filter shape shown in Figure 8-5. The filter is a bandpass filter with a nominal gain of G_0 at its center frequency, f_0. If the input noise is constant across the filter shape, the output noise power can be determined by integrating the gain of the filter:

$$P_N = N_0 \int_0^\infty G(f)\, df \qquad (8\text{-}6)$$

where

N_0 = power density of the input noise (volts²/Hz)

$G(f)$ = power gain of the filter

[3] This model has the annoying property that the noise amplitude approaches infinity near 0 Hz.

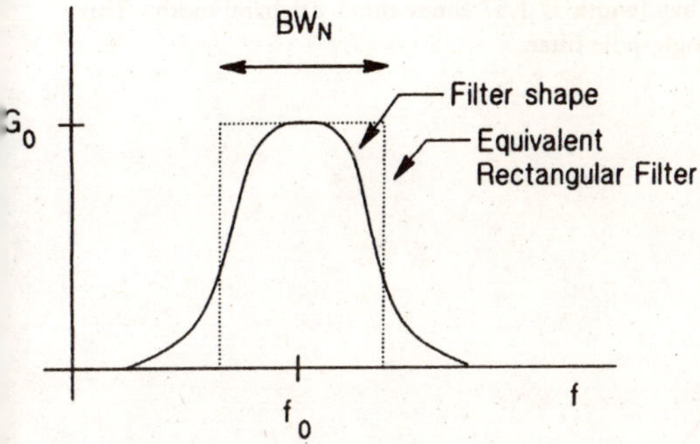

Figure 8-5. The noise equivalent bandwidth of a filter is defined by a rectangular filter that passes the same amount of white noise as the original filter.

Now suppose that an ideal rectangular filter with gain G_0 is used instead. What bandwidth will this filter need to be to produce the same noise power at the output? The output power with such a filter is given by

$$P_N = N_0 G_0 BW_N \tag{8-7}$$

where BW_N is the bandwidth of the ideal rectangular filter.

Therefore,

$$BW_N = \frac{1}{G_0}\int_0^{\infty} G(f)\, df \tag{8-8}$$

The rectangular filter bandwidth, BW_N, is called the ***noise equivalent bandwidth*** of the filter. If the noise equivalent bandwidth of a filter is known, the exact filter shape is not needed to perform noise calculations, as long as the input noise is constant over the bandwidth of the filter. Note that this definition of bandwidth is NOT the same as some of the other classical definitions such as the 3 dB and 6 dB bandwidth.

Example 8.1

What is the noise equivalent bandwidth of a single-pole low-pass filter? The filter is shown in Figure 8-6, with cutoff frequency, f_c.

A single-pole low-pass filter has the voltage transfer function

$$H(f) = \frac{f_c}{f_c + jf}$$

Taking the magnitude of $H(f)$ and squaring to get the power gain gives

$$G(f) = |H(f)|^2 = \frac{f_c^2}{f_c^2 + f^2}$$

The nominal gain $G_0 = 1$ and the noise equivalent bandwidth is

$$BW_N = \int_0^{\infty} \frac{f_c^2}{f_c^2 + f^2}\, df = \frac{\pi}{2} f_c = 1.57\, f_c$$

Therefore, the noise equivalent bandwidth is 1.57 times the 3 dB bandwidth. This relationship is valid only for a single-pole filter.

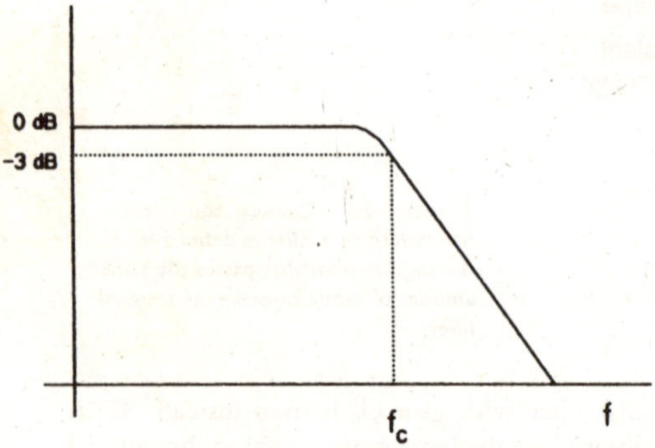

Figure 8-6. The transfer function of a single-pole low-pass filter.

8.6 NOISE UNITS AND DECIBEL RELATIONSHIPS

As previously stated, the power spectral density of noise has the units of volts^2/Hz. The volts^2 implies that this is a measure of power. The PSD can be converted to a voltage measurement by taking the square root, with the units changed to volts/$\sqrt{\text{Hz}}$. Spectrum analyzers may display the measured results this way, which is convenient when the user is working in terms of voltage.

Alternatively, it is often convenient to refer to the noise level as normalized to 1 Hz and expressed in decibel form. At this point, we will introduce the possibility of a resistance other than 1 Ω.

$$\text{Noise (dBm, 1 Hz)} = 10 \log[N_0/(Z_0 \times 1 \text{ mW})] \tag{8-9}$$

where

$$N_0 = \text{power spectral density (in volts}^2\text{/Hz)}$$

$$Z_0 = \text{resistance (impedance) of the system}$$

The 1 Hz noise level can be converted to other (noise equivalent) bandwidths, assuming that the noise density remains constant across all bandwidths of interest, by the following equations.

$$\text{Noise (dBm)} = 10 \log[BW_N\, N_0/(Z_0 \times 0.001)] \tag{8-10}$$

$$= 10 \log(BW_N) + 10 \log[N_0/(Z_0 \times 0.001)] \tag{8-11}$$

The dBm (1 Hz) value is increased by 10 log (BW_N) to obtain the new noise power adjusted to the new bandwidth. To go from one bandwidth directly to another, a correction factor can be computed

$$K_{dB} = 10 \log(BW_2/BW_1) \tag{8-12}$$

To convert from BW_1 to BW_2 add K_{dB} to the noise value associated with BW_1. Note that the bandwidth is treated with a factor of 10 in decibel form similar to power (and not voltage). This is because noise power is proportional to bandwidth, given the foregoing assumptions. The reader is urged to be careful in applying these equations since the stated assumption of constant noise power density across the bandwidths of interest must be true.

Example 8.2

Given a noise power density of 2×10^{-12} volts²/Hz and a resistance of 50 Ω, what is the noise power present in a noise equivalent bandwidth of 1 kHz? What is the noise voltage (in the same bandwidth)? Express the noise level in dBm (1 Hz). Convert the dBm (1 Hz) value to a 1 kHz bandwidth.

The noise power in 1000 Hz is

$$P_N\,(1\text{ kHz}) = (1000\text{ Hz})\,(2 \times 10^{-12}\text{ volts}^2/\text{Hz})/(50\ \Omega)$$
$$= 40\text{ pW}$$

In terms of voltage,

$$V_N = \sqrt{2 \times 10^{-12}\text{ volts}^2/\text{Hz} \times 1000\text{ Hz}}$$
$$= 44.7\ \mu\text{V}$$

$$P_N\,(1\text{ Hz}) = 10 \log(2 \times 10^{-12}\text{ volts}^2/\text{Hz}/(50\ \Omega \times 0.001\text{ mW}))$$
$$= -104\text{ dBm (1 Hz)}$$

To convert to a 1 kHz bandwidth, add 10 log (1000/1) = 30 dB:

$$P_N\,(1\text{ kHz}) = -104\text{ dBm (1 Hz)} + 30\text{ dB} = -74\text{ dBm}$$

8.7 NOISE MEASUREMENT

Since the level of noise at the analyzer detector is affected by the resolution bandwidth, the noise level on the analyzer display depends on the resolution bandwidth setting. Narrowing the resolution bandwidth reduces the displayed noise level and widening the bandwidth increases the noise level. Such a measurement is usually uncalibrated due to the unknown noise equivalent bandwidth of the resolution bandwidth filter and the unknown characteristics of the detector. Since spectrum analyzer resolution bandwidth filters are designed to be swept quickly, the filter shape is not very steep. Typically, the noise equivalent bandwidth is wider than the 3 dB bandwidth by 15% to 20%.

Most spectrum analyzer detectors are designed to detect sinusoids, which introduces an error when measuring random noise. However, for a particular detector, correction factors can be determined to account for the error in random noise measurements. Also, most spectrum analyzers have an amplifier with a logarithmic gain characteristic, which causes the noise peaks to be amplified less than the smaller noise values, introducing an additional error in the measurement. For a spectrum analyzer with a peak (envelope) detector and a logarithmic amplifier before the detector, the error correction factor is typically 2 to 2.5 dB. The correctly calibrated noise reading is given by

$$\text{Noise (dBm, 1 Hz)} = \text{spectrum analyzer reading (dBm)} + K_{det} - 10\log(BW_N) \qquad (8\text{-}13)$$

where K_{det} is the error correction factor for the detector and log amp.

Since the measurement is inherently noisy, it is usually desirable to use a large amount of video filtering or averaging to smooth out the noise reading.

If relative (not absolute) noise measurements are required, the correction factor can be eliminated. Also, it may be reasonable to assume that the noise equivalent bandwidths of the resolution bandwidth filters have a constant relationship to the resolution bandwidth. Thus, a factor of 2 change in resolution bandwidth would imply the same change in noise equivalent bandwidth.

8.8 AUTOMATIC NOISE LEVEL MEASUREMENT

The level of random noise can be measured with most any spectrum analyzer, but the calculations required to produce a calibrated measurement may be difficult. The noise equivalent bandwidth of the analyzer's resolution bandwidth filter and the characteristics of the detector must be known and the measured result must be adjusted accordingly. Most modern spectrum analyzers provide an automatic and calibrated means of measuring the spectral density of random noise. When this feature is invoked, the analyzer takes a large number of independent readings (perhaps 100 or so) and averages them together to reduce the variation in the noise measurement. The result is automatically corrected for the noise equivalent bandwidth of the resolution bandwidth filter, the logarithmic IF gain, and the characteristics of the detector. Finally, the measurement is normalized to a 1 Hz bandwidth.

Since the noise is a random variable, it would take an infinite number of measurements to produce a perfect measurement. However, a large number of samples is sufficient to limit the error in the measurement to a reasonable amount.

The noise measurement can be performed using any resolution bandwidth and the analyzer will still normalize the result to a 1 Hz bandwidth. The user must be careful not to pick too wide a bandwidth, since the noise is assumed to be white over the resolution bandwidth. The analyzer cannot discern a noise spectral density shape which varies over its resolution bandwidth. Also, any discrete spectral lines which fall inside the resolution bandwidth will be lumped into the measurement and treated as noise in the calculations, introducing an error into the noise measurement.

8.9 NOISE FLOOR

To the user of an analyzer, the noise present in a measurement (either from the circuit or signal under test or the internal noise of the analyzer) shows up as a noise trace on the analyzer display. The noise is usually relatively constant with frequency but may be worse at certain frequencies (particularly low frequencies, due to $1/f$ noise). Although the use of a narrower resolution bandwidth forces the measured noise to be lower, the spectral density of the noise remains unchanged.

When measuring spectral lines, the instrument user may narrow the resolution bandwidth as needed to lower the noise level. Thus, small signals can still be measured in the presence of noise. The situation is different when measuring random noise—the internal noise of the analyzer must be lower than the noise being measured. Reducing the resolution bandwidth does not help, since it will reduce the noise being measured along with the internal noise of the analyzer. Again, in terms of spectral density, the analyzer noise must be less than the noise being measured.

Obviously, reducing the bandwidth of the resolution bandwidth filter reduces the amount of noise in the measurement system. Most often the noise is undesirable and should be removed from the measurement so the narrowest possible predetection filter is used. Unfortunately, there are other considerations in the choice of predetection filter bandwidth, most notably measurement speed. Narrow filters cannot be swept quickly, so the time required to complete a measurement increases.

8.10 CORRECTION FOR NOISE FLOOR

If the measured noise is much larger than the internal noise of the analyzer, no significant error will be introduced. However, as the external noise level approaches the internal noise level, the measurement will be in error. Table 8-1 summarizes this

TABLE 8-1 ERROR DUE TO ANALYZER INTERNAL NOISE

Measured noise level in dB relative to internal noise	Error in measured noise, dB
20	0.04
15	0.14
10	0.46
9	0.58
8	0.75
7	0.97
6	1.26
5	1.65
4	2.20
3	3.02
2	4.33
1	6.87

effect. The left column corresponds to the noise level as measured by the analyzer while the right column indicates the amount of error in that measurement due to the analyzer's internal noise. The error is always positive (i.e., the measured value is larger than the actual noise). To obtain the actual noise level, the error (in dB) should be subtracted from the measured value. Note that even with measured noise levels as large as 10 dB above the analyzer noise floor, an error of 0.46 dB is introduced. When the actual noise equals the analyzer's noise, the measurement will appear to be 3 dB above the noise floor (with an error of 3 dB).

Example 8.3

A spectrum analyzer with a noise floor of −140 dBm (1 Hz) shows a measured value of −135 dBm (1 Hz). What is the actual noise level being measured?

The measured value is 5 dB above the internal noise floor. From Table 8-1, this produces an error of 1.65 dB. The actual noise level being measured is

$$-135 \text{ dBm} - 1.65 \text{ dB} = -136.65 \text{ dBm (1 Hz)}$$

8.11 PHASE NOISE

Phase noise is an important measure of the spectral purity of a sine wave, usually associated with synthesized (phase-locked) oscillators. In the time domain, phase noise is exhibited as a jitter in the zero crossings of the waveform (Figure 8-7). For a

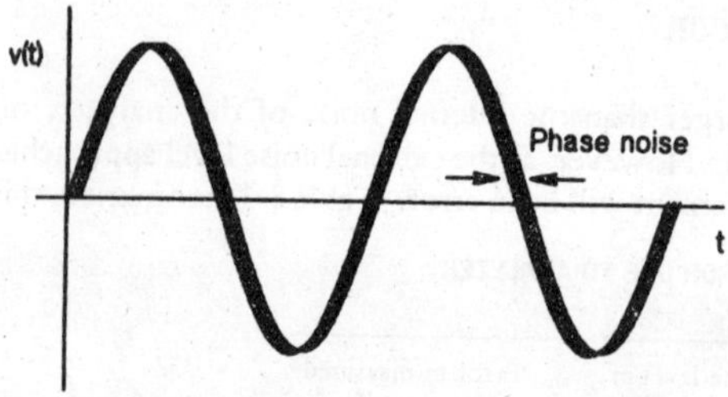

Figure 8-7. In the time domain, phase noise causes a jitter in the zero crossings of a waveform.

high-quality oscillator design, the phase noise will usually not be discernable in the time domain. In the frequency domain, the phase noise shows up as noise sidebands on the carrier (Figure 8-8).

A pure sine wave can be represented by the following equation:

$$v(t) = V_0 \sin 2\pi f_0 t \tag{8-14}$$

where

V_0 = zero-to-peak amplitude

f_0 = carrier frequency

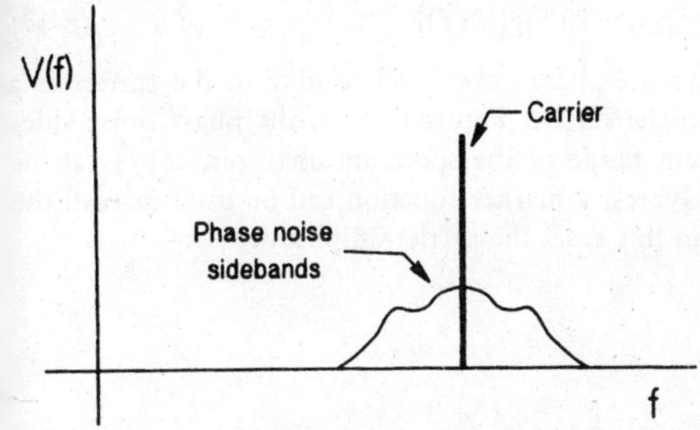

Figure 8-8. In the frequency domain, phase noise appears as noise sidebands on both sides of the carrier.

A sine wave which exhibits both amplitude and frequency fluctuations is given by

$$v(t) = [V_0 + a(t)] \sin[2\pi f_0 t + \phi(t)] \tag{8-15}$$

where

$$a(t) = \text{amplitude noise}$$
$$\phi(t) = \text{phase noise}$$

Notice that this noise process resembles the amplitude and angle modulation processes, but with the modulation "source" being random noise mechanisms in the system. The amplitude noise may be significant, even in a high-quality oscillator design. However, in many systems, the amplitude noise is removed when the signal passes through an amplitude limiting device such as a mixer. As shown in Chapter 6, combined amplitude and angle (phase) modulation results in an asymmetrical set of sidebands. We will assume that the amplitude noise is much smaller than the phase noise and then verify this assumption by comparing the upper and lower noise sidebands with the spectrum analyzer.

The phase term, $\phi(t)$, could include both long-term and short-term phase or frequency fluctuations. Long-term effects are usually specified in terms of frequency drift (and not phase noise) while the short-term effects are characterized as phase noise.

Phase noise in the frequency domain can be expressed as

$$\mathcal{L}(f) = \frac{V_N \text{ (1 Hz BW)}}{V_C} \tag{8-16}$$

where

V_N (1 Hz BW) = RMS noise level in a 1 Hz bandwidth at a frequency f Hz away from the carrier

V_C = RMS voltage of the carrier

$\mathcal{L}(f)$ is often expressed in terms of decibels,

$$\mathcal{L}(f)_{dB} = 20 \log[\mathcal{L}(f)] \quad (8\text{-}17)$$

The resulting plot of $\mathcal{L}(f)$ shows the phase noise level relative to the carrier as a function of frequency away from the carrier (Figure 8-9). If the phase noise sidebands are within the measurement range of the spectrum analyzer, $\mathcal{L}(f)$ can be measured directly. On most analyzers, a marker function can be used to read the noise level relative to an offset, in this case, the carrier amplitude.

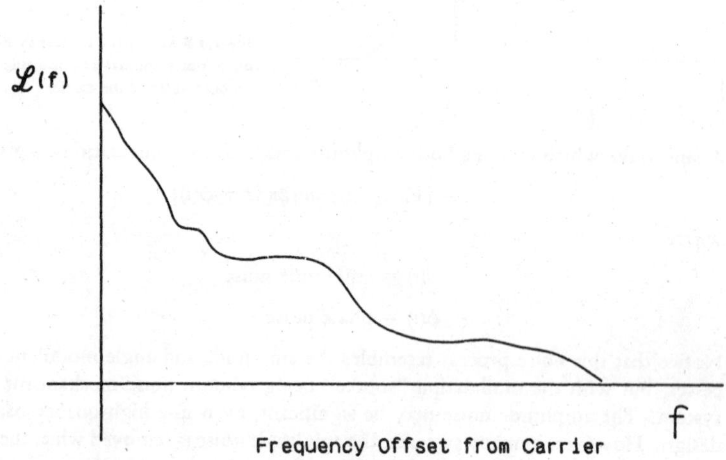

Figure 8-9. Phase noise is usually plotted as a function of frequency away from the carrier.

A problem exists as the frequency offset, f, approaches zero. The definition of phase noise becomes in question as we approach the carrier frequency. For instance, a frequency offset of 0.001 Hz (period = 17 minutes) is usually categorized as a frequency drift and not phase noise. Also, our ability to measure the noise at such a small frequency offset becomes increasingly difficult. The resolution bandwidth must be narrow enough to reject the large carrier power while still measuring the phase noise.[4]

In order for the spectrum analyzer to measure the phase noise of a sine wave directly, a few conditions must be met. Obviously, the noise floor of the spectrum analyzer must be low enough that the analyzer's internal noise is not larger than the measured phase noise. A more subtle phenomenon occurs due to the spectrum analyzer's local oscillator purity. Recall from Chapter 5 that the analyzer's local oscillator mixes with the input signal to produce a new signal at the analyzer inter-

[4] Hybrid analyzer systems employing both FFT and heterodyne techniques have been developed to provide narrowband analysis at high frequency in order to measure the close-in phase noise of high-frequency oscillators.

mediate frequency. We usually think of this process as mixing the input signal down to the intermediate frequency. Certainly the phase noise of the input signal will also be mixed down, but less obvious is that the phase noise on the local oscillator will also appear at the intermediate frequency. So a close-in measurement of phase noise on an input signal is a measurement of the combined phase noise of the input signal and the local oscillator. If the input signal were completely free of phase noise, it could be used to measure the internal phase noise of the spectrum analyzer. In order to get a good measurement of an input signal's phase noise, the analyzer's local oscillator must have lower phase noise than the signal being measured. Figure 8-10 shows the phase noise of an oscillator as directly measured by a spectrum analyzer.

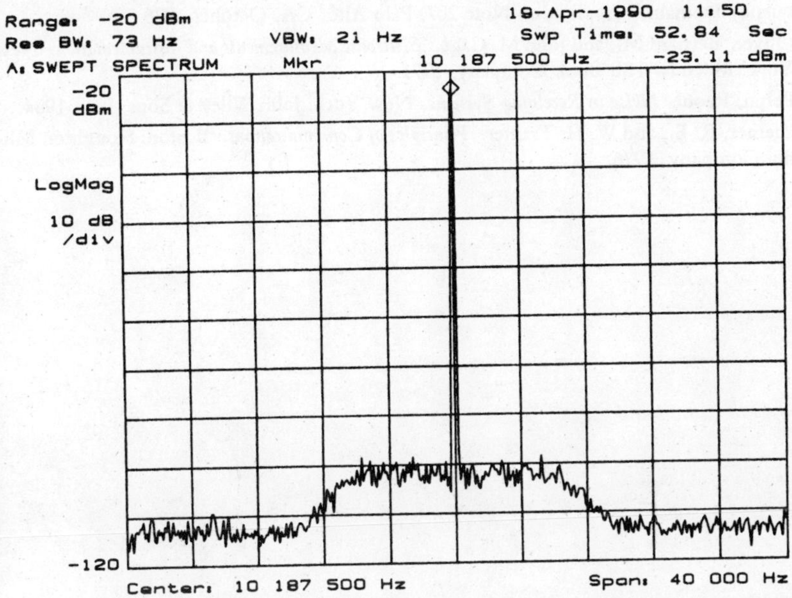

Figure 8-10. A spectrum analyzer measurement showing the close-in phase noise of an oscillator which appears as a noise pedestal.

While we have just shown that the phase noise of an oscillator can sometimes be measured directly with a spectrum analyzer, in many cases the internal phase noise and broadband noise floor of the spectrum analyzer is not good enough to perform the measurement accurately. High-quality phase noise measurements usually require external quadrature mixing and low-noise amplification of the detected result.[5]

[5] See Hewlett-Packard (1976) and Feher (1987) for more advanced phase noise measurement techniques.

REFERENCES

1. Cooper, George R., and Clare D. McGillem. *Probabilistic Methods of Signal and System Analysis.* New York: Holt, Rinehart and Winston, Inc., 1971.
2. Feher, Kamilo. *Telecommunications Measurements, Analysis, and Instrumentation.* Englewood Cliffs, NJ: Prentice Hall, Inc., 1987.
3. Engelson, Morris. *Modern Spectrum Analyzer Theory and Applications.* Dedham, MA: Artech House, 1984.
4. Hewlett-Packard Company. "Spectrum Analysis . . . Noise Measurements," Application Note 150-4, Palo Alto, CA, April 1974.
5. Hewlett-Packard Company. "Understanding and Measuring Phase Noise in the Frequency Domain," Application Note 207, Palo Alto, CA, October 1976.
6. Oliver, Bernard M., and John M. Cage. *Electronic Measurements and Instrumentation.* New York: McGraw-Hill Book Company, 1971.
7. Pettai, Raoul. *Noise in Receiving Systems.* New York: John Wiley & Sons, Inc., 1984.
8. Ziemer, R. E., and W. H. Tranter. *Principles of Communications.* Boston: Houghton Mifflin Company, 1976.

9

Pulse Measurements

Pulsed waveforms are an important class of signals in such systems as radar and digital radio. Pulsed signals can present a more difficult measurement problem than continuous waveforms. With a small-resolution bandwidth, the displayed spectrum has discrete spectral lines. With wide-resolution bandwidths, these line spectra are smeared together and the spectrum appears to be continuous. Under such measurement conditions, the settings of the spectrum analyzer greatly affect the measured results.

The principles associated with the pulsed waveform (or pulse train) are also applicable to pulsed radio frequency signals. The envelope of the spectrum is the same and depends on the pulse width, but the spectrum is centered on the RF carrier frequency.

9.1 SPECTRUM OF A PULSED WAVEFORM

As shown in Chapter 3, the Fourier transform of a single pulse has a $(\sin x)/x$ shape (Figure 9-1):

$$V(f) = \tau \frac{\sin[2\pi f(\tau/2)]}{2\pi f(\tau/2)} \tag{9-1}$$

The nulls of the spectrum occur at multiples of $1/\tau$. The amplitude of the spectrum is proportional to the pulse width. This makes sense in that the wider the pulse, the more energy present in it.

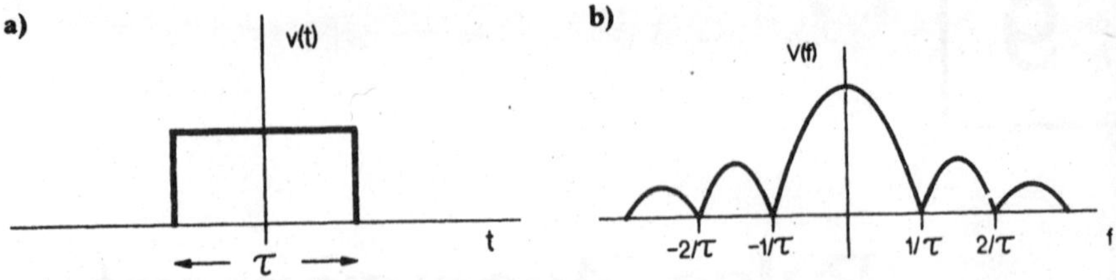

Figure 9-1. (a) A pulse in the time domain. (b) The corresponding spectrum in the frequency domain.

A swept spectrum analyzer is not capable of measuring a transient event such as a single pulse. However, an FFT spectrum analyzer can produce the spectrum of such a signal as long as it is within the bandwidth of the analyzer.

A pulse train is produced by repeating the pulse periodically (Figure 9-2a). Since the waveform is periodic, it can be expanded into a Fourier series to determine the harmonic content of the waveform. As listed in Table 3-1, the Fourier series for this waveform is

$$x(t) = \frac{\tau}{T} + \frac{2\tau}{T}\sum_{n=1}^{\infty}\frac{\sin(\pi n\tau/T)}{(\pi n\tau/T)}\cos(2\pi nt/T) \tag{9-2}$$

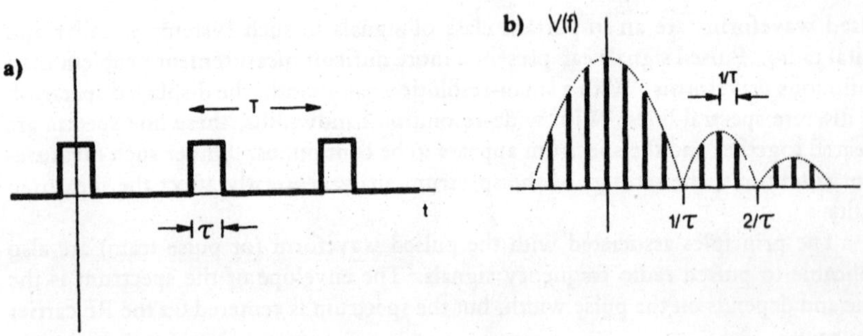

Figure 9-2. (a) A repetitive pulse train in the time domain. (b) The pulse train spectrum in the frequency domain.

The waveform has a DC component of τ/T, which is just the average value of the waveform. The harmonics of the signal will fall at multiples of the waveform frequency, which is $1/T$ (Figure 9-2b). The period of the waveform is also known as the *pulse repetition frequency* (PRF). The overall shape or envelope of the harmonics takes on the (sin x)/x characteristic, which is the same shape as the Fourier transform of the single pulse. As shown in Figure 9-2b, the envelope of the spectrum has nulls at integer multiples of $1/\tau$.

The amplitude of the spectrum of the pulse train is proportional to the time that the pulse is high. The greater amount of time that the pulse is high, the greater the power in the waveform. More specifically, the amplitude of the spectrum depends on the duty cycle of the waveform, which is the ratio of the pulse width to the waveform period:

$$\text{duty cycle} = \frac{\tau}{T} \tag{9-3}$$

The overall shape of the pulse train spectrum is determined by the width of the pulse, while the PRF determines the spacing of the spectral lines. Figure 9-3 illustrates the phenomenon. In Figure 9-3a, with $\tau/T = 1/4$, the spectral lines are widely spaced. If the PRF is decreased with τ remaining constant (Figure 9-3b), the spectral

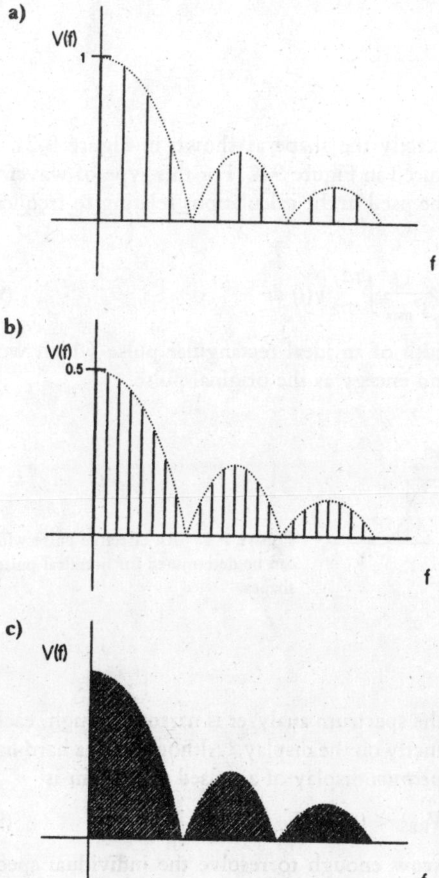

Figure 9-3. The PRF determines the spacing of the spectral lines within the $(\sin x)/x$ envelope. (a) Widely spaced spectral lines (high PRF). (b) Closely spaced spectral lines (moderate PRF). (c) Continuous spectrum (low PRF).

lines move closer together while the shape of the spectrum remains the same. Note that the amplitude of the spectrum decreases, consistent with the decrease in duty cycle of the waveform. (A factor of 2 decrease in duty cycle corresponds to a factor of 2 decrease in the amplitude of the spectrum.)

If the PRF is made very small, the spectral lines get very close together and begin to approximate a continuous spectrum (Figure 9-3c). In reality, the spectral lines are always distinct for repetitive waveforms, but as the spacing between the harmonics gets smaller than the resolution bandwidth of the spectrum analyzer, the spectrum will appear to be continuous. The amplitude of the spectrum continues to be proportional to the duty cycle of the waveform. Note that as the PRF approaches zero, corresponding to the waveform period approaching infinity, the time domain and frequency domain representations of the signal revert back to being those of a single pulse.

9.2 EFFECTIVE PULSE WIDTH

Many pulsed waveforms are not exactly the shape as shown in Figure 9-2a. An example of such a waveform is pictured in Figure 9-4. For this type of waveform we define an *effective pulse width* to be used in the calculations relating to frequency spectrum.

$$\tau_{\text{eff}} = \frac{1}{V_{\max}} \int_{-T/2}^{T/2} v(t)\, dt \qquad (9\text{-}4)$$

The effective pulse width is the width of an ideal rectangular pulse which would have the same maximum voltage and energy as the original pulse.

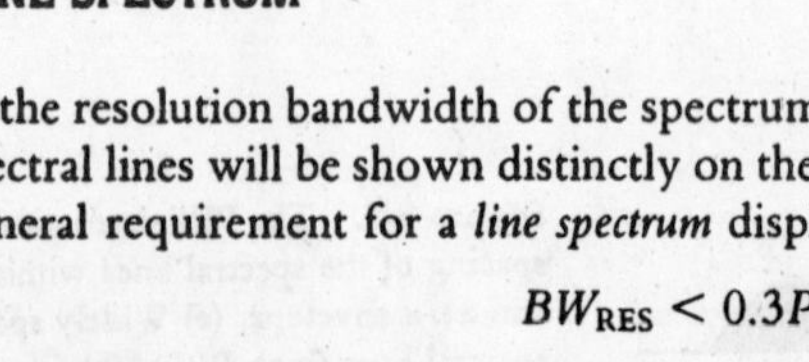

Figure 9-4. An effective pulse width can be determined for nonideal pulse shapes.

9.3 LINE SPECTRUM

When the resolution bandwidth of the spectrum analyzer is narrow enough, each of the spectral lines will be shown distinctly on the display. Although not a hard limit, the general requirement for a *line spectrum* display of a pulsed waveform is

$$BW_{\text{RES}} < 0.3PRF \qquad (9\text{-}5)$$

With the resolution bandwidth narrow enough to resolve the individual spectral lines, the spectrum measurement is fairly conventional, with the display being a close representation of the signal's spectrum. (Compare this with the pulse spectrum case discussed later.) Changing the measurement span widens or narrows the displayed spectrum as appropriate and changing the sweep time does not affect the shape of the spectrum.[1]

Example 9.1

A pulse waveform has a period of 10 μsec and a duty cycle of 10%. What is the maximum resolution bandwidth that will cause a line spectrum to be displayed?

The waveform has a period of 10 μsec, so

$$PRF = 1/T = 1/10\ \mu\text{sec} = 100\ \text{kHz}$$

The maximum resolution bandwidth is determined by

$$BW_{\text{RES}} < 0.3PRF = (0.3)(100\ \text{kHz}) = 30\ \text{kHz}$$

A typical spectrum analyzer measurement resulting in a line spectrum is shown in Figure 9-5.

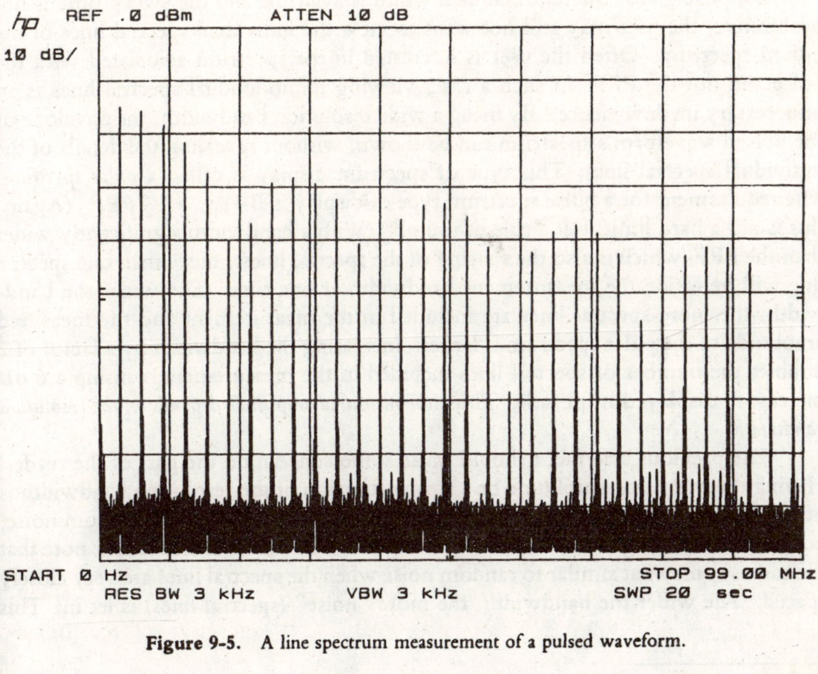

Figure 9-5. A line spectrum measurement of a pulsed waveform.

[1] As long as the sweep limitations of the resolution bandwidth are not violated.

Example 9.2

Estimate the pulse repetition frequency and the effective pulse width of the signal shown in Figure 9-5.

The spectral lines are spaced approximately every 3 MHz, so the $PRF = 3$ MHz. The first null of the spectrum envelope occurs on the sixth spectral line from the left, which is approximately 18 MHz.

$$\tau = 1/18 \text{ MHz} = 56 \text{ nsec}$$

9.4 PULSE SPECTRUM

It may seem desirable to always measure a pulsed waveform using spectrum analyzer settings which cause a line spectrum to be displayed. However, such a display may not always be possible or desirable. When the PRF is very small, the spacing of the spectral lines is also small and a suitably small resolution bandwidth may not be available. Even if such a bandwidth setting is available, the required sweep time may result in an unacceptably slow measurement.

Supposing that the required bandwidth is available and the sweep time is not prohibitive, the user may still not want to view the individual spectral lines of the pulsed spectrum. Often the user is interested in the spectrum associated with the pulse and not the PRF. In such a case, viewing the individual spectral lines is an unnecessary inconvenience. By using a wide resolution bandwidth, the envelope of the pulsed waveform's spectrum can be shown, without revealing the details of the individual spectral lines. This type of spectrum display is called a *pulse spectrum*. The requirement for a pulse spectrum type of display is $BW_{RES} > 1.7PRF$. (Again. this is not a hard limit, but a rule of thumb.) With a bandwidth significantly wider than the PRF (which is also the spacing of the spectral lines), more than one spectral line will be inside the measurement bandwidth at one time. The wider the bandwidth, the more spectral lines are included in the measurement and the measured amplitude of the pulse spectrum is larger. Increasing the bandwidth by a factor of 2 doubles the number of spectral lines included in the measurement, causing a 6 dB increase in displayed amplitude. Thus, *the measured amplitude depends on the resolution bandwidth*.

The previous statement should cause some concern on the part of the reader! Having the measured amplitude be a strong function of the resolution bandwidth is not a desirable feature.[2] One might expect this type of behavior with random noise, but not when the signal has discrete spectral lines. On closer examination, note that the case is somewhat similar to random noise when the spectral lines are very closely spaced. The wider the bandwidth, the more "noise" (spectral lines) is let in. This

---

[2] Unless one likes the freedom of adjusting the measuring instrument until the desired answer appears.

type of signal is sometimes categorized as "impulse noise," which implies a large number of closely spaced spectral lines.

The resolution bandwidth must not be too large; otherwise the envelope of the pulsed spectrum may become washed out. The resolution bandwidth must remain small compared to $1/\tau$, which defines the spacing of the nulls in the spectrum envelope. To summarize both constraints, the resolution bandwidth must be larger than the PRF, but significantly smaller than $1/\tau$.

$$1.7PRF < BW_{\text{RES}} < 0.1/\tau \tag{9-6}$$

With a swept spectrum analyzer, the sweep time can interact with the PRF to produce discrete spectral lines. If the sweep time is set much greater than the period of the pulse train, the pulse spectrum is continuous. With faster sweep times, the on/off rate of the pulse train can show up as spectral lines. For example, if the sweep time is 100 msec and the pulse waveform repeats every 5 msec, spectral lines will occur at every 5 msec during the sweep, which corresponds to every 1/20 of the frequency span. If the frequency span is 200 MHz wide, these spectral lines would appear every 10 MHz. Changing the sweep time to 50 msec will cause the responses to appear at every 1/10 of the frequency span, which would appear to be spaced every 20 MHz. Clearly, both cases are misleading as to the actual spectral content of the signal. When individual spectral lines are visible on the display, they no longer represent the actual spacing of the harmonics in frequency, but instead occur every 1/PRF seconds during the analyzer's sweep. Increasing the sweep time to much greater than 1/PRF will eliminate this effect and cause the spectrum to appear as a continuous $(\sin x)/x$ function. A useful guideline is

$$\text{sweep time} \geq \frac{100}{PRF} \tag{9-7}$$

which will produce at least 100 spectral lines in the spectrum.

Example 9.3

A pulse waveform has a period of 1 msec and a pulse width of 500 nsec. Determine the limitations on sweep time and resolution bandwidth for a pulse spectrum measurement.

The $PRF = 1/(1\text{ msec}) = 1\text{ kHz}$, $\tau = 500\text{ nsec}$

$$1.7\,PRF < BW_{\text{RES}} < 0.1/\tau$$

$$1.7\,(1\text{ kHz}) < BW_{\text{RES}} < 0.1/(500\text{ nsec})$$

$$1.7\text{ kHz} < BW_{\text{RES}} < 200\text{ kHz}$$

$$\text{sweep time} \geq \frac{100}{PRF} = 100/1\text{ kHz} = 100\text{ msec}$$

Figure 9-6 shows a typical measurement of a pulsed waveform resulting in a pulse spectrum.

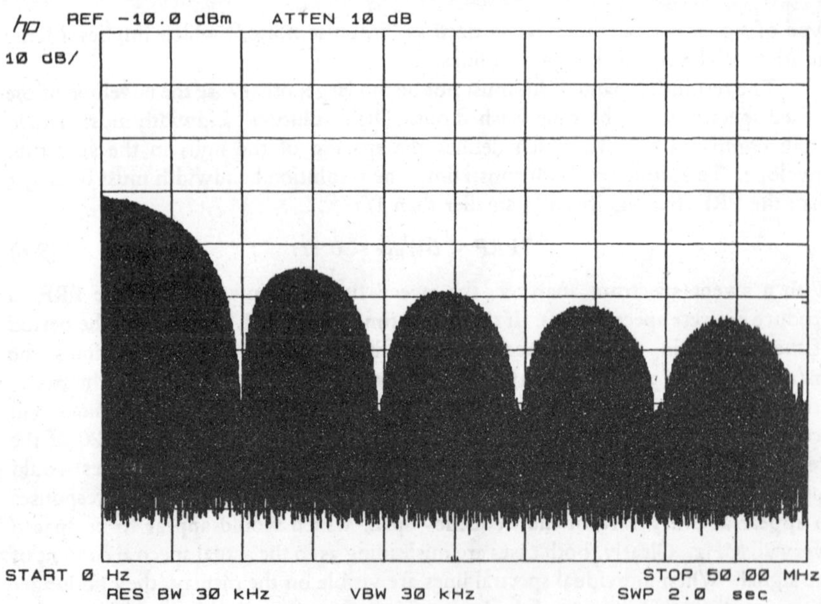

Figure 9-6. A pulse spectrum measurement of a pulsed waveform.

Example 9.4

Determine the effective pulse width of the signal shown in Figure 9-6.

The first null of the pulse spectrum occurs at approximately 10 MHz, so the effective pulse width is τ = 1/10 MHz = 100 nsec.

9.5 PULSED RF

Another signal which is closely related to the pulse train is the *pulsed sinusoid or pulsed RF* (Figure 9-7a). This type of signal can be derived by pulse modulating a radio frequency carrier (i.e., using the pulse train to turn the carrier on and off.) Radar signals are one common example of pulsed RF. The modulation property from Table 3-3 can be used to derive the pulsed RF spectrum from our previous results. The modulation property is described by the following transform pair:

$$x(t)\cos(2\pi f_0 t) \leftrightarrow 1/2[X(f-f_0) + X(f+f_0)] \qquad (9\text{-}8)$$

$X(f)$, the spectrum of the modulating signal $x(t)$ appears centered on the carrier frequency, f_0. Since the two-sided transform is used, the modulating spectrum occurs at both $\pm f_0$. In the case of pulsed RF, the modulating signal is the pulse train,

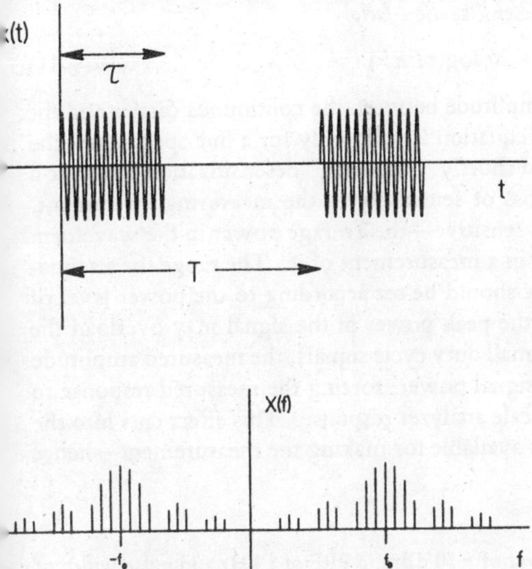

Figure 9-7. (a) A pulsed RF signal in the time domain. (b) The frequency spectrum of a pulsed RF signal.

so the $(\sin x)/x$ shaped spectrum is no longer centered on the origin but is centered on the carrier frequency (Figure 9-7b).

Since the pulsed RF case can be directly related to the baseband pulse train, the principles derived relative to the pulse train are also valid for pulsed RF. For example, when measuring a pulsed RF signal with a spectrum analyzer, the display may show discrete spectral lines or may show a continuous "pulse" spectrum, depending on the pulse repetition frequency.

9.6 PULSE DESENSITIZATION

The RMS value of a pulsed RF signal is directly proportional to the duty cycle of the waveform. The extreme cases are when the RF carrier is left "on" all of the time and when the RF carrier is always "off." In between, the RMS value depends on the duty cycle of the waveform, τ/T. The RMS value of the pulsed RF waveform is given by

$$V_{\text{RMS}} = V_{\text{CAR}}\ \tau/T = V_{\text{CAR}}\ \tau PRF \qquad (9\text{-}9)$$

where V_{CAR} is the RMS voltage of a constant carrier. This equation can be expressed in decibel form as

$$V_{\text{RMS(dB)}} = V_{\text{CAR(dB)}} + 20\log(\tau PRF) \qquad (9\text{-}10)$$

It is customary to define the *pulse desensitization factor*

$$\alpha_L = 20 \log(\tau PRF) \tag{9-11}$$

which represents the difference in amplitude between the continuous carrier and the pulsed RF signal (in decibels). This equation is valid only for a line spectrum—the pulse spectrum case will be discussed shortly. The term "desensitization" may be a poor choice since it may imply a loss of sensitivity in the measuring instrument. The instrument is not really any less sensitive—the average power in the waveform decreases, which should be reflected in a measurement of it. The range (or attenuator setting) of the spectrum analyzer should be set according to the power level of the continuous carrier. Otherwise, the peak power in the signal may overload the input circuitry of the analyzer. For small duty cycle signals, the measured amplitude will be much smaller than the peak signal power, forcing the measured response to be considerably lower than the full-scale analyzer response. This effect cuts into the dynamic range of the analyzer that is available for making the measurement—hence the term "pulse desensitization."

Example 9.5

A pulsed RF signal has a peak power of −10 dBm, a PRF of 1 kHz and pulse width of 10 μsec. What is the amplitude of the main lobe of the spectrum? If the spectrum is to be measured with 40 dB of dynamic range, what is the required dynamic range of the spectrum analyzer (assuming that full scale on the spectrum analyzer corresponds to −10 dBm)?

The power of a continuous carrier is −10 dBm.

$$\begin{aligned} V_{RMS(dB)} &= V_{CAR(dB)} + 20 \log(\tau PRF) \\ &= -10 \text{ dBm} + 20 \log(10\ \mu\text{sec } 1 \text{ kHz}) \\ &= -10 \text{ dBm} - 40 \text{ dB} = -50 \text{ dBm} \end{aligned}$$

The amplitude of the main lobe is −50 dBm.

Another 40 dB below the main lobe is −90 dBm. Therefore, the spectrum analyzer must measure −90 dBm with full scale equal to −10 dBm, which requires 80 dB of dynamic range.

In the case of the pulse spectrum, the situation is different. In addition to the pulse width, the measured amplitude also depends on the resolution bandwidth of the spectrum analyzer. The pulse desensitization factor for the pulse spectrum case is defined as

$$\alpha_p = 20 \log(\tau BW_{IMP}) \tag{9-12}$$

The resolution bandwidth of a spectrum analyzer is normally specified in terms of a 3 dB bandwidth, which is useful in most cases, but is not appropriate when analyzing pulsed signals. Instead, we introduce a new concept of bandwidth called the *effective impulse bandwidth*, BW_{IMP}. This bandwidth is the bandwidth of an ideal rectangular filter which has a pulse response equivalent to the actual IF filter.

Alternatively, we can say that the typical IF filter behaves like it is wider than the normal 3 dB bandwidth when the input signal is a pulse. Mathematically, we can state

$$BW_{\text{IMP}} = K \times BW_{\text{RES}} \tag{9-13}$$

where K is a factor relating the resolution bandwidth and the impulse bandwidth. A typical value for K is 1.5, assuming a typical synchronously tuned IF filter. For a pulse spectrum, reduced τ or BW_{IMP} decreases the amplitude of the measured response. Reduced τ causes the average power in the signal to decrease, while reduced BW_{IMP} leaves the average power in the signal unchanged but reduces the amount of energy present in the analyzer's IF bandwidth. Either way, the measurement is "desensitized" in the sense that the measured reading will be lower.

REFERENCES

1. Adam, Stephen F. *Microwave Theory and Applications.* Englewood Cliffs, NJ: Prentice-Hall, Inc., 1969.
2. Engelson, Morris. *Modern Spectrum Analyzer Theory and Applications.* Dedham, MA: Artech House, 1984.
3. Hewlett-Packard Company. "Spectrum Analysis . . . Pulsed RF," Application Note 150-2, Palo Alto, CA, November 1971.

10

Averaging and Filtering

Many analyzer measurements have considerable amounts of noise present in them. Except in those cases in which the noise is to be measured, the noise can be considered undesirable. Two basic techniques are used to reduce the noise—filtering and averaging.[1] Filtering usually takes the form of an analog filter, but it can also be implemented in digital form, while averaging is almost always done digitally. The two concepts are closely related and are treated here in a unified manner. Both filtering and averaging can be classified as either *predetection* (before the detector) or *postdetection* (after the detector). Predetection averaging/filtering reduces the noise present in a measurement while postdetection averaging/filtering reduces the amount of fluctuation in the noise.

10.1 PREDETECTION FILTERING

In the spectrum or network analyzer block diagram, filtering can be broken down into two types—predetection and postdetection—depending on whether the filter resides before or after the detector, as shown in Figure 10-1. In most swept analyzers, the detector is implemented as a distinct circuit block, although digital signal processing techniques can be used instead. There is no detector circuit in an FFT analyzer—instead the magnitude detection is done by computing the magnitude of

[1] It can be argued that averaging is a special case of filtering.

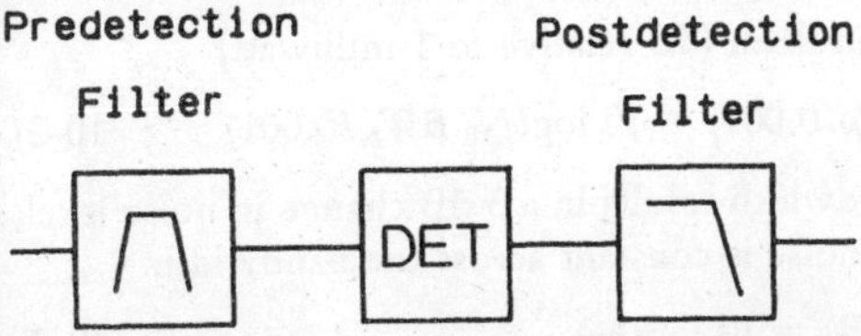

Figure 10-1. Predetection filtering takes place in front of the detector while postdetection filtering is performed behind the detector.

the complex frequency domain data provided by the FFT algorithm. Whether there is or is not an actual detector circuit, the concept remains the same.

Noise

There is always some noise present in the front end and IF sections of an analyzer. This noise may come from the signal or network being measured or may be generated internal to the analyzer since the analyzer's circuits contribute noise. Noise present at the input of the detector degrades the measurement depending on the size of the noise relative to the amplitude of the signal. As shown in Figure 10-2, the wider the predetection bandwidth, the more noise that gets included in the measurement at the detector. Because the resolution bandwidth of the analyzer is usually narrow, the noise can be considered constant or white across its passband. The noise power is

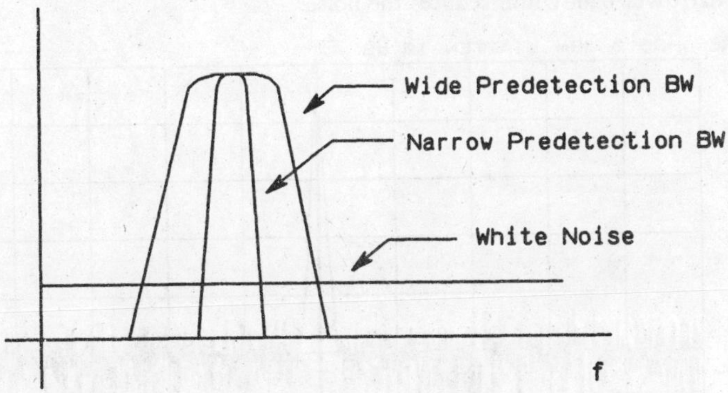

Figure 10-2. A wide predetection filter lets more noise into the measurement than does a narrow predetection filter.

$$P_N = N_0 \, BW_N \tag{10-1}$$

where

N_0 = noise spectral density (watts/Hz)

BW_N = the noise equivalent bandwidth of the predetection filter (Hz)

The noise power may be expressed in dBm (dB relative to 1 milliwatt).

$$\text{Noise (dBm)} = 10 \log(P_N/0.001) = 10 \log(N_0\ BW_N/0.001) \qquad (10\text{-}2)$$

Note that a factor of 2 change in bandwidth results in a 3 dB change in noise level, assuming the spectral density of the noise is constant across the bandwidth.

$$\Delta P_N(\text{dB}) = 10 \log(k_B) \qquad (10\text{-}3)$$

where

ΔP_N = change in noise power

k_B = ratio of the two noise equivalent bandwidths

k_B	ΔP_N
2	3.01 dB
5	6.99 dB
10	10.00 dB

Figure 10-3 shows the effect on the analyzer display due to changing the resolution bandwidth. With a wider bandwidth, the noise on the display is higher while a narrower bandwidth reduces the noise.

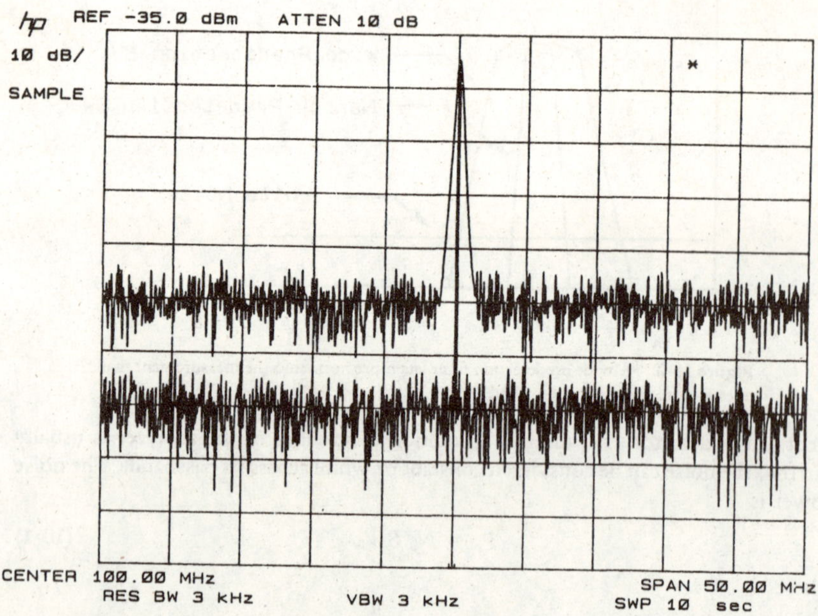

Figure 10-3. The upper trace was measured with a 300 kHz resolution bandwidth and lower trace was measured with a 3 kHz resolution bandwidth. The displayed noise level is 20 dB lower with the narrower bandwidth.

Example 10.1

A spectrum analyzer measures the noise at a particular frequency to be −70 dBm using a resolution bandwidth of 1 kHz. What will the noise reading be using a resolution bandwidth of 300 Hz? Assume that the noise is white noise and that the bandwidths given are noise equivalent bandwidths.

The noise will be reduced by the ratio of the bandwidths, expressed in dB.

$$\text{Noise (dBm)} = -70\text{ dBm} + 10\log(300/1000) = -75.2\text{ dBm}$$

10.2 PREDETECTION FILTERS

Predetection filters are easily identified in the block diagrams of traditional swept analyzers. The narrowest filter in the signal path before the detector is effectively the predetection filter. This filter's bandwidth is usually selectable and is indicated on the front panel as resolution bandwidth or IF bandwidth.

In the case of the FFT analyzer, the predetection bandwidth is not a distinct filter in the block diagram. Instead, it is the effective bandwidth resulting from the use of the FFT. To a rough approximation, the predetection bandwidth will be the frequency span divided by the number of displayed points. To be more exact, this number must be adjusted depending on the time window function, with the actual bandwidth usually being somewhat larger. Since the number of displayed points is usually constant, the only choice available to the user is selection of frequency span. Most FFT analyzers will indicate the effective predetection bandwidth for a given measurement setting, either via the operator's manual or via the display. For noise measurements it is important to always use the noise equivalent bandwidth which is not necessarily the same as the 3 dB bandwidth (see Chapter 8).

10.3 POSTDETECTION FILTERING

Filters that reside after the detector in the signal processing chain are called postdetection filters. On swept analyzers, postdetection filters are usually called video filters. Postdetection filtering is not capable of reducing the noise level since the noise has already been detected. However, it can reduce the variations in the noise, exposing previously obscured signals which are near the noise floor. Also, if noise is being measured, postdetection filtering helps stabilize the measurement. Notice that in Figure 10-3, there is considerable variation in the amplitude of the noise, independent of which resolution bandwidth is used.

One can think of the output of the detector (over a short period of time) as being a constant DC value with some noise superimposed on it (Figure 10-4a). The DC level represents the amount of energy present within the predetection bandwidth in front of the detector. This energy could be made up of discrete spectral lines or noise or both. The noise on the DC level is the statistical variation in the

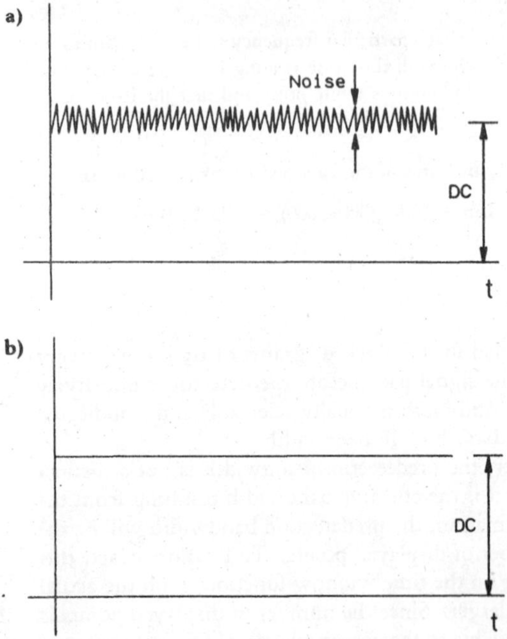

Figure 10-4. (a) The output of the detector consists of a constant DC level plus some noise. (b) Low-pass filtering the detector output removes the noise without altering the DC level.

predetection energy, which is caused by the noise in the measurement. A low-pass filter applied to the detector output can reduce the variation in the detector output (Figure 10-4b), giving a more stable, noise-free output. This does not, however, reduce the DC level. Thus, postdetection filtering can reduce the variation in the detector's output but does not affect the average output level.

It is important to distinguish between predetection and postdetection noise. Predetection noise can be reduced by narrower predetection filtering, thereby reducing the level out of the detector. The predetection noise will be detected and will contribute to the absolute level at the output of the detector. Postdetection noise is the variation in the predetection noise. This variation can be reduced by appropriate filtering, but the DC level which represents predetection noise cannot be reduced by postdetection filtering.

To understand the effect of postdetection filtering on a typical measurement, consider the typical analyzer display shown in Figures 10-5a and b. With a wide postdetection filter, the variance in the noise is quite large. With a narrower postdetection filter, the variance is reduced considerably. Note that the average value of the measured noise remains the same, only the variation in the noise is different. While postdetection filtering does not lower the average noise level, the reduction in the variance does reduce the peak noise level and may expose low-level signals that cannot be observed with a wider postdetection bandwidth.

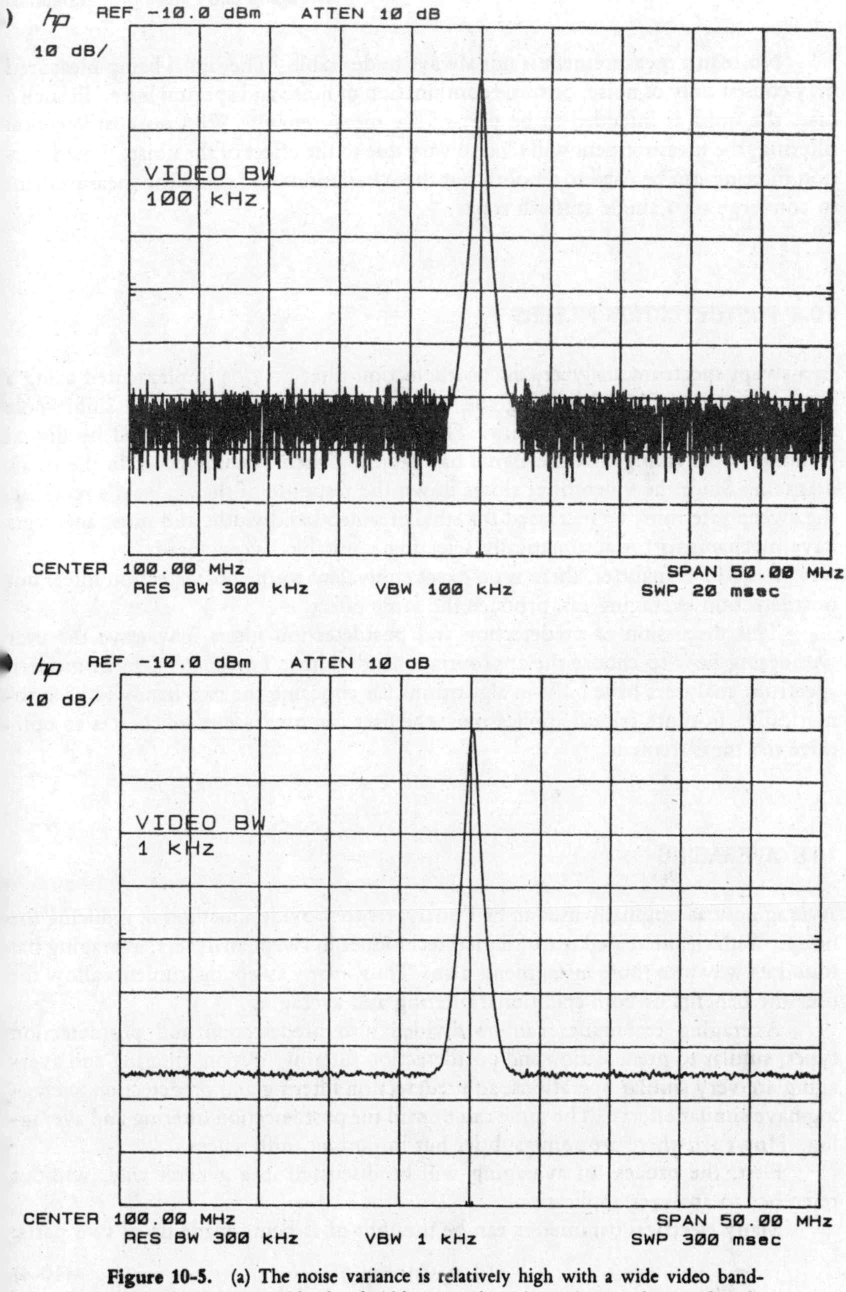

Figure 10-5. (a) The noise variance is relatively high with a wide video bandwidth. (b) A narrow video bandwidth causes the noise variance to be significantly reduced.

Noise in a measurement is not always undesirable. The signal being measured may consist only of noise, or some combination of noise and spectral lines. In such a case, the noise is intended to be part of the measurement. With no postdetection filtering, the measurement will tend to vary due to the effect of the noise. Postdetection filtering can be used to smooth out these variations and cause the measurement to converge on a single smooth trace.

10.4 POSTDETECTION FILTERS

In a swept spectrum analyzer, the postdetection filter may be implemented using a low-pass analog filter following the detector. This filter is usually a single-pole filter, often a simple RC network. The filter can also be implemented by digital techniques providing that the signal has been digitized at some point in the block diagram. Since the video filter slows down the response of the analyzer's receiver, the sweep rate must be increased for smaller video bandwidths and most analyzers have mechanisms for automatically selecting a suitable sweep speed.

In an FFT analyzer, there is no exact equivalent to the postdetection filter, but postdetection averaging can produce the same effect.

This discussion of predetection and postdetection filters may leave the user wondering how to choose the appropriate bandwidths. Fortunately, most modern spectrum analyzers have built-in algorithms for choosing the two bandwidths automatically. In more critical applications, the user can override these choices to optimize the measurement.

10.5 AVERAGING

Averaging was originally used in FFT analyzers to provide a method of reducing the noise. With the increased use of digital techniques in swept analyzers, averaging has found its way into those instruments also. Thus, many swept instruments allow the user the benefits of both traditional filtering and averaging.

Averaging techniques can be divided into predetection and postdetection types, similar to predetection and postdetection filtering. Again, filtering and averaging are very similar operations, so predetection filtering and predetection averaging have similar effects. The same can be said for postdetection filtering and averaging. However, there are some subtle, but important, differences.

First, the process of averaging will be discussed in a general way, without reference to analyzer applications.

Many electrical parameters can be thought of as being made up of two parts:

$$x(t) = s(t) + n(t) \tag{10-4}$$

where

$$x(t) = \text{the measured value}$$

$$s(t) = \text{the desired signal to be measured}$$

$$n(t) = \text{the noise contaminating the signal}$$

Noise and signals contaminated by noise must be treated on a statistical basis. The variance, σ^2, is defined as

$$\sigma^2 = E[x^2] - E^2[x] \tag{10-5}$$

where $E[\ ]$ indicates the expected value.

The variance is the square of the standard deviation, σ. As the name implies, the variance is a measure of how much a noisy parameter varies away from its average value. If the noise has zero mean, then the average value of $x(t)$ equals $s(t)$, the desired signal.

Usually when a measured parameter is averaged, the signal portion of $x(t)$ will be retained while the noise portion, $n(t)$, will be decreased. This assumes that the signal portion is correlated with the sample rate, producing the same value on each sample. Similarly, the noise is assumed to be uncorrelated to the sample rate and will vary in value with each sample. Any portion of $x(t)$ that is correlated to the sample rate will tend to be retained after averaging. Any portion that is uncorrelated will tend to be averaged out.

10.6 VARIANCE RATIO

Averaging can be considered as a process with an input, $x(t)$, and an output, $y(t)$, as shown in Figure 10-6. Both the input and the output have corresponding variances, σ_x^2 and σ_y^2. By averaging, the variance of the measured signal is reduced and $y(t)$ is a better approximation to the desired signal, $s(t)$. The *variance ratio* (VR) is used as a figure of merit for the averaging process.

$$VR = \frac{\sigma_y^2}{\sigma_x^2} \tag{10-6}$$

where

$$\sigma_x^2 = \text{variance of the unaveraged signal}$$

$$\sigma_y^2 = \text{variance of the averaged signal}$$

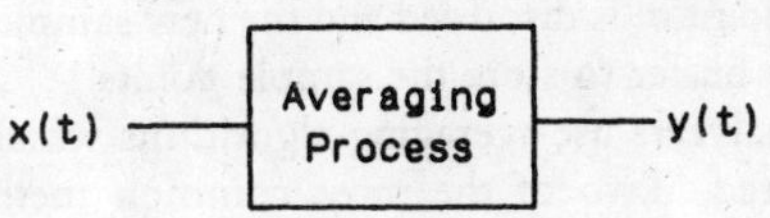

Figure 10-6. Averaging produces an output signal, $y(t)$, which has a lower variance than the input signal, $x(t)$.

The variance of a signal is associated with its noise power (not its voltage). The *standard deviation,* which is the square root of the variance, should be used to analyze noise in terms of voltage. Since the variance ratio is power related, it can be converted to decibel form by

$$VR_{dB} = 10\log(VR) \tag{10-7}$$

Example 10.2

For a given averaging process, the variance of the averaged output is 0.2 times the variance of the input. If the noise power at the input is −45 dBm, what is the noise power at the output of the averaging process?

$$\sigma_y^2 = 0.2\,\sigma_x^2$$

$$VR = 0.2$$

$$VR_{dB} = 10\log(0.2) = -6.99\text{ dB}$$

$$\text{Noise (dBm)} = -45\text{ dBm} - 6.99\text{ dB} = -51.99\text{ dBm}$$

10.7 GENERAL AVERAGING

In general, averaging is accomplished by sampling a signal at discrete points in time, weighting the samples according to some algorithm, and adding them together. Mathematically, this can be expressed as

$$y_N = w_1x_1 + w_2x_2 + w_3x_3 + \cdots + w_Nx_N \tag{10-8}$$

where

$w_1, w_2, w_3, \ldots, w_N$ = weighting factors

$x_1, x_2, x_3, \ldots, x_N$ = last N samples of $x(t)$

y_N = current averaged output

The variance ratio for the averaging process can be shown to be just the sum of the squares of the weighting factors.

$$VR = \sum_{n=1}^{N} w_n^2 \tag{10-9}$$

Such a general approach to averaging requires that N samples of $x(t)$ be always stored with the averaged output, y_N computed by summing the weighted x_n values. As a new x_n sample is acquired, the oldest x_n is discarded and the new sample saved. (This requires some type of memory buffer to store the sample points.)

Most spectrum and network analyzers use averaging algorithms which minimize the amount of memory required. Two of the most common methods of weighting the samples will be described here. These weighting functions greatly reduce the storage requirements (and, to a certain extent, the computational complexity) while providing significant noise improvement.

10.8 LINEAR WEIGHTING

The most obvious way to weight the data is to weight them all equally. After all, what makes one sample point any more valid than any other point? This type of averaging is probably what most engineers think of when the term "averaging" is used.

The variance ratio for linear averaging is

$$VR = \frac{1}{N} \quad \text{(linear averaging)} \tag{10-10}$$

where N is the number of samples averaged. Thus, for N measurements averaged together ("N averages"), the noise power is reduced by a factor N and the noise voltage is reduced by a factor of the square root of N.

When linear weighting is used in instrumentation averaging, the final averaged result (with all N measurements averaged together) cannot be displayed until all N measurements have been acquired. Many instruments will display the intermediate results of the averaging process so the user has some measurement information without having to wait for all N acquisitions. Linear weighting is usually implemented as a terminating average. That is, after N measurements are taken, averaged, and displayed, the data are discarded and any subsequent averaging starts over with completely new data.

For a given number of samples, linear weighting provides the best possible variance ratio. The more samples that are averaged together, the better the noise reduction at the expense of longer measurement time.

10.9 EXPONENTIAL WEIGHTING

The weighting function may be an exponential function with the most recent sample weighted the highest and previous samples weighted exponentially less. Although an exponential weighting function would seem to be computationally complex, it can be implemented with a simple algorithm. The averaged output is computed by summing the input sample multiplied by a factor of $1/k$ and the previous result multiplied by $1 - (1/k)$. A single accumulation register (memory location) is used to hold the y_{n-1} (previous) value. In a typical analyzer application, an accumulation register is required for each displayed frequency point or bin.

$$y_n = (1/k)x_n + (1 - (1/k))y_{n-1} \tag{10-11}$$

Exponential averaging is not usually implemented as a terminating average, since there is no number N that defines the number of samples to be acquired. Instead, the averaging algorithm can continue to run indefinitely. A newly acquired sample is first weighted heavily and then the weighting factor for that sample gradually decreases as additional samples are taken. This type of algorithm has the ability to track changes in the measured value.

The variance ratio, assuming that the averaging process has been running for a very long time (much more than k samples), is

$$VR = \frac{1}{2k - 1} \quad \text{(exponential averaging)} \tag{10-12}$$

The exponential weighting function produces a step response very similar to a single-pole low-pass filter (Figure 10-7). If the input of the averaging process starts at zero and abruptly changes to a constant value, the output of the process rises exponentially and asymptotically approaches the final value of the input. For k's of interest, the time constant of this system, T is given by

$$T = k + 1/2 \tag{10-13}$$

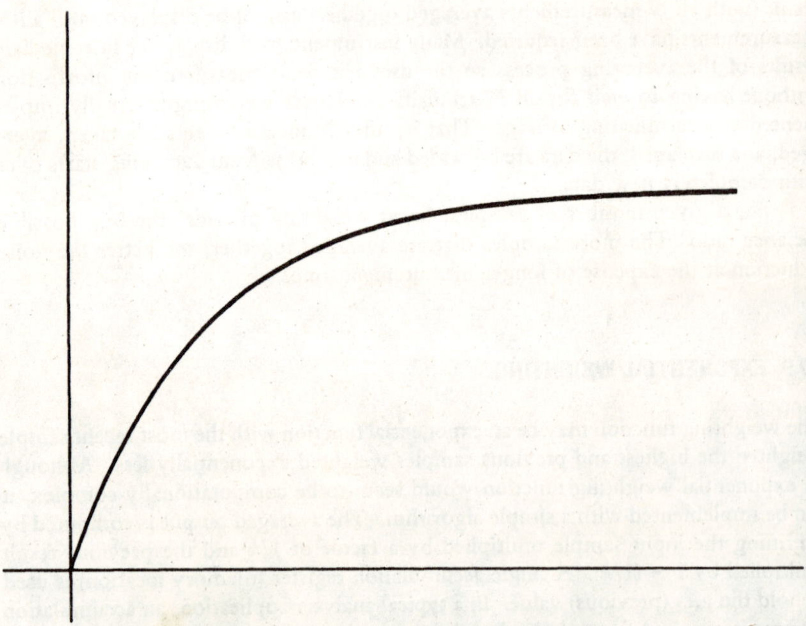

Figure 10-7. The step response of an averaging process with exponential weighting is an exponential function.

Thus, the time constant of the system is approximately k. After k samples the step response will reach 63% of the final value, just as one would expect in a single-pole analog system. In a purely exponential average, the user should keep this behavior in mind and wait several time constants for the measurement to settle out. Selecting a large value for k will provide the maximum amount of noise reduction, but at the expense of increased measurement time and slower response to changes.

The exponential average has an initialization problem, in that if the y_{n-1} accumulation register starts out set to zero it will take k samples to get 63% of the way to the final value. It will take even longer before the averaging process produces an output close to the true answer. One solution is to load the first sample into the y_{n-1} register and then let the averaging algorithm run. This immediately gets the averaged output close to the correct answer (depending on how "good" the first sample is). Unfortunately, this also causes the first sample to be weighted much heavier than the others, which is most noticeable with large k's since the subsequent samples are weighted very lightly.

A better solution to the initialization problem is to start the averaging algorithm with k small and load the first sample into the y_{n-1} register. As the averaging algorithm progresses, the k value is automatically increased, eventually stopping at the value selected by the user. This technique has the effect of producing a linear or near-linear weighting on the early samples. Later, when the k value reaches the maximum imposed by the user, the averaging algorithm reverts to a pure exponential. This type of weighting combines the advantages of linear weighting and exponential weighting. It provides the good variance improvement of a linear average in the early samples but with the exponential's advantage of ability to track changes. Thus, the modified exponential weighting algorithm is commonly used in electronic instruments.

10.10 AVERAGING IN SPECTRUM AND NETWORK ANALYZERS

The previous discussion has centered on the general principles of averaging along with the common weighting functions. Now we will focus on how these weighting functions are applied to the sampled data in a spectrum or network analyzer.

First, it must be made clear in what dimension the averaging is taking place. *Trace-to-trace averaging* is accomplished by taking a sample at one particular frequency or bin and averaging it with samples *at the same frequency* from other traces or sweeps. This represents most of the averaging algorithms found in spectrum and network analyzers.

Adjacent-point averaging is accomplished by averaging several data points together from the same sweep. For example, the nth bin might be computed by the linear average of the nth − 1, nth, and nth + 1 bins. This type of averaging is used to implement smoothing functions.

Scalar Data

Next, the particular type of data that is to be averaged must be considered. For most spectrum analyzers,[2] there is only one type of data to be averaged—scalar data produced by sampling the output of the detector. Since these data exist after the detector, only postdetection averaging is possible. The behavior of scalar (postdetection) averaging is very similar to postdetection (video) filtering. The average noise level remains the same, but the variance of the noise is reduced. More precisely, the output of the averaging process approaches the mean of the signal plus the noise.

Vector Data

Vector network analyzers and FFT spectrum analyzers have two types of data which can be averaged—scalar data and vector data. Vector data are represented by a complex number in a real + *j* imaginary form. The vector data appear in front of the detector and contain the magnitude and phase information necessary for vector network measurements. During detection, the vector data are converted into scalar magnitude information, the same as the spectrum analyzer case.

A vector representing a complex signal can be plotted on the complex plane (Figure 10-8a). The vector has a real part and an imaginary part, which can be converted into magnitude and phase form, if necessary. Noise present in the measurement will add vectorally to the signal vector (Figure 10-8b). Assuming that the noise is random in amplitude and phase, the noise vector will modify the signal vector, creating a signal-plus-noise vector which varies in amplitude and phase (Figure 10-8c). Applying an averaging algorithm (independently) to the real part and to the imaginary part of the noisy vector tends to average out the noise portion while retaining the signal. With sufficient averaging, the noise portion will approach zero, leaving only the signal. Thus, vector averaging produces the mean of the desired signal.

Vector averaging is predetection averaging, since it operates on the complex data before the detector and reduces the amount of noise seen by the detector.

10.11 AVERAGING VERSUS FILTERING

One important, but subtle, difference between averaging and filtering in analyzers is the dimension in which the averaging/filtering takes place. Since filtering occurs in the IF section, the analyzer is filtering as it sweeps. Thus, filtering is done across the frequency axis of the display (similar to adjacent-point averaging). In order for the filtering to not distort the measurement by smearing the trace, the sweep time must

---

[2] Swept spectrum analyzers usually have only scalar (magnitude) data available, but FFT spectrum analyzers usually have vector data.

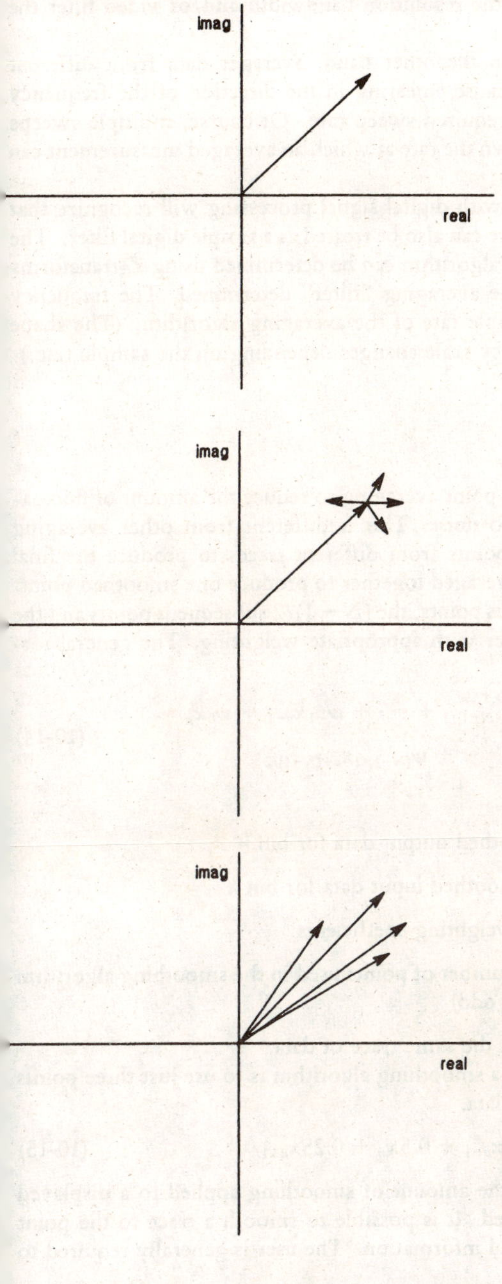

Figure 10-8. (a) The vector representation of a complex signal. (b) Random noise adds vectorally with the signal. (c) The original signal is varied in both magnitude and phase.

not be too small. The narrower the resolution bandwidth and/or video filter the slower the sweep must be.

Trace-to-trace averaging, on the other hand, averages data from different sweeps together. This does not cause smearing in the direction of the frequency axis, and this does *not* affect the required sweep rate. Of course, multiple sweeps must be acquired which slows down the rate at which an averaged measurement can be completed.

Engineers who are familiar with digital signal processing will recognize that what has been called averaging here can also be treated as a simple digital filter. The transfer function of the averaging algorithm can be determined using Z-transforms and the frequency response of the averaging "filter" determined. The frequency response is dependent on the sample rate of the averaging algorithm. (The shape remains constant, but the frequency scale changes depending on the sample rate.)

10.12 SMOOTHING

Smoothing functions use adjacent-point averaging to reduce the amount of fluctuation in the measured trace due to noise. This is different from other averaging techniques which combine data points from different traces to produce the final result. N points of the trace are averaged together to produce one smoothed point. (N is odd.) The $(N-1)/2$ previous points, the $(N-1)/2$ subsequent points and the current point are summed together with appropriate weighting. The general formula for smoothing is

$$\begin{aligned} y_n = {} & w_{-(N-1)/2}x_{n-(N-1)/2} + \cdots + w_{-1}x_{n-1} + w_0x_n \\ & + w_1x_{n+1} + \cdots + w_{(N-1)/2}x_{n+(N-1)/2} \end{aligned} \qquad (10\text{-}14)$$

where

y_n = smoothed output data for bin n

x_k = unsmoothed input data for bin k

$w_{-(N-1)/2}$ through $w_{(N-1)/2}$ = the weighting coefficients

N = the number of points used in the smoothing algorithm (N is odd)

The data points are all taken from the same trace of data.

A simple implementation of a smoothing algorithm is to use just three points ($N = 3$) in the smoothing of the data.

$$y_n = 0.25x_{n-1} + 0.5x_n + 0.25x_{n+1} \qquad (10\text{-}15)$$

If the user is given control over the amount of smoothing applied to a displayed trace, good judgment must be used. It is possible to smooth a trace to the point where it provides little or no useful information. The user is generally required to

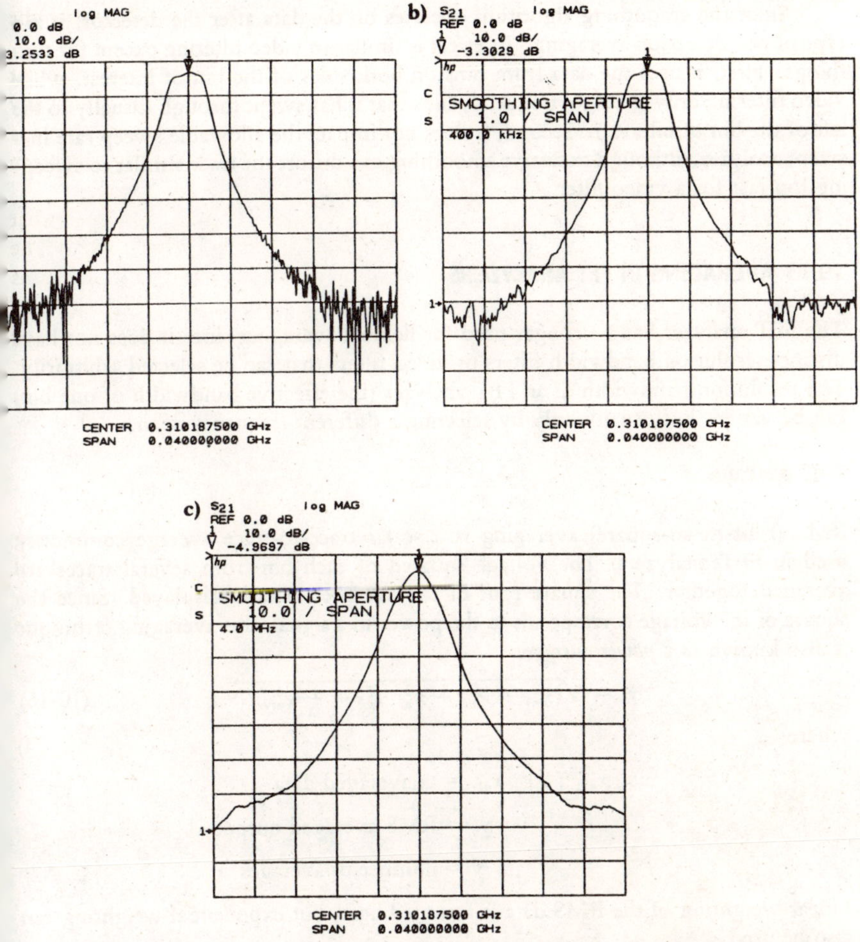

Figure 10-9. (a) A network measurement with no smoothing. (b) The same measurement with some smoothing. (c) The same measurement with excessive smoothing.

select the amount of smoothing which reduces the noise without significantly changing the shape of the trace.[3] Figure 10-9 shows the transfer function of a network with varying amounts of smoothing.

---

[3] In other words, select the amount of smoothing which makes the trace look "good."

Since the smoothing algorithm operates on the data after the detector, it is a type of postdetection averaging. Its effect is similar to video filtering except for two things. First, it uses the data from bins on both sides of the bin of interest, while video filtering only averages frequency bins that it has swept through (usually to the left of the bin of interest). Second, it does not impact the allowable sweep rate in a swept analyzer, although excessive smoothing can distort the trace similar to sweeping too fast for a video filter.

10.13 AVERAGING IN FFT ANALYZERS

The FFT analyzer has a stronger need for flexible averaging since it does not have distinct resolution bandwidth filters or video filters that can be selected arbitrarily. The resolution bandwidth of an FFT analyzer (the effective bandwidth of one bin) can be varied, but usually only by selecting a different frequency span.

RMS Average

RMS (root-mean-square) averaging is a scalar trace-to-trace average commonly used in FFT analyzers. The voltage squared of each bin from several traces are averaged together. The square root of the averaged data is displayed. Since the square of the voltage corresponds to the power in a signal, this averaging technique is also known as a *power average.*

$$y_n = \sqrt{(x_{n1}^2 + x_{n2}^2 + x_{n3}^2 + \cdots + x_{nN}^2)/N} \tag{10-16}$$

where

$$x_{n1}, x_{n2}, \ldots, x_{nN} = \text{unaveraged data}$$

$$y_n = \text{RMS averaged output}$$

$$N = \text{number of averages}$$

Linear weighting of the RMS data is assumed here, but exponential weighting can also be used.

RMS averaging is a postdetection process and therefore reduces the variance of the noise, but not the absolute noise level (Figure 10-10). It is very similar to video filtering, but operates as a trace-to-trace average. Either linear or exponential weighting can be applied to this form of averaging.

Vector Averaging

FFT analyzers usually offer vector averaging, although it may be called something different. The terms *time average* and *linear average* (not to be confused with linear weighting of other types of averaging) are often used.

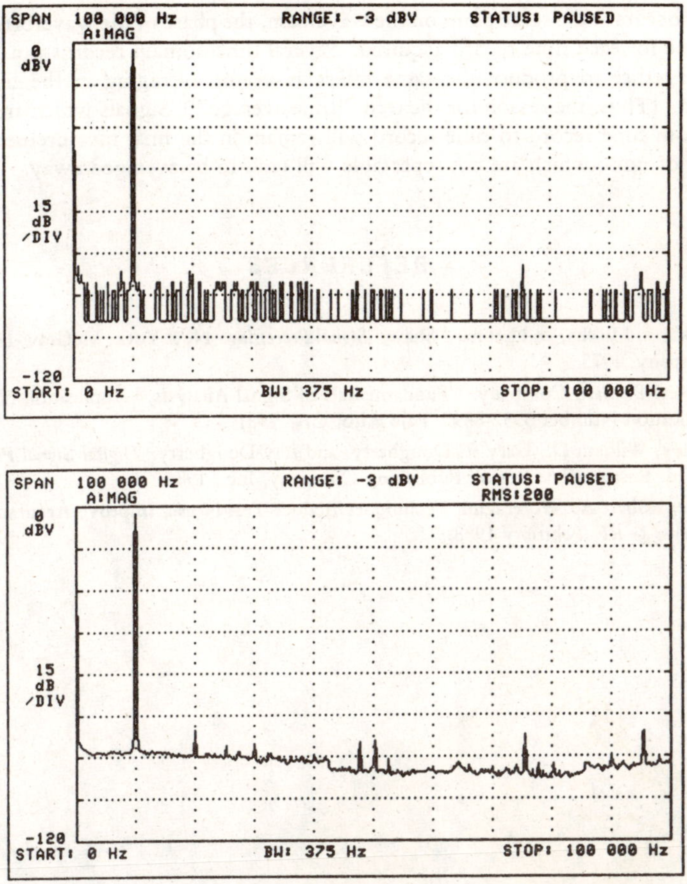

Figure 10-10. (a) FFT spectrum measurement with no averaging. (b) The same measurement with RMS averaging.

As previously stated, vector averaging is not usually found in spectrum analyzers because the phase of the vector data is not consistent from trace to trace. However, in an FFT analyzer this limitation can be overcome, even for spectrum measurements. The FFT analyzer's data starts out in the form of time domain data. If the sampled waveform in the time record is repeatable from record to record, the (vector) frequency domain data will have a repeatable phase from trace to trace and vector averaging can be used.

A triggering circuit, similar to the triggering circuit found in an oscilloscope, is employed to start the collection of the time domain data. Assuming that the trigger always occurs at the same point on the waveform, the phase of the waveform will be the same for each time record acquired. Several time domain records can be averaged together to produce the same effect as vector averaging in the frequency domain. (Thus, the reason for the term "time average.") Signals which are repeatable from time record to time record will remain in the final measurement while signals or noise which are not repeatable will tend to be averaged away.

REFERENCES

1. Schwartz, Mischa, and Leonard Shaw. *Signal Processing*. New York: McGraw-Hill Book Company, 1975.
2. Hewlett-Packard Company. "Fundamentals of Signal Analysis," Application Note 243, Publication Number 5952-8898, Palo Alto, CA. 1981.
3. Stanley, William D., Gary R. Dougherty, and Ray Dougherty. *Digital Signal Processing*, 2nd ed. Reston, VA: Reston Publishing Company, Inc., 1984.
4. Witte, Robert A. "Averaging Techniques Reduce Test Noise, Improve Accuracy." *Microwaves & RF* (February 1988).

11

Transmission Lines

Transmission lines are commonly used to connect test and measurement instruments to circuits under test. Transmission lines are used to control the effects of inductance and capacitance which are unavoidable in high-frequency systems. Coaxial transmission lines are the most common and provide the best shielding of the signals being measured.

Measurement error can be introduced due to impedance mismatch at either end of a transmission line. These errors must be understood and minimized in order to ensure an accurate measurement.

11.1 THE NEED FOR TRANSMISSION LINES

When connecting DC circuits, the major concern is the resistance of the wires. According to Ohm's law, a drop in voltage will occur when a current flows through a wire with nonzero resistance. Inductance and capacitance are not a concern for strictly DC voltages and currents.

For circuits with AC voltages and currents, the inductance and capacitance of wires start to come into play. A typical wire exhibits self-inductance and has some capacitance to other nearby conductors. The higher the frequency, the more significant the effect of inductive and capacitive reactance. Uncontrolled, these reactive effects can distort signals by loading the driving circuit and causing reflections on the wire. Transmission lines are used to solve these problems by controlling the inevitable inductance and capacitance.

Signals do not travel down a wire infinitely fast. It takes a finite amount of time for a signal to propagate from one place to another. For circuits and systems that have short connections (relative to the wavelength of interest), these effects are usually ignored. As the frequency of the signal and/or length of the wire is increased, the delays along the wire become significant. As the signal propagates down the wire, it may encounter variations in the impedance that it sees. The signal will be fully or partially reflected at each of these impedance discontinuities.

Reflections on a wire can cause the impedance looking into the wire to be uncontrolled, which can present an unknown or undesirable impedance to the driving circuit. When terminated properly, transmission lines provide a known impedance at each end of the line. This allows the system to be designed for maximum power transfer, with the signal source loaded by an impedance equal to its output impedance.

11.2 DISTRIBUTED MODEL

The unavoidable inductance and capacitance associated with lengthy wires are used to advantage in a transmission line. The reactances are controlled such that a signal traveling down the line sees a constant impedance. A circuit model for an arbitrarily small section of transmission line is shown in Figure 11-1a. As alluded to earlier, the transmission line contains some series inductance (L) and some capacitance (C) between the two conductors. Also included in the circuit model is a series resistance (R) and a shunt conductance (G) associated with losses incurred in the transmission line. All the circuit parameters are normalized to unit length (e.g., the inductance is in henries/meter).

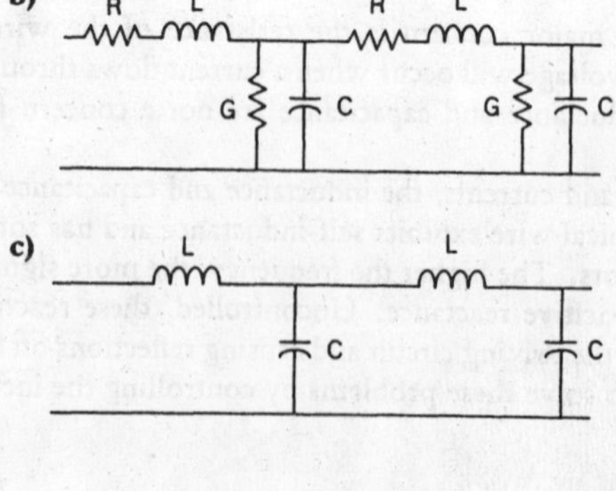

Figure 11-1. (a) The circuit model for a small section of transmission line. (b) The circuit model for a transmission line. (c) The circuit model for a lossless transmission line.

The circuit model represents an extremely small section of transmission line. A finite-length line is modeled as a large number of these sections cascaded end to end (Figure 11-1b). If the line is assumed to be lossless ($R = G = 0$), the resistive components are removed from the model (Figure 11-1c).[1]

The capacitance per unit length and the inductance per unit length depend on the physical construction of the transmission line. The dielectric constant of the material between the conductors and the physical geometry of the transmission line both affect the electrical properties of the line.

11.3 CHARACTERISTIC IMPEDANCE

The impedance looking into the end of an infinitely long lossless transmission line is called the *characteristic impedance*, Z_0. (The transmission line is specified here as infinitely long to sidestep any impedance changes due to reflections from the other end of the line.)

$$Z_0 = \sqrt{L/C} \quad \text{(lossless line)} \tag{11-1}$$

11.4 PROPAGATION VELOCITY

Electromagnetic waves in free space propagate at the speed of light. Inside a transmission line there is usually a dielectric material which lowers the propagation velocity. Thus, the propagation velocity of a transmission is given by

$$v_p = k_v c \tag{11-2}$$

where

k_v = velocity factor

c = velocity of light in free space (approximately 3×10^8 meters/sec)

The velocity factor simply expresses the propagation factor as a percent of free space light velocity. The velocity factor has a value between 0 and 1 depending on the dielectric material in the transmission line. The cable manufacturer will usually specify the propagation velocity in the form of velocity factor, often expressed in percent. Typically, k_v ranges from 60% to 90%.

11.5 GENERATOR, LINE, AND LOAD

First, consider the generator and load shown in Figure 11-2. The generator produces a 1 volt step and has an output impedance equal to Z_0. The generator is

---

[1] Transmission line losses will be ignored in this chapter unless otherwise specified.

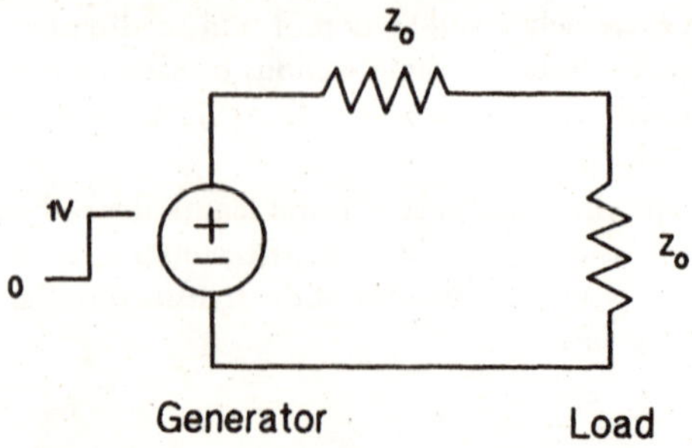

Figure 11-2. A Z_0 generator drives a Z_0 load.

connected to the Z_0 load by very short wires, and therefore there are no transmission line effects. At the same instant the generator voltage changes from 0 volts to 1 volt, the voltage across the load resistor changes from 0 volts to 0.5 volts. Note that the load voltage is one-half of the generator voltage due to the voltage divider effect.

Z_0 Load

If a transmission line is inserted between the generator and the load, the situation changes (Figure 11-3a). When the generator voltage changes from 0 volts to 1 volt, a forward-going (incident) voltage is created at the generator end of the transmission line. Since the generator sees the Z_0 impedance of the line, this incident voltage is equal to one-half of the generator voltage. This voltage moves down the transmission line at the propagation velocity until it meets the load. Since the load impedance is equal to the characteristic impedance of the line, no reflections occur. The incident voltage is "absorbed" by the load.

There is a time delay in the system, as the voltage wave travels down the transmission line (Figure 11-3b). This is unlike the previous example where the wires are so short that the load voltage instantaneously follows the generator voltage. Notice that the final value of the load voltage is the same in both cases. After the transmission line effects settle out, the DC voltages should be consistent.

The system shown in Figure 11-3 has the transmission line matched at both ends. That is, the impedances that the transmission line sees at the generator end and the load end are both Z_0. This eliminates any possible reflections and is usually the desirable case in instrumentation use. However, the generator and the load impedances may not be Z_0, so other cases must be considered.

Non-Z_0 Load

Suppose the Z_0 load is replaced by a load which is some other value (Figure 11-4a). As in the Z_0 load case, the incident voltage of 0.5 volts appears at the generator end of the transmission line. The incident voltage is not affected by the change in load impedance since the generator initially sees only the Z_0 impedance of the line. The 0.5 volt step propagates down the line and eventually reaches the load. The load is

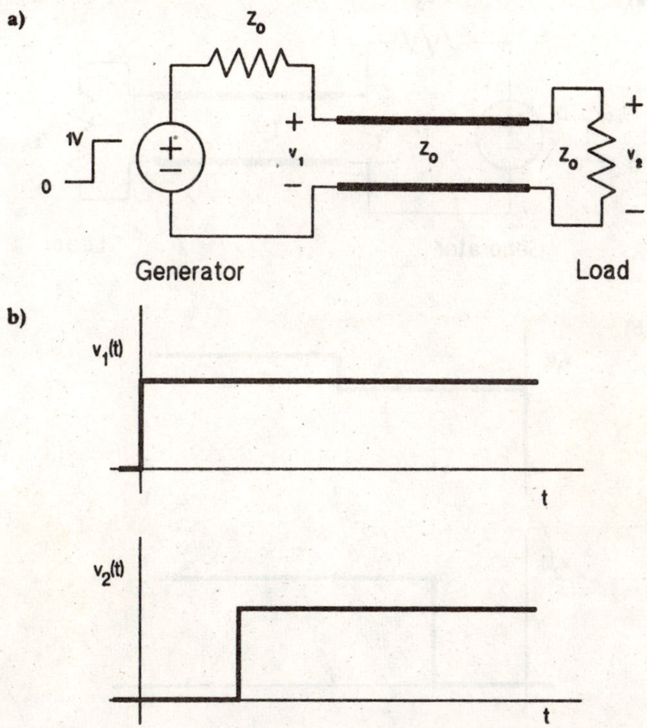

Figure 11-3. (a) A Z_0 generator drives a Z_0 load using a transmission line. (b) It takes a finite amount of time for the voltage to travel down the transmission line.

not matched to the Z_0 line, so some of the forward-going voltage is reflected back toward the generator. The reflected voltage is given by

$$V_R = \Gamma V_I \tag{11-3}$$

where

$$V_R = \text{reflected voltage}$$

$$V_I = \text{incident voltage}$$

$$\Gamma = \text{reflection coefficient}$$

The absolute value of Γ cannot exceed 1, since the reflected voltage cannot be larger than the incident voltage. The value of Γ can vary between −1 and +1, inclusive.[2]

[2] The reflection coefficient is introduced here as a scalar quantity, but the definition will be expanded to include complex values.

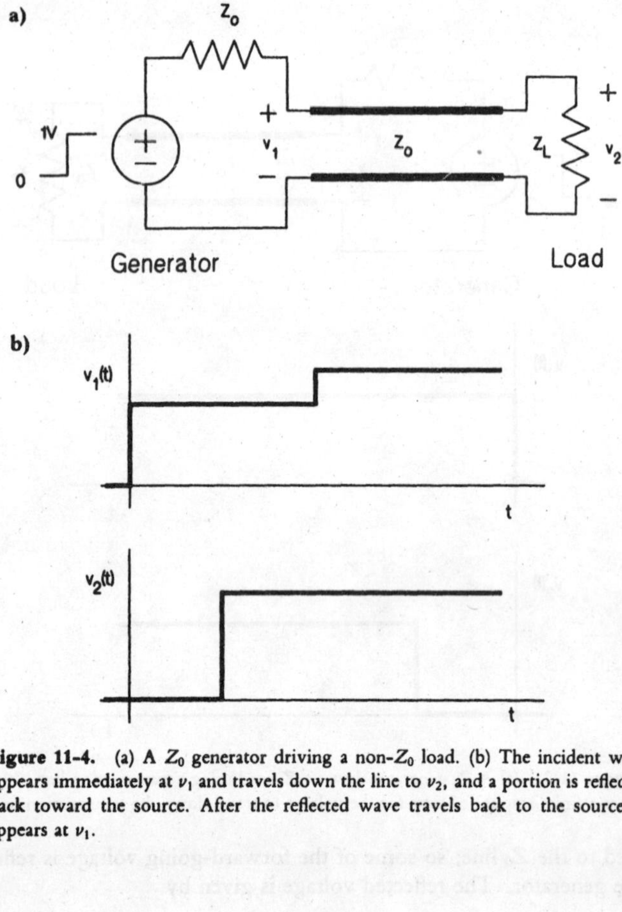

Figure 11-4. (a) A Z_0 generator driving a non-Z_0 load. (b) The incident wave appears immediately at v_1 and travels down the line to v_2, and a portion is reflected back toward the source. After the reflected wave travels back to the source, it appears at v_1.

For the case shown, Γ can be computed by

$$\Gamma = \frac{Z_L - Z_0}{Z_L + Z_0} \tag{11-4}$$

Also,

$$Z_L = Z_0 \frac{1 + \Gamma}{1 - \Gamma} \tag{11-5}$$

The reflected voltage, V_R, propagates back down the line toward the generator. The voltage at any point on the line is the sum of the incident and reflected voltages taking into account how far the two waves have traveled. The line is initially at volts (because the generator has presumably been at 0 volts for some time). As the incident wave travels down the line, the line becomes charged to V_I. Then the reflected wave starts back down the line moving from the load toward the generator. As the wave passes any given point, the voltage on the line at that point goes from V_I to $V_I + V_R$. When the reflected wave reaches the generator, it sees a Z_0 impedance (of the generator) and no additional reflections occur. Had the generator impedance been other than Z_0, additional reflections would occur.

Example 11.1

Determine the incident and reflected voltages for the case shown in Figure 11-5. What is the final value of the load voltage?

The incident wave $V_I = 4(50)/(50 + 50) = 2$ volts, the reflected wave $V_R = \Gamma V_I$, and $\Gamma = [(30 - 50)/(30 + 50)] = -0.25$. So $V_R = (-0.25)(2) = -0.5$ volts.

The final value of the load voltage is

$$V_L = V_I + V_R = 1.5 \text{ volts}$$

Note that this answer agrees with a simple DC analysis, ignoring the transmission line:

$$V_L = (4)[30/(50 + 30)] = 1.5 \text{ volts}$$

Let's describe what happens. The line is initially at 0 volts. When the voltage source steps to 4 volts, a 2 volt incident wave propagates down the line. When the incident voltage reaches the load, a −0.5 volt reflected wave starts its way back. As the reflected wave propagates back, the transmission line voltage becomes 1.5 volts. Finally, the reflected wave is absorbed when it reaches the source, since the source is matched to Z_0.

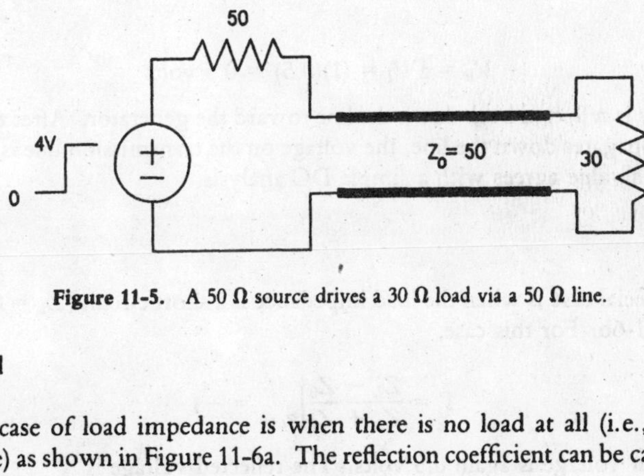

Figure 11-5. A 50 Ω source drives a 30 Ω load via a 50 Ω line.

Open Load

A special case of load impedance is when there is no load at all (i.e., an infinite impedance) as shown in Figure 11-6a. The reflection coefficient can be calculated for this case:

$$\Gamma = \left.\frac{Z_L - Z_0}{Z_L + Z_0}\right|_{Z_L=\infty} = 1 \tag{11-6}$$

Thus, all of the incident voltage is reflected back toward the generator. The incident voltage is once again 0.5 volts, which propagates down the line until it encounters the load.

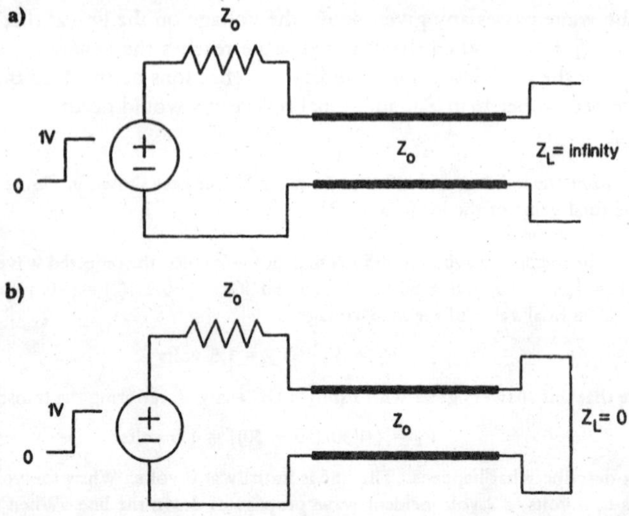

Figure 11-6. (a) When the line is terminated in an open circuit, the reflection coefficient is 1. (b) When the line is terminated in a short circuit, the reflection coefficient is −1.

$$V_R = \Gamma V_I = (1)(0.5) = 0.5 \text{ volts} \quad (11\text{-}7$$

So 0.5 volts is reflected back down the line toward the generator. After the reflecte voltage propagates down the line, the voltage on the transmission line is $V_I + V_R =$ 1 volt. This value agrees with a simple DC analysis.

Short Load

Another special case is when the load impedance is a short circuit ($Z_L = 0$) as show in Figure 11-6b. For this case,

$$\Gamma = \left.\frac{Z_L - Z_0}{Z_L + Z_0}\right|_{Z_L=0} = -1 \quad (11\text{-}8$$

The incident voltage is again 0.5 volts. The reflected voltage is

$$V_R = \Gamma V_I = (-1)(0.5) = -0.5 \text{ volts}$$

The incident voltage of 0.5 volts propagates down the line to the load where th negative of it is reflected back toward the generator. Since $V_I + V_R = 0$, the ne result is that the voltage returns to zero since the incident and reflected voltage cancel. This is required for the result to make any sense—the final DC voltag across a short circuit must be zero.

11.6 IMPEDANCE CHANGES

So far we have given examples where a generator drives a line which is connected to a load impedance. Now we will expand the concept of reflection coefficient to include the case where a voltage is incident at the junction of two different impedances. When the incident voltage encounters an impedance change, part of the incident voltage is reflected back and part of it travels on through (Figure 11-7).

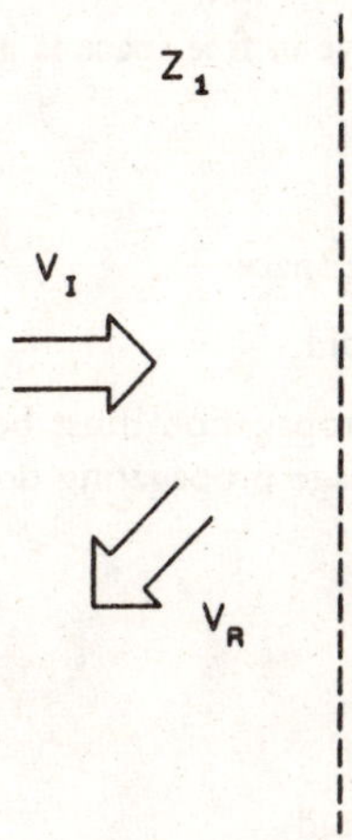

Figure 11-7. When a traveling wave encounters an impedance change, a portion of it is reflected while a portion of it is transmitted through the mismatch.

The two impedances involved (Z_1 and Z_2) might be two transmission lines with differing characteristic impedance or perhaps a slight impedance mismatch due to connector imperfections. Whatever the cause of the impedance change, it will result in a reflected wave.

Rewriting the expression for the reflection coefficient

$$\Gamma = \frac{Z_2 - Z_1}{Z_2 + Z_1} \tag{11-9}$$

The portion of the voltage wave which is reflected is

$$V_R = \Gamma V_I \tag{11-10}$$

A portion of the incident voltage may be transmitted on through the impedance mismatch, but modified by the amount reflected. The transmitted voltage is equal to

$$V_T = (1 + \Gamma) V_I \tag{11-11}$$

11.7 SINUSOIDAL VOLTAGES

A common electrical signal in spectrum and network measurement systems is th[e] sinusoid. Thus, we will expand our discussion of transmission lines to include thi[s] type of waveform.

Wavelength

The wavelength of a sinusoidal electromagnetic wave in free space is given by

$$\lambda = c/f \tag{11-12}$$

where

c = velocity of light in free space

f = frequency of the sinusoid

However, in a transmission line the velocity of propagation must be taken int[o] account. Thus, the wavelength of a sinusoidal voltage propagating down a transmission line is

$$\lambda = v_p/f = k_v\, c/f \tag{11-13}$$

where

v_p = propagation velocity

f = frequency of the sinusoid

k_v = velocity factor

c = velocity of light in free space

The slower the propagation velocity, the shorter the wavelength.

Example 11.2

What is the wavelength of a sinewave which has a frequency of 146.52 MHz in transmission line with a 66% velocity factor?

$$\lambda = k_v\, c/f = 0.66(3 \times 10^8)/(146.52 \times 10^6)$$

$$= 1.35 \text{ meters}$$

11.8 COMPLEX REFLECTION COEFFICIENT

Sine waves are normally characterized by their magnitude and phase. In order t[o] accommodate this representation, the concept of reflection coefficient is expanded t[o] allow for the reflection coefficient as a complex number. The reflection coefficient [is] often shown as a magnitude and a phase angle:

$$\Gamma = \rho\angle\theta \tag{11-14}$$

Both ρ and Γ are referred to as the reflection coefficient, but ρ is a scalar quantity while Γ is a complex number.

The previously introduced definition of the reflection coefficient (reflected voltage over the incident voltage) is modified to allow complex (vector) voltages.

$$\Gamma = \frac{|V_R|\angle\theta_R}{|V_I|\angle\theta_I} = \frac{|V_R|}{|V_I|}\angle\theta_R - \theta_I \tag{11-15}$$

$$\Gamma = \rho\angle\theta_R - \theta_I \tag{11-16}$$

The magnitudes of the incident and reflected voltages do not change with position on the transmission line and therefore ρ does not change with position. The phase of the complex reflection coefficient does change as the position changes.

The complex reflection coefficient due to an impedance mismatch (as shown in Figure 11-7) is computed using the values of the complex impedances.

$$\Gamma = \frac{Z_2 - Z_1}{Z_2 + Z_1} \tag{11-17}$$

11.9 RETURN LOSS

Another commonly measured quantity in radio frequency systems is *return loss*. The return loss of a particular system is the scalar reflection coefficient expressed in decibels.

$$RL = -20\log(\rho) \tag{11-18}$$

The minus sign in the equation causes the decibel form to indicate the amount of loss from the incident wave to the reflected wave, hence the name return loss. It is a measure of how large the reflected wave is relative to the incident wave. For example, if the return loss is 30 dB, then a 0 dBm incident wave causes a −30 dBm reflected wave. The return loss of a system can range from 0 to ∞ dB, 0 dB being the case where all of the incident wave is reflected and ∞ occurring when none of the incident wave is reflected.

Example 11.3

A sine wave generator with an output impedance of 50 Ω drives a load impedance of 30 + j20 ohms through a 50 Ω transmission line. What is the value of the reflection coefficient and return loss at the load?

$$\Gamma = \frac{Z_L - Z_0}{Z_L + Z_0} = \frac{(30 + j20) - 50}{(30 + j20) + 50} = \frac{28.28\angle 135^\circ}{82.46\angle 14.0^\circ}$$

$$\Gamma = 0.342\angle 121^\circ \text{ or } -0.176 + j0.293$$

$$\rho = |\Gamma| = 0.342$$

$$RL = -20\log(\rho) = 9.32 \text{ dB}$$

11.10 STANDING WAVES

When a sinusoidal signal first propagates down a transmission line, the sinusoidal voltage moves toward the load somewhat like a step waveform moves down the line. When the incident voltage encounters the load, a portion of the incident wave is reflected, according to the complex reflection coefficient (Figure 11-8). This reflected wave travels back toward the generator and may also be reflected again when the generator is reached, depending on the generator's impedance. All of this is very similar to the behavior of the step waveform.

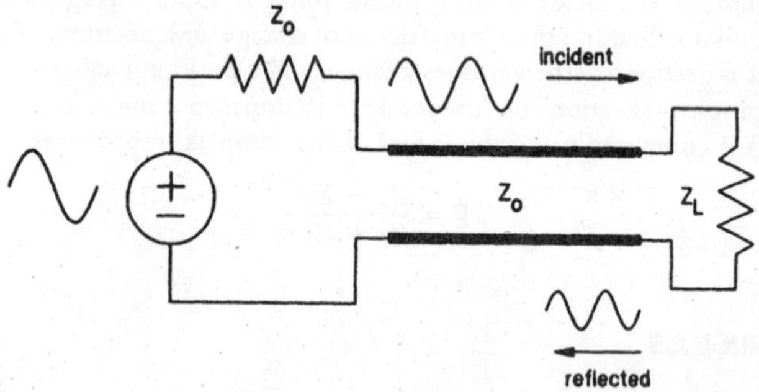

Figure 11-8. Sinusoidal signals also experience reflections on a transmission line.

Something different happens with a sinusoidal signal. The incident voltage and the reflected voltage are both sine waves. When they intersect, going up and down the transmission line, an interference pattern is set up (Figure 11-9). The envelope of the sinusoidal voltage will remain in a constant shape called a *standing wave,* as shown in the figure. The envelope (or magnitude) of the voltage varies with distance down the line but the voltage at each point on the line varies sinusoidally.

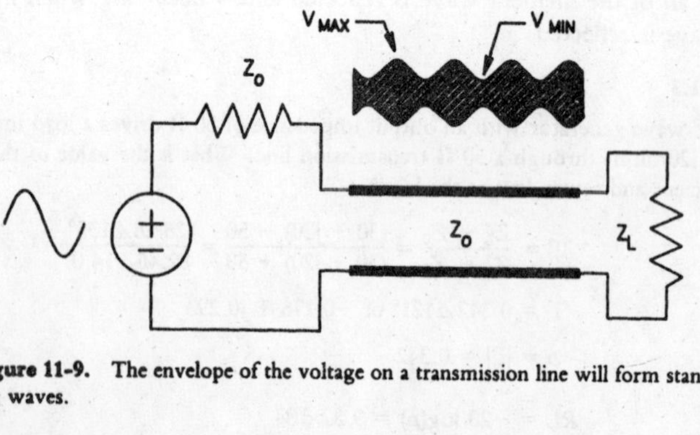

Figure 11-9. The envelope of the voltage on a transmission line will form standing waves.

The *voltage standing wave ratio (VSWR)* or simply the *standing wave ratio (SWR)* is the ratio of the maximum and minimum of the envelope.

$$SWR = V_{MAX}/V_{MIN} \tag{11-19}$$

The SWR is always greater than or equal to one, with 1.0 being the case where no mismatch occurs. In this case, V_{MAX} is equal to V_{MIN} since no reflections occur. A properly loaded transmission line is often called a "flat" line, referring to the lack of standing waves.

The maximum of the envelope occurs when the incident and reflected voltages add constructively. Similarly, the minimum of the envelope occurs where the incident and reflected voltages add destructively.

$$V_{MAX} = |V_I| + |V_R| \tag{11-20}$$

and

$$V_{MIN} = |V_I| - |V_R| \tag{11-21}$$

Thus, the SWR can be determined from the incident and reflected voltages as well as the scalar reflection coefficient.

$$SWR = \frac{|V_I| + |V_R|}{|V_I| - |V_R|} \tag{11-22}$$

$$SWR = \frac{1 + \rho}{1 - \rho} \tag{11-23}$$

Also,

$$\rho = \frac{SWR - 1}{SWR + 1} \tag{11-24}$$

Example 11.4

A 50 Ω sine wave generator drives a 100 Ω load through a 50 ohm transmission line. What are the values of the reflection coefficient and standing wave ratio? If the incident wave has a peak voltage of 4 volts, determine the minimum and maximum envelope voltages occurring on the line.

$$\Gamma = \frac{Z_L - Z_0}{Z_L + Z_0} = \frac{100 - 50}{100 + 50} \approx 0.33$$

$$\rho = |\Gamma| = 0.33$$

$$SWR = \frac{1 + \rho}{1 - \rho} = 2$$

$$= \frac{|V_I| + |V_R|}{|V_I| - |V_R|}$$

so

$$|V_R| = \frac{|V_I|\,(SWR - 1)}{(SWR + 1)} = \frac{4(2 - 1)}{(2 + 1)} = 1.33 \text{ volts}$$

The maximum and minimum envelope voltages are

$$V_{MAX} = |V_I| + |V_R| = 4 + 1.33 = 5.33 \text{ volts}$$

$$V_{MIN} = |V_I| - |V_R| = 4 - 1.33 = 2.67 \text{ volts}$$

Example 11.5

What are the values of reflection coefficient, return loss, and SWR for the special cases of a shorted load and an open load?

Shorted Load:

$$\Gamma = \frac{Z_L - Z_0}{Z_L + Z_0} = \frac{0 - Z_0}{0 + Z_0} = -1$$

$$\rho = 1, \ RL = -20 \log(1) = 0 \text{ dB}$$

$$SWR = \frac{1 + \rho}{1 - \rho} = \infty$$

Open Load:

$$\Gamma = \frac{Z_L - Z_0}{Z_L + Z_0} = \frac{\infty - Z_0}{\infty + Z_0} = 1$$

$$\rho = 1, \ RL = -20 \log(1) = 0 \text{ dB}$$

$$SWR = \frac{1 + \rho}{1 - \rho} = \infty$$

Note that both load conditions result in an infinite SWR and 0 dB return loss, although the sign of the complex reflection coefficient is different.

A table of reflection coefficient, return loss, and standing wave ratio is shown in Table 11-1.

11.11 INPUT IMPEDANCE OF A TRANSMISSION LINE

When an incident wave first encounters a transmission line it sees an impedance of Z_0. As it propagates down the line, a portion of the incident wave may be reflected back toward the generator end of the line. When this wave encounters the generator end, it will affect the voltage at that end of the line. This also means that the impedance seen looking into the end of the line will depend on the magnitude and phase of the reflections, and it will no longer be simply Z_0. For the situation shown in Figure 11-10, the input impedance looking into the line is[3]

$$Z_{IN} = Z_0 \frac{Z_L + j\, Z_0 \tan \theta_L}{Z_0 + j\, Z_L \tan \theta_L} \tag{11-25}$$

[3] This equation is adapted from Hayt (1974).

TABLE 11-1. TABLE OF REFLECTION COEFFICIENT, RETURN LOSS AND STANDING WAVE RATIO

Reflection coefficient	Return loss	Standing wave ratio
1.00	0.00	∞
0.90	0.92	19.00
0.80	1.94	9.00
0.70	3.10	5.67
0.60	4.44	4.00
0.50	6.02	3.00
0.40	7.96	2.33
0.30	10.46	1.86
0.20	13.98	1.50
0.10	20.00	1.22
0.09	20.92	1.20
0.08	21.94	1.17
0.07	23.10	1.15
0.06	24.44	1.13
0.05	26.02	1.11
0.04	27.96	1.08
0.03	30.46	1.06
0.02	33.98	1.04
0.01	40.00	1.02
0.00	∞	1.00

where θ_L represents the distance away from the load, expressed as an angle (degrees or radians).

The angle representing the distance from the load may be found from the physical distance.

$$\theta_L = \frac{360\, d}{\lambda} \tag{11-26}$$

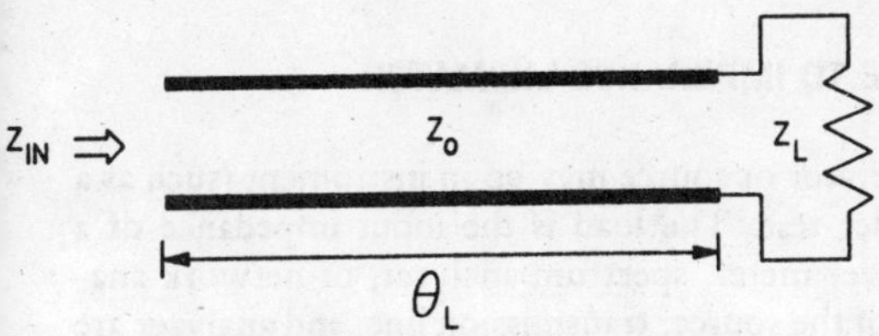

Figure 11-10. Z_{IN} is the impedance looking into the end of a transmission line with a load Z_L at the end of it.

where

$$d = \text{distance from the load}$$
$$\lambda = \text{wavelength}$$
$$\theta_L = \text{in units of degrees}$$

The velocity factor must be accounted for when computing the wavelength.

Matched System

For the special case of a perfectly matched system, $Z_L = Z_0$, the Z_{IN} equation reduces to

$$Z_{IN} = Z_0 \left. \frac{Z_L + j\, Z_0 \tan \theta_L}{Z_0 + j\, Z_L \tan \theta_L} \right|_{Z_L = Z_0} = Z_0 \qquad \text{(11-27)}$$

Since there are no reflections in a perfectly matched system, the impedance looking into the end of the line just equals Z_0.

Example 11.6

For $f = 50$ MHz, what is the impedance looking into the end of a 1 meter length of 50 Ω transmission line with a 70% velocity factor, terminated in 25 Ω?

$$\lambda = k_v\, c/f = 0.70(3 \times 10^8)/(50 \times 10^6) = 4.2 \text{ meters}$$

$$\theta_L = 360\, d/\lambda = 360(1)/4.2 = 85.7°$$

$$Z_{IN} = Z_0 \frac{Z_L + j\, Z_0 \tan \theta_L}{Z_0 + j\, Z_L \tan \theta_L}$$

$$= 50 \frac{25 + j\, 50 \tan(85.7°)}{50 + j\, 25 \tan(85.7°)}$$

$$= 98.9\ \Omega\ \angle 6.4° \text{ or } 98.3 + j\ 11.0\ \Omega$$

Not only is the input impedance larger than either Z_0 or Z_L, the impedance has an imaginary component while Z_0 and Z_L are both real. (This illustrates the transformer action possible with a transmission line, as the 25 Ω load impedance is transformed up to approximately 100 Ω.)

11.12 MEASUREMENT ERROR DUE TO IMPEDANCE MISMATCH

In a measurement situation, the generator or source may be an instrument (such as a signal generator) or the device under test. The load is the input impedance of a measuring instrument such as a power meter, spectrum analyzer, or network analyzer. Often the output impedance of the source, transmission line, and analyzer are all nominally Z_0 (Figure 11-11). However, the exact value of each of these imped-

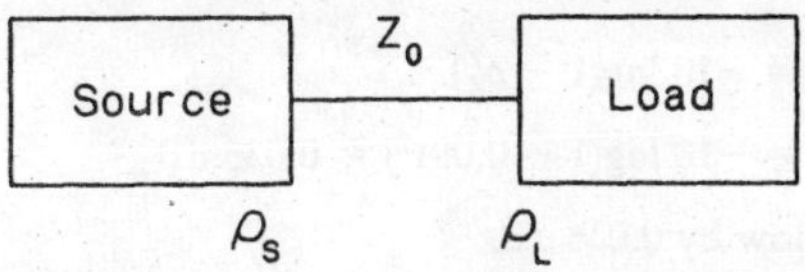

Figure 11-11. Measurement diagram for discussing mismatch loss and mismatch uncertainty.

ances may vary somewhat, causing slight mismatches in the system and errors in the measurement.

CASE 1. PERFECT SOURCE, IMPERFECT LOAD

As shown in previous examples, the incident voltage from the source will propagate to the load and a portion of it will be reflected back. The reflected voltage will travel back to the source and will be absorbed there if the source is perfectly matched. The major portion of the incident wave is transmitted into the load, which is to say it is measured by the instrument.

This mismatch effect will be examined from a power transfer point of view. The incident wave traveling toward the load has a power associated with it which is

$$P_I = V_I^2/Z_0 \tag{11-28}$$

The power reflected back from the load is

$$P_R = V_R^2/Z_0 = (\rho_L V_I)^2/Z_0 \tag{11-29}$$

The power delivered to the load must be the difference between these two powers:

$$P_L = P_I - P_R = \frac{V_I^2}{Z_0}(1 - \rho_L^2) = P_I(1 - \rho_L^2) \tag{11-30}$$

Ideally, all the incident power would be delivered to the load, but the mismatch causes some amount of power loss. The *mismatch loss*[4] is the power transfer to the load, relative to the incident power, expressed in dB:

$$\text{mismatch loss} = -10\log(P_L/P_I) = -10\log(1 - \rho_L^2) \tag{11-31}$$

[4] Beatty (1964) further refines this definition, calling it "Z_0 mismatch loss" to differentiate it from other possible definitions.

Example 11.7

A power meter with an *SWR* of 1.2 is used to measure the power at the end of a transmission line. How much error will be introduced in the measurement due to the mismatch at the power meter?

First, we need to compute the reflection coefficient from the *SWR*.

$$\rho = \frac{SWR - 1}{SWR + 1} = \frac{1.2 - 1}{1.2 + 1} = 0.091$$

The mismatch loss is given by

$$\text{mismatch loss} = -10 \log(1 - \rho_L^2)$$
$$= -10 \log(1 - 0.091^2) = 0.036 \text{ dB}$$

The power meter will read too low by 0.036 dB.

The mismatch loss has been derived in terms of a perfect source driving an imperfect load. The same effect exists for the case where the load impedance is exactly Z_0 and the source has a non-Z_0 impedance. The mismatch loss is computed the same way, but using the source's reflection coefficient (not the load's).

CASE 2. IMPERFECT SOURCE, IMPERFECT LOAD

Now consider the case where the source also has an output impedance which is not exactly Z_0. In addition to the error described in Case 1, there will be an additional error introduced due to source mismatch. The mismatch loss at the source is described mathematically as $(1 - \rho_S^2)$, where ρ_S is the source reflection coefficient. Note the similarity of this term to the load mismatch equation of Case 1.

There is another source of error when both the load and source are not perfectly matched. When the reflection from the load travels back to the source, instead of being absorbed at the source, it is reflected back again to the load. It adds constructively or destructively at the load, depending on the phase of the signal. This double reflection adds another term to the equation, which will be derived shortly. Additional reflections also occur, with each reflected wave being smaller than the previous one. For source and load impedances reasonably close to Z_0, the additional reflections are much smaller than the first load/source round trip reflection. Since the traveling waves in the system may add vectorally, they will be analyzed as voltage waveforms and then converted to power.

The incident wave, V_S, leaves the source and travels to the load. At the load end, there is an incident voltage present, V_I, which gets reflected back to the source which reflects it again to the load, which produces a second incident wave, $\rho_S\rho_L V_I$. There are really two waves incident at the load—the first direct wave from the source and the doubly reflected wave. Mathematically, we can express this as

$$V_I = V_S \pm \rho_S\rho_L V_I \tag{11-32}$$

Solving for V_I/V_S,

$$\frac{V_I}{V_S} = \frac{1}{1 \pm \rho_S\rho_L} \tag{11-33}$$

The scalar reflection coefficient is used here since we usually don't know the phase of the reflection and after the signal travels an unknown length of cable the phase relationship is lost anyway. The uncertainty in the sign of the reflected

term indicates that we don't know whether the reflected wave will add constructively or destructively.

Taking the square of Equation 11-33 and combining it with the mismatch losses at the source and load gives the complete power transfer function.

$$\frac{P_L}{P_S} = \frac{(1 - \rho_S^2)(1 - \rho_L^2)}{(1 \pm \rho_S\rho_L)^2} \qquad \textbf{(11-34)}$$

The numerator of the equation indicates the effect of the mismatch loss. These two terms are deterministic in the sense that a given reflection coefficient will cause a corresponding loss in power to the load. On the other hand, the denominator represents a *mismatch uncertainty*, with the uncertainty in the power transfer bounded by taking the + or − sign in the denominator. The actual power transfer can fall anywhere in between these two extremes. From a measurement point of view, we concentrate on the maximum and minimum power transfer, which represents the maximum and minimum error that can be incurred due to impedance matching problems.

Taking the decibel form of the equation allows it to be easily broken up into the individual error mechanisms.

$$10 \log\left(\frac{P_L}{P_S}\right) = 10 \log(1 - \rho_S^2) + 10 \log(1 - \rho_L^2) + 20 \log(1 \pm \rho_S\rho_L) \qquad \textbf{(11-35)}$$

Example 11.8

Determine the worst case error, expressed in decibels, due to mismatches when a source with a 10 dB return loss drives a lossless transmission line connected to a power meter having a return loss of 20 dB.

$$\rho_S = 10^{(RL/-20)} = 10^{(10/-20)} = 0.32$$

$$\rho_L = 10^{(RL/-20)} = 10^{(20/-20)} = 0.1$$

The mismatch loss due to the source is

$$10 \log(1 - \rho_S^2) = 10 \log(1 - 0.32^2) = -0.469 \text{ dB}$$

The mismatch loss due to the load is

$$10 \log(1 - \rho_L^2) = 10 \log(1 - 0.1^2) = -0.0436 \text{ dB}$$

The mismatch uncertainty due to the double reflection is

$$20 \log(1 \pm \rho_S\rho_L) = 20 \log[1 \pm (0.32)(0.1)] = -0.282 \text{ dB}, +0.274 \text{ dB}$$

The total error is bounded by

$$-0.469 - 0.0436 + 0.274 = -0.239 \text{ dB}$$

$$-0.469 - 0.0436 - 0.282 = -0.795 \text{ dB}$$

11.13 LINE LOSSES

Lossless transmission lines have been assumed thus far, which is a good approximation for many situations. If high-quality cables are used, frequencies are low, and cable length is short, other error mechanisms in the measurement system will dominate. The longer the cable becomes and the higher the frequency, the more attention needs to be paid to the cable loss. At microwave frequencies, the loss of even high quality cables can be severe.

Manufacturers normally specify the loss in their cables in dB, often in dB per hundred feet.

11.14 COAXIAL LINES

Transmission lines are available in a variety of physical configurations, but coaxial transmission lines are normally used in measurement applications below 1 GHz (Figure 11-12). The center conductor is surrounded by a dielectric material, which is surrounded by the outer conductor shield. The characteristic impedance is given by

$$Z_0 = \frac{138}{\sqrt{\varepsilon}} \log(D/d) \qquad (11\text{-}36)$$

where

D = inner diameter of the shield

d = outer diameter of the center conductor

ε = dielectric constant of the dielectric material

For air, the dielectric constant is equal to 1.

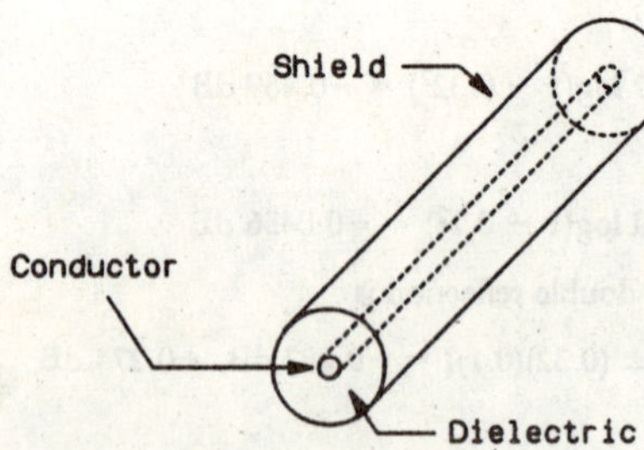

Figure 11-12. The coaxial transmission line is commonly used in measurement applications.

The coaxial structure inherently provides shielding from external electromagnetic fields and can result in transmission lines which can be moved and flexed somewhat without causing the characteristic impedance to change. Not all coaxial lines are flexible since some of the highest-quality lines, often with air dielectric, are fabricated with rigid or semirigid outer conductors.

The most common impedances in high-frequency measurement applications are 50 and 75 Ω. The 50 Ω impedance is most common due to its good compromise between power handling capability and loss. The 75 Ω impedance is found more in telecommunications applications where its low-loss characteristics are important.

REFERENCES

1. Adam, Stephen F. *Microwave Theory and Applications.* Englewood Cliffs, NJ: Prentice-Hall, Inc., 1969.
2. Beatty, Robert W. "Insertion Loss Concepts." *Proceedings of the IEEE* (June 1964).
3. Hayt, William H., Jr. *Engineering Electromagnetics,* 3rd ed. New York: McGraw-Hill Book Company, 1974.
4. Hayward, W. H. *Introduction to Radio Frequency Design.* Englewood Cliffs, NJ: Prentice-Hall, Inc., 1982.
5. Hewlett-Packard Company. "High Frequency Swept Measurements," Application Note 183, December 1978.
6. Laverghetta, Thomas S. *Practical Microwaves.* Indianapolis, IN: Howard W. Sams & Co., Inc., 1984.
7. *Reference Data for Radio Engineers,* 6th ed. Indianapolis, IN: Howard W. Sams & Co., 1981.

12

Measurement Connections

Connecting an instrument to a device under test (DUT) invariably involves disturbing that device. When making precision measurements, it is desirable to minimize loading and other effects so that the measurement is not corrupted by the measuring instrument. Probes, attenuators, impedance matching devices, and filters are used to couple the signal of interest into the instrument in the most efficient and accurate manner.

12.1 THE LOADING EFFECT

Any attempt to measure a voltage in a circuit will change that voltage. Consider the circuit shown in Figure 12-1. The circuit under test is modeled as a voltage source with some internal impedance, Z_S. The open circuit voltage of the circuit is V_S since no current can flow through Z_S under open circuit conditions. When a load impedance, Z_L is connected to the circuit, the situation changes. By the voltage divider relationship

$$V_L = V_S\, Z_L/(Z_S + Z_L) \tag{12-1}$$

Unless Z_S is equal to zero or Z_L is equal to infinity, V_L will always be less than V_S.

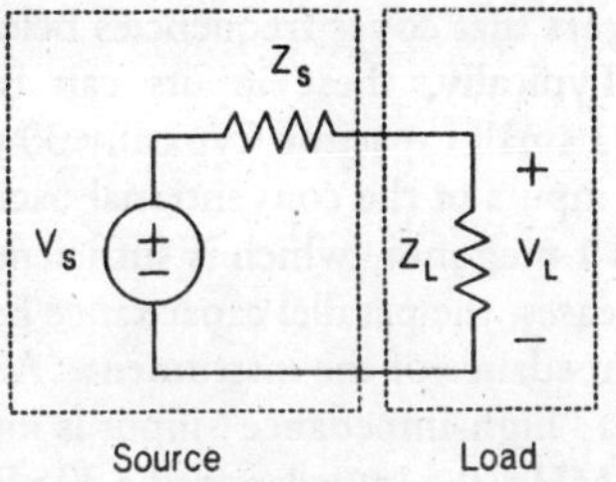

Figure 12-1. A finite impedance voltage source is attached to a resistive load.

12.2 MAXIMUM VOLTAGE AND POWER TRANSFER

Some electronic systems are designed to maximize the voltage transfer in the system. In order to maximize the voltage transfer, Z_L should be made much larger than Z_S. Thus, Z_S is made as small as possible and Z_L as large as possible. If Z_L is infinite, the desired result of $V_L = V_S$ occurs.

Other electronic systems, especially those operating at frequencies above 10 MHz, are designed to maximize the power transfer in the system. The power dissipated in the load impedance of Figure 12-1 is given by the equation

$$P_L = \frac{V_S^2 Z_L}{(Z_S + Z_L)^2} \tag{12-2}$$

It can be shown that the power in the load is maximized when[1]

$$Z_L = Z_S^* \quad (\text{*indicates complex conjugate}) \tag{12-3}$$

If the impedances are real, then the complex conjugate designation can be dropped and power is maximized when both of the impedances (resistances) are the same.

The impedances are typically chosen to be the same as the characteristic impedance, Z_0 of a transmission line. If all input and output impedances are equal to this value, the system will inherently terminate the transmission lines properly, minimizing reflections.

12.3 HIGH-IMPEDANCE INPUTS

It is often desirable to measure a given voltage in a circuit with minimal loading. In most measurement situations, Z_S is predetermined since it is a function of the circuit being measured and Z_L must be made large relative to Z_S. This is most easily accomplished at low frequencies and is increasingly more difficult as the frequency is increased.

---

[1] This assumes that Z_S is nonzero and Z_L is to be chosen. Otherwise $Z_S = 0$ would be a good choice.

Spectrum and network analyzers that cover frequencies below 30 MHz often provide high-impedance inputs. Typically, these inputs can be modeled by a 1 megohm resistor in parallel with a small capacitor (typically 30 pF). This type of input is very similar to the high-Z inputs of the conventional oscilloscope. At low frequencies, the input impedance is 1 megohm, which is sufficiently large for most applications. As the frequency increases, the parallel capacitance becomes dominant and reduces the equivalent input impedance of the instrument. An instrument user must be careful to not assume that a "high-impedance" input is high impedance for all frequencies. For example, at 10 MHz, the impedance of a 30 pF capacitor is only 530 Ω. Besides causing a reduction in amplitude, the high-impedance input can cause a phase shift due to the parallel capacitance.

12.4 HIGH-IMPEDANCE PROBES

Standard oscilloscope probes can be used with high-impedance analyzer inputs to provide convenient probing of circuit nodes (Figure 12-2). A 1X or 1 : 1 probe has no intentional attenuation and is basically equivalent to connecting the instrument input to the circuit under test with a shielded cable. The circuit model is shown in Figure 12-3.

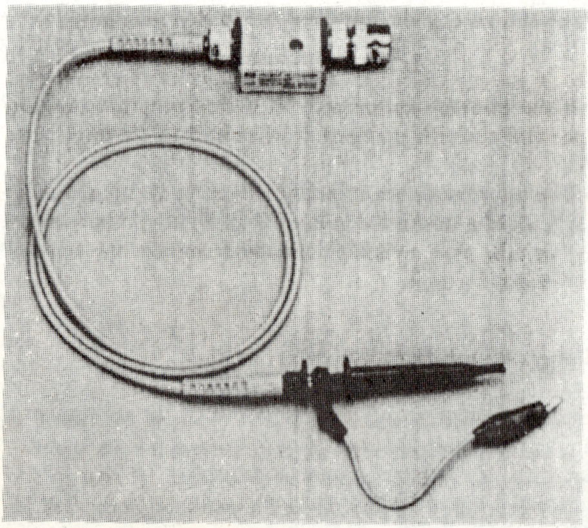

Figure 12-2. A typical 10 : 1 high-impedance oscilloscope probe. Photo courtesy of Hewlett-Packard Company.

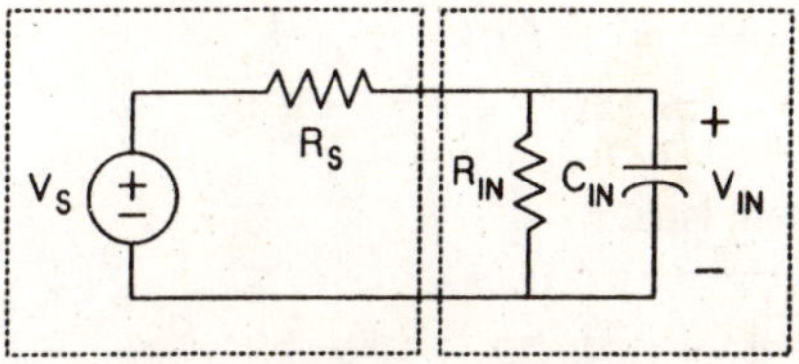

Figure 12-3. The circuit under test and a high-impedance instrument input produce a single-pole low-pass transfer function.

The measured voltage is

$$V_{IN} = V_S \frac{R_{IN}}{R_{IN} + R_S} \frac{1}{1 + j2\pi f C_{IN}(R_{IN} \| R_S)} \tag{12-4}$$

The input capacitance, C_{IN} creates a pole in the transfer function, causing V_{IN} to decrease at high frequencies. The magnitude of the transfer function is reduced by 3 dB at $f = 1/[2\pi(R_{IN}\|R_S)C_{IN}]$. Note that this frequency (essentially the bandwidth of the system) depends on R_{IN}, C_{IN}, and R_S. Normally, R_{IN} is much larger than R_S, so R_S dominates. While C_{IN} is part of the measuring instrument, R_S is the equivalent output impedance of the circuit under test. Thus, the impedance of the node being measured will influence the bandwidth of the measurement.

Attenuating Probes

The bandwidth-limiting effect of the analyzer's input capacitance can be compensated for at the price of some attenuation of the input signal. An attenuating probe includes a resistor and capacitor in the signal path (Figure 12-4).[2] The voltage delivered to the analyzer input is

$$V_{IN} = V_S \frac{R_{IN}(j2\pi f R_P C_P + 1)}{R_{IN}(j2\pi f R_P C_P + 1) + R_P(j2\pi f R_{IN} C_{IN} + 1)} \tag{12-5}$$

If $R_P C_P = R_{IN} C_{IN}$ then the equation reduces to

$$V_{IN} = V_S \frac{R_{IN}}{R_{IN} + R_P} \tag{12-6}$$

Under this condition, the effect of the input capacitance is canceled and other parameters such as cable capacitance will limit the probe bandwidth. The loading on the device under test is decreased, since the DUT sees a higher-probe impedance (smaller capacitance). For a 10X or 10 : 1 probe, R_P is chosen to be 9 times R_{IN}; V_{IN} is one-tenth of V_S. Any particular model of probe is designed for a certain range of

[2] This is a simplified probe circuit. Practical probe circuits may be arranged differently and may have additional circuit components.

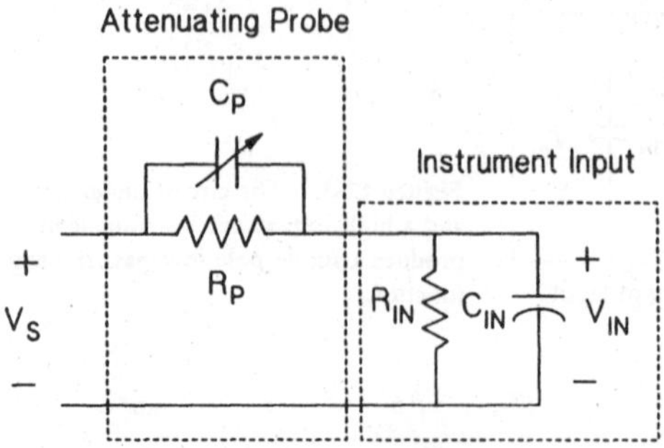

Figure 12-4. An attenuating probe will cancel out the effect of the instrument's input capacitance.

input capacitance and since input capacitance will vary with the design of the instrument, the probe must be selected to match the input.

C_P is made variable to allow the user to match precisely the probe to the input. When used with an oscilloscope, the probe is compensated (tweaked) by optimizing the pulse response of the system. In spectrum and network analyzer applications, a probe can be compensated by adjusting it for the flattest possible frequency response, using a tracking generator or signal generator with flat frequency response.

The 10:1 probe is the most common attenuating probe, supplying 20 dB of attenuation. Other attenuation factors are possible, with each trading off increased signal attenuation for increased system bandwidth.

12.5 Z_0 IMPEDANCE INPUTS

At higher frequencies (say, above 10 MHz), stray capacitance and other effects seriously degrade the performance of high-impedance inputs. Although high-impedance inputs may be present on high-frequency analyzers, for quality measurements a Z_0 input impedance is used. Most electronic systems operating in this frequency range generally use low input and output impedances, typically 50 or 75 Ω.

Spectrum and network analyzers are most often offered in these Z_0 input impedances. The goal is not so much to provide maximum power to the analyzer, but that the system is designed to be loaded in Z_0 and therefore such a load must be provided during the measurement. Many circuits such as filters, amplifiers, attenuators, and oscillators need to see a Z_0 load in order to function properly. The analyzer is usually connected to the device under test via a Z_0 impedance transmission line, so a Z_0 input properly terminates the line.

Although the analyzer may function as a Z_0 load, it usually is not capable of handling large power levels. The input voltage and/or power to the analyzer must not exceed its recommended rating. A *power attenuator* or *attenuating coupler* is designed to handle large signal levels and may be used to reduce the power present at the analyzer's input.

12.6 INPUT CONNECTORS

A variety of connector types are found on network and spectrum analyzer inputs, depending on the accuracy and frequency range of the instrument. A connector must introduce very little impedance mismatch, in order to allow an accurate measurement. A connector's impedance varies depending on the frequency of operation. Connectors which are perfectly acceptable at low frequencies may perform miserably at 1 GHz. Connector repeatability is also important, since it will limit the repeatability of the measurement. This also has implications about the quality of the instrument calibration since connector repeatability errors will occur during the calibration procedure.

For analyzers with upper frequency limits less than 40 MHz, the BNC (bayonet Neill Concelman) is almost universal. The bayonet-style locking mechanism provides a quick and convenient means of attaching and removing the connectors. The BNC connector's return loss degrades at higher frequencies, but the BNC fills the role of general-purpose connector for noncritical inputs and outputs regardless of the instrument's frequency range. BNC's are available in both 50 and 75 Ω versions.

For good-quality radio frequency measurements, the type N (Neill) connector is widely used. The type N connector, found in both 50 and 75 Ω versions, is much larger than the BNC and works well up into the 10 GHz frequency range. The threaded coupling mechanism provides good repeatability and is often used below microwave frequencies for this reason. The 50 Ω and 75 Ω versions are not the same, and accidentally mixing them will cause damage to the connectors.

At microwave frequencies, the choice of connector is even more critical. Examples of connectors used in this frequency range are the APC-7, the APC-3.5, the SMA, and the SMB.

12.7 Z_0 TERMINATIONS

In many measurement situations, it is important that all ports of the DUT are properly terminated. This means that the device must see the correct (usually Z_0) impedance at the port. A Z_0 termination is usually just a good-quality resistor conveniently packaged with the appropriate connector. A feedthrough termination is one that has a connector at both ends. Such a connector may be used to connect a high-impedance input to a DUT (Figure 12-5). The instrument input impedance is presumably much higher than Z_0 and the device under test sees roughly a Z_0 impedance (at low frequencies).

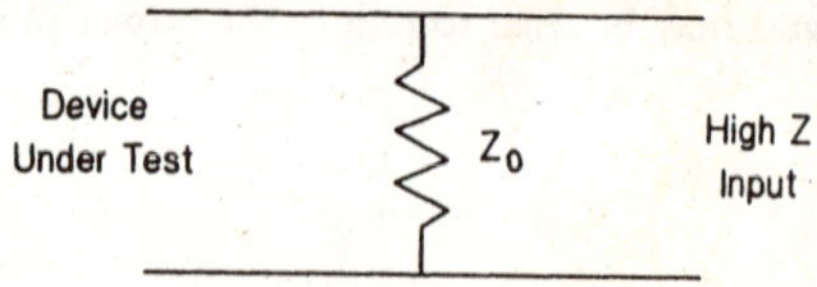

Figure 12-5. A feedthrough termination can be used to connect a Z_0 device with a high-impedance input.

12.8 POWER SPLITTERS

Power splitters are used to provide a common signal to multiple ports, either DUTs or instrument inputs. Most power splitters are two-way splitters (providing two outputs), but three-way splitters are also used.

Three-resistor power splitters can be configured in one of two different ways (Figure 12-6). These two circuits are totally equivalent from an external point of view. If each port is loaded by a Z_0 resistor to ground, then the impedance looking into each port of the power splitter is Z_0. (This is a simple example of a circuit that must be properly terminated at each port.)

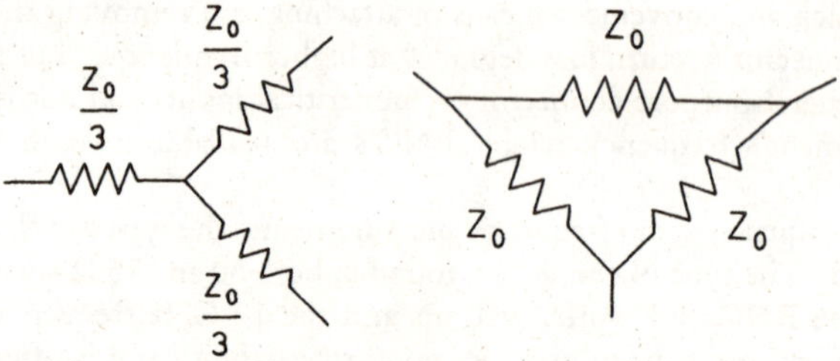

Figure 12-6. Two types of three-resistor power splitters.

The two-resistor power splitter is shown in Figure 12-7. This type of power splitter should be used in leveling and ratio applications. With a two-resistor power splitter, both output ports receive the same incident voltage. (See Chapter 14 for further discussion of ratio applications.)

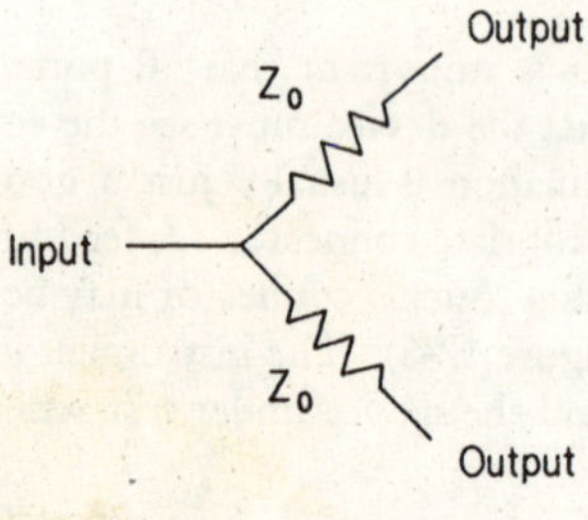

Figure 12-7. The two-resistor power splitter.

12.9 ATTENUATORS

Attenuators (sometimes called *pads*) are used to reduce the signal level in a measurement system. This may be required in order to bring a large signal within the measuring range of an instrument or to control distortion by reducing the signal level. Also, improvement in the impedance match (return loss) may be achieved by using attenuators.

Fixed attenuators provide only one, constant amount of attenuation, while variable attenuators can be adjusted (often in discrete steps). High-quality attenuators are commercially available, but sometimes it is necessary or convenient for the instrument user to construct them.

High-Impedance Attenuators

If an instrument with a high-input impedance is used and the device driving the attenuator has a low impedance, a simple voltage divider can be used as an attenuator (Figure 12-8a). The high-impedance input of the instrument ensures that little or

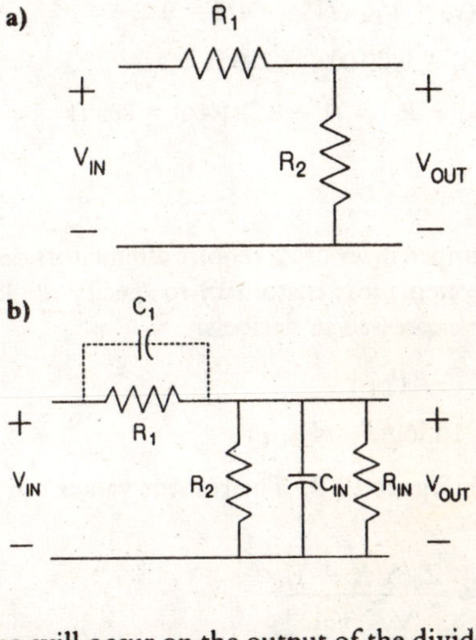

Figure 12-8. (a) A voltage divider circuit can be used as an attenuator for high-impedance systems. (b) A capacitor can be used to compensate the attenuator, improving the frequency response.

no loading will occur on the output of the divider. However, the effect of the input capacitance should be considered as the frequency increases. A compensated high-impedance attenuator can compensate for any stray capacitance across R_2 or instrument input capacitance by placing a capacitor in parallel with R_1 (Figure 12-8b). This is the same technique used in an attenuating probe, described earlier in the chapter. The capacitor value must satisfy the equation

$$R_1C_1 = C_{IN}(R_2 \| R_{IN})$$

The lack of a Z_0 or other reference impedance and the usage of the common voltage divider suggest that the attenuator be specified in terms of voltage gain or loss. The voltage gain[3] of the circuit is given by

$$G_V = V_{OUT}/V_{IN} = R_2/(R_1 + R_2) \tag{12-7}$$

The sum of R_1 and R_2 should be chosen so that it is much larger than the output impedance of the device driving the voltage divider. Then, R_1 and R_2 can be computed:

$$R_2 = G_V(R_1 + R_2) \tag{12-8}$$

$$R_1 = (1 - G_V)(R_1 + R_2) \tag{12-9}$$

Example 12.1

Design a high-impedance attenuator to produce a 1 volt RMS signal at the output when the input signal is 5 volts RMS. The loading on the driving circuitry must be at least 1 kΩ.

The load on the driving circuitry is $R_1 + R_2$, so choose $R_1 + R_2$ equal to 1 Ω. The voltage gain of the attenuator is $G_V = V_{OUT}/V_{IN} = 1/5 = 0.2$.

$$R_2 = G_V(R_1 + R_2) = 0.2(1000) = 200\ \Omega$$

$$R_1 = (1 - G_V)(R_1 + R_2) = (1 - 0.2)(1000) = 800\ \Omega$$

Z_0 Attenuators

Devices that have input and output impedances of Z_0 require attenuators designed to match the Z_0 impedance. In such systems it is customary to specify the loss of the attenuator as a power ratio, usually expressed in decibels.

$$K = P_{IN}/P_{OUT} \tag{12-10}$$

$$K_{dB} = 10 \log(P_{IN}/P_{OUT}) \tag{12-11}$$

The T attenuator circuit is shown in Figure 12-9. The resistor values can be determined from the following equations:

$$R_1 = \frac{Z_0(\sqrt{K} - 1)}{\sqrt{K} + 1} \tag{12-12}$$

$$R_2 = \frac{2Z_0\sqrt{K}}{K - 1} \tag{12-13}$$

Another Z_0 attenuator configuration, the π attenuator, is shown in Figure 12-10. The resistor values are determined from the following equations.

---

[3] Of course, the attenuator will always be lossy ($V_{OUT} < V_{IN}$), and the gain will be less than unity.

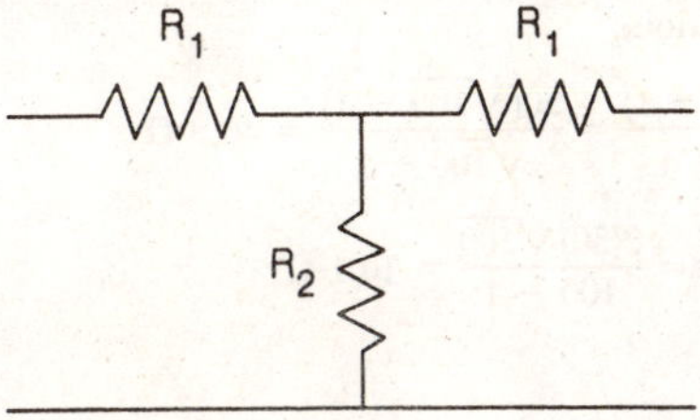

Figure 12-9. The *T* attenuator circuit.

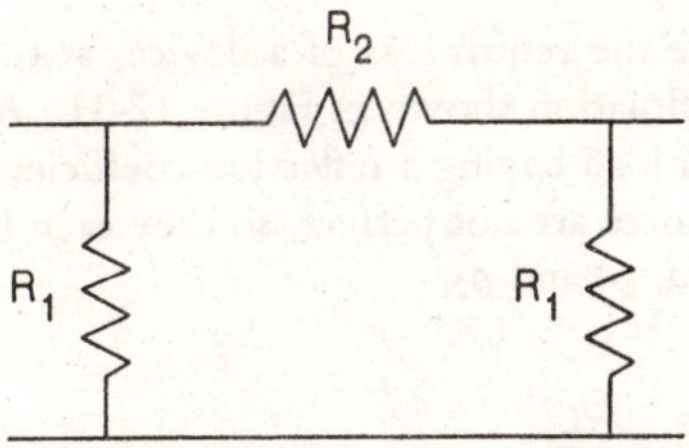

Figure 12-10. The π attenuator circuit.

$$R_1 = Z_0 \frac{\sqrt{K} + 1}{\sqrt{K} - 1} \tag{12-14}$$

$$R_2 = Z_0 \frac{K - 1}{2\sqrt{K}} \tag{12-15}$$

The two attenuator configurations shown are equivalent, but for a particular application, one configuration may result in more reasonable component values. Note that both Z_0 attenuator configurations are symmetrical from one port to another. Therefore, they provide the same attenuation in both directions.[4]

Example 12.2

In a particular 50 Ω system, a −10 dBm signal must be attenuated to −30 dBm. Design an attenuator to accomplish this.

First compute the loss required. In dB,

$$K_{dB} = -10 \text{ dBm} - (-30 \text{ dBm}) = 20 \text{ dB}$$

In power ratio form,

$$K = 10^{(K_{dB}/10)} = 100$$

---

[4] Many attenuators are symmetrical, but not all attenuators have this characteristic (consider the voltage divider previously discussed).

Using the T attenuator configuration,

$$R_1 = \frac{Z_0(\sqrt{K} - 1)}{\sqrt{K} + 1} = \frac{50(\sqrt{100} - 1)}{\sqrt{100} + 1} = 40.9\ \Omega$$

$$R_2 = \frac{2Z_0\sqrt{K}}{K - 1} = \frac{2(50)\sqrt{100}}{100 - 1} = 10.1\ \Omega$$

12.10 RETURN LOSS IMPROVEMENT

Attenuators can be used to improve the return loss of a device, at the expense of reduced signal level. Consider the situation shown in Figure 12-11. An attenuator with power loss K is connected to a load having a reflection coefficient of ρ_L. The attenuator input and output impedances are not perfect, so they each have a reflection coefficient associated with them, ρ_1 and ρ_2.

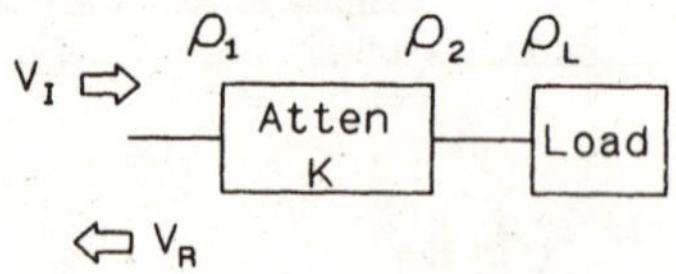

Figure 12-11. The return loss of a device can be improved by adding an attenuator.

Without the attenuator in the system, an incident voltage, V_I produces a reflected voltage,

$$V_R = \rho_L V_I \tag{12-16}$$

With the attenuator connected, V_I is attenuated by $\sqrt{K}$, the loss in the attenuator. (Since K is the ratio of input and output power, the voltage is reduced by $\sqrt{K}$.) This makes the voltage incident on the device equal to $V_I/\sqrt{K}$. A reflected voltage is produced which is equal to $\rho_L V_I/\sqrt{K}$. Assuming a symmetrical attenuator, the reflected voltage is attenuated by $\sqrt{K}$ on its way back. The reflected voltage seen at the attenuator port is

$$V_R = \rho_L V_I/K \tag{12-17}$$

The attenuator input and output impedances also cause another set of reflections. A reflected voltage is caused by ρ_1 as V_I is incident on the attenuator input. Also, when the incident voltage makes its way to the load and a reflected voltage is produced due to ρ_L, that reflected voltage may be reflected again by ρ_2. This reflection will be ignored in the analysis, since with a good-quality attenuator, ρ_2 will be small and any subsequent reflection from the load will be even smaller. Also, its effect will be reduced by $\sqrt{K}$ as it passes back through the attenuator. So combining the main reflection from the load and the attenuator input reflection,

$$\begin{aligned} V_R &= \rho_1 V_I + \rho_L V_I/K \\ V_R &= V_I(\rho_1 + \rho_L/K) \end{aligned} \tag{12-18}$$

Since we don't know whether the reflections will add in phase or out of phase, we simply added them together to produce a worst case reflection.

The reflection coefficient looking into the input of the attenuator is

$$\rho_a = V_R/V_I = \rho_1 + \rho_L/K \tag{12-19}$$

Thus, the reflection coefficient equals the reflection coefficient of the attenuator plus the load reflection coefficient reduced by the power loss factor, K. With a perfect attenuator, ρ_1 is zero and the equation reduces to

$$\rho_a = \rho_L/K \tag{12-20}$$

This is a good approximation for estimating the improvement in return loss achievable with an attenuator. For a high-quality attenuator and a load with a poor reflection coefficient, this is a reasonable approximation. The ρ_1 term in equation 12-19 serves to remind us that the improved return loss will never be better than the inherent return loss of the attenuator.

By expressing the reflection coefficient in decibel form, the return loss can be determined.

$$\begin{aligned} \mathrm{RL}_a &= -20\log(\rho_a) = -20\log(\rho_L/K) \\ &= -20\log(\rho_a) + 20\log(K) \end{aligned} \tag{12-21}$$

Since K is a power ratio, we will rewrite the equation

$$\begin{aligned} \mathrm{RL}_a &= -20\log(\rho_L) + 2[10\log(K)] \\ \mathrm{RL}_a &= \mathrm{RL}_L + 2K_{\mathrm{dB}} \end{aligned} \tag{12-22}$$

Thus, the return loss is improved by twice the attenuation (expressed in dB). The penalty for such an improvement is that the signal level to the load is reduced.

Example 12.3

What are the return loss and the reflection coefficient (ρ) at the output of a perfect 10 dB attenuator connected to signal generator having a return loss of 8 dB? Does the answer change if the attenuator has a return loss of 20 dB?

Perfect Attenuator:

$$\text{return loss} = RL = 8 + 2(10) = 28 \text{ dB}$$

The reflection coefficient is $\rho = 10^{(-RL/20)} = 0.040$.

Attenuator with 20 dB Return Loss:
Since the return loss predicted by the ideal analysis is 28 dB, the attenuator's 20 dB return will clearly limit the overall return loss. Adding the reflection coefficient of the attenuator to the reflection coefficient of the ideal case will produce the overall reflection coefficient.

The reflection coefficient of the attenuator is $\rho_1 = 10^{(-20/20)} = 0.1$. Since

$$\rho_a = \rho_1 + \rho_{a(\mathrm{IDEAL})} = 0.1 + 0.04 = 0.14$$

therefore,

$$RL_a = -20 \log(0.14) = 17 \text{ dB}$$

Note that the overall return loss is somewhat worse than the attenuator return loss.

Although the previous analysis used an attenuator connected to a load, the same principles apply for improving the return loss of a source. The source's return loss will be improved by twice the loss of the attenuator, except where limited by the attenuator match. Again, the disadvantage of such an improvement is the reduced signal level available at the attenuator output.

12.11 THE CLASSICAL ATTENUATOR PROBLEM

When a device under test is placed between a single-ended source and a single-ended detector such as a spectrum or network analyzer, a significant error can be introduced. This effect is due to nonzero cable shield impedance and occurs only at low frequency (less than 100 kHz). As the frequency increases, the cable acts more like a transmission line, and the shield impedance is less critical.

Consider the circuit model shown in Figure 12-12. A signal source is connected to an attenuator via a coaxial cable. The output of the attenuator is connected to the input of an analyzer with another cable. R_{C1} and R_{C2} represent the cable ground impedances. The input of the analyzer is initially assumed floating, with impedance R_G between its input ground and chassis ground. To illustrate the problem, the attenuation of the attenuator is infinite and no signal should be present at the input to the analyzer.

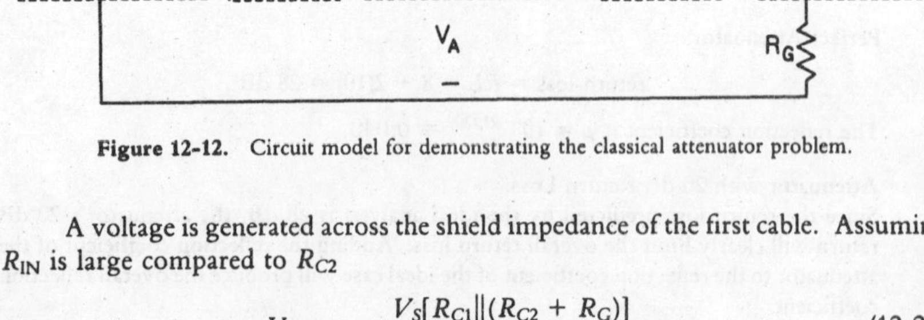

Figure 12-12. Circuit model for demonstrating the classical attenuator problem.

A voltage is generated across the shield impedance of the first cable. Assuming R_{IN} is large compared to R_{C2}

$$V_a = \frac{V_S[R_{C1}\|(R_{C2} + R_G)]}{R_S + R_{A1} + [R_{C1}\|(R_{C2} + R_G)]} \tag{12-23}$$

This voltage is in turn transferred onto R_{C2} and the instrument input. Again, if R_{IN} is much larger than R_{C2}

$$V_{IN} = \frac{V_a R_{C2}}{R_{C2} + R_G} \qquad (12\text{-}24)$$

With infinite attenuation, V_{IN} should be zero. But as shown, a small voltage is present at the instrument input. To drive V_{IN} to zero and reduce this error, without corrupting the measurement, R_G can be made large.

To put the situation in perspective, a plot of the measured loss in an attenuator in a typical measurement setting is shown in Figure 12-13. With $R_G = 0$ (the input is grounded), the attenuator loss is measured as low as 90 dB. With R_G large (the input is isolated or floating), the attenuator loss is measured correctly as 120 dB.

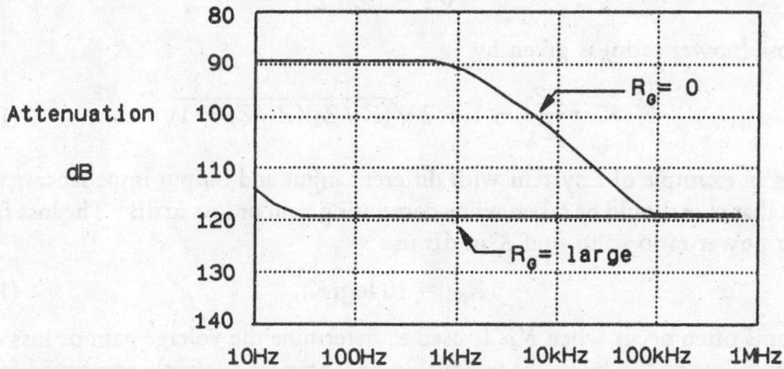

Figure 12-13. The measured attenuation of a 120 dB attenuator with ($R_G = 0$) and without (R_G = large) the classical attenuator problem.

The classical attenuator problem applies to any low-frequency situation where a large amount of attenuation is encountered, including measurements of devices such as filters. Analyzers that measure in this frequency range often supply two forms of defense against the problem. One is to isolate or float the input relative to chassis ground at all frequencies. In this case, a switch is often supplied to allow the user to conveniently select a grounded or floated input. The other technique is to isolate the input from the chassis, but only at low frequencies. For high frequencies, where the classical attenuator problem doesn't exist, the input is grounded.

12.12 IMPEDANCE MATCHING DEVICES

Spectrum and network analyzers are normally offered with standard input impedances, typically 50 or 75 Ω. Sometimes the analyzer does not have the same impedance as the device under test and the user may need a matching network to eliminate mismatch problems.

Minimum Loss Pads

An attenuator can be used to provide an impedance change, at the expense of some signal loss. Such an attenuator is called an impedance matching attenuator or impedance matching pad. There is a class of impedance matching pads which incur the theoretical minimum amount of loss called *minimum loss pads.*

The circuit for a minimum loss pad which matches impedances Z_1 and Z_2 is shown in Figure 12-14. Z_1 must be greater than Z_2. The resistor values can be computed from

$$R_1 = Z_1\sqrt{1 - (Z_2/Z_1)} \tag{12-25}$$

$$R_2 = \frac{Z_2}{\sqrt{1 - (Z_2/Z_1)}} \tag{12-26}$$

The loss (power ratio) is given by

$$K = \frac{2Z_1}{Z_2} - 1 + 2\sqrt{(Z_1/Z_2)(Z_1/Z_2 - 1)} \tag{12-27}$$

This is an example of a system with different input and output impedances, which means that care should be taken when computing gain or loss in dB. The loss factor, K, is a power ratio so to find K in dB use

$$K_{dB} = 10 \log(K) \tag{12-28}$$

Problems often occur when K_{dB} is used to determine the voltage gain or loss of the minimum loss pad. Unless the unequal impedances are explicitly accounted for, the results will be in error.

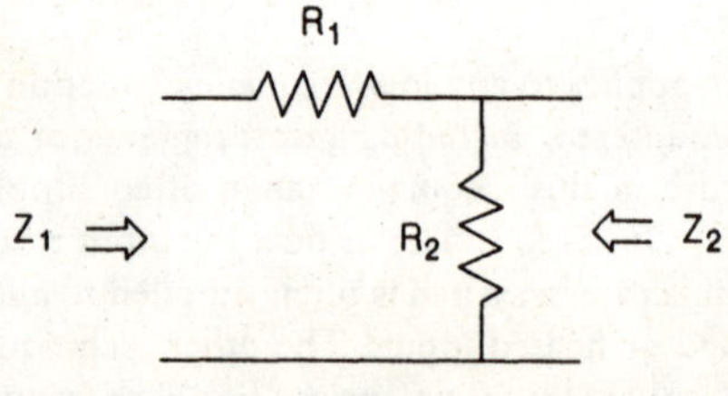

Figure 12-14. A minimum loss pad is used to match unequal impedances. Z_1 must be greater than Z_2.

Example 12.4

Compute the values for a minimum loss pad that will convert 50 Ω to 75 Ω. What is the (power) loss of the pad? What is the ratio of the input and output voltage when the pad is correctly terminated?

$$Z_1 = 75 \quad \text{and} \quad Z_2 = 50$$

$$R_1 = Z_1\sqrt{1 - (Z_2/Z_1)} = 75\sqrt{1 - (50/75)} = 43.3\ \Omega$$

$$R_2 = \frac{Z_2}{\sqrt{1 - (Z_2/Z_1)}} = \frac{50}{\sqrt{1 - (50/75)}} = 86.6\ \Omega$$

The loss is

$$K = \frac{2(75)}{50} - 1 + 2\sqrt{(75/50)(75/50 - 1)}$$

$$= 3.73$$

$$K_{dB} = 10 \log(3.73) = 5.7 \text{ dB}$$

$$K = P_1/P_2 = (V_1^2/Z_1)/(V_2^2/Z_2)$$

$$V_1/V_2 = \sqrt{Z_1K/Z_2} = \sqrt{75(3.73)/50} = 2.37$$

Transformers

Transformers can be used to match impedances for measurement purposes. A transformer consists of two separate coils that share the same core. The coupling of the magnetic fields of the coils causes a voltage on one coil to induce a voltage on another coil. Since the coupling mechanism depends on a changing magnetic field, a transformer works only for AC signals and not DC.

The ideal transformer is a two-port device with the following voltage and current relationships (Figure 12-15a).

$$V_2 = nV_1 \tag{12-29}$$

$$I_2 = I_1/n \tag{12-30}$$

where n is the turns ratio of the transformer.

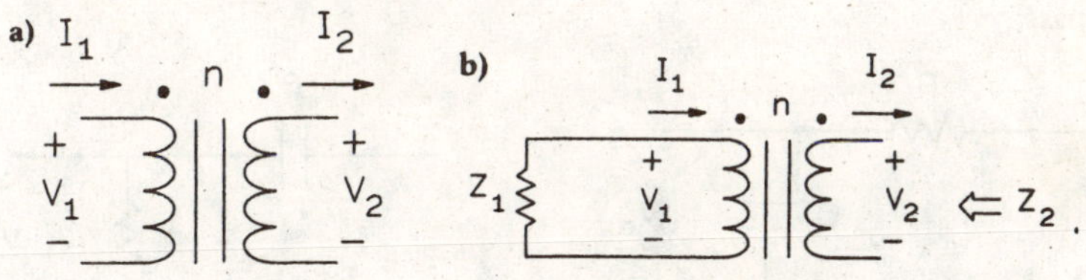

Figure 12-15. (a) The ideal transformer. (b) A transformer can be used to produce an impedance change.

The voltage is changed according to the turns ratio, while the current is transformed inversely proportional to the turns ratio. If an impedance, Z_1, is connected to the V_1, I_1 port of the transformer, the impedance looking into the V_2, I_2 port will be given by the following equation (Figure 12-13b):

$$Z_2 = \frac{V_2}{-I_2} = \frac{nV_1}{-I_1/n} = n^2(-V_1/I_1) = n^2Z_1 \tag{12-31}$$

So the impedance is transformed by the square of the turns ratio.

A particular transformer is usually optimized for some particular frequency range, and the user should consider this when selecting one for measurement use.

Ideally, the transformer is lossless—the power in will equal the power out. But in reality there will be some loss in the transformer. Such a loss can be characterized and normalized out of the measurement.

Another use of transformers in electronic measurements is to provide DC isolation between the device under test and the measuring instrument. Since there is no direct connection between the two ports of the transformer, transformers are isolated for DC. In other words, a transformer can be used to convert a grounded input instrument into one with floating inputs.

12.13 MEASUREMENT FILTERS

It is sometimes desirable to condition a signal's frequency content before it enters the instrument input. For example, an undesirable out-of-band signal might be large enough to cause distortion in the analyzer. Often, a simple filter can remove the offending signal or signals. High-quality filters are commercially available for many different applications, but they can also be built by the instrument user. Electronic filter design has consumed the pages of many other books and only a few low-pass and high-pass topologies will be discussed here.

High-Impedance Filters

Two filters suitable for use in situations with high-impedance instrument inputs are shown in Figure 12-16. Both of these are single-pole filters—one low pass and one high pass.

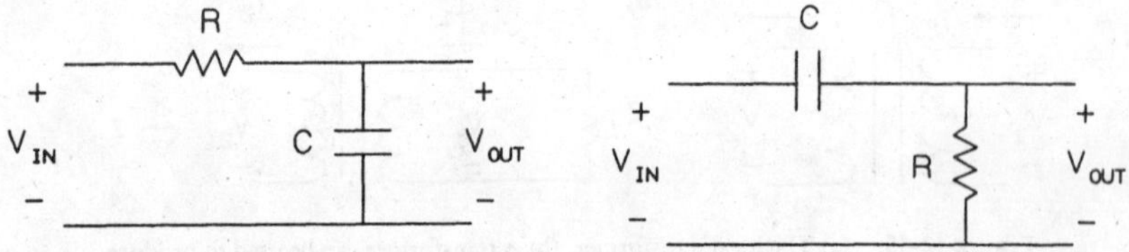

Figure 12-16. (a) A low-pass filter for high-impedance inputs. (b) A high-pass filter for high-impedance inputs.

The low-pass transfer function is

$$V_{OUT}/V_{IN} = \frac{1}{1 + j(f/f_{3dB})} \tag{12-32}$$

and the high-pass transfer function is

$$V_{OUT}/V_{IN} = \frac{j(f/f_{3db})}{1 + j(f/f_{3dB})} \tag{12-33}$$

where f_{3dB} is the frequency at which the response is reduced by 3 dB. For both filters,

$$f_{3dB} = \frac{1}{2\pi RC} \tag{12-34}$$

Being a single-pole filter, the transfer function rolls off at the rate of 20 dB per decade above or below the 3 dB frequency (depending on whether it's a high-pass or a low-pass design).

Z_0 Filters

If the previous filters are used in a Z_0 system (say, 50 Ω), the loading of the 50 Ω impedance on the filter output would distort the response of the filter. A different approach is needed, one which takes into account the Z_0 loading of such a system. In fact, these filters are required to be loaded in Z_0 to obtain the desired response.

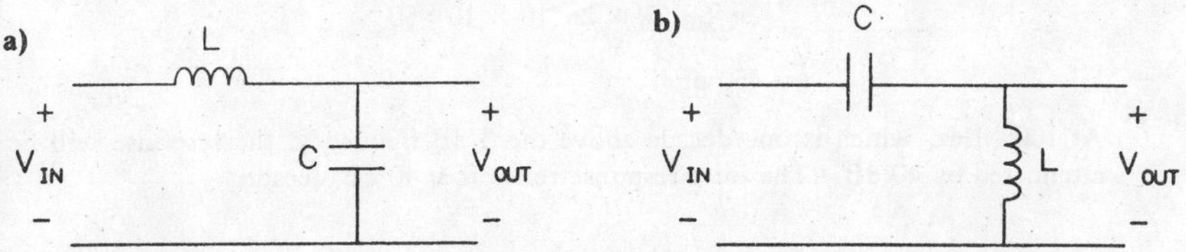

Figure 12-17. (a) A low-pass filter for Z_0 systems. (b) A high-pass filter for Z_0 systems.

Two filter networks, a high pass and a low pass, are shown in Figure 12-17. The values for the low-pass version are computed from the following equations:[5]

$$L = \frac{\sqrt{2}\, Z_0}{2\pi f_{3dB}} \qquad C = \frac{\sqrt{2}}{2\pi f_{3dB}\, Z_0} \tag{12-35}$$

For the high-pass filter, the equations are

$$L = \frac{Z_0}{\sqrt{2}\,(2\pi f_{3dB})} \qquad C = \frac{1}{\sqrt{2}\,(2\pi f_{3dB})\, Z_0} \tag{12-36}$$

The low-pass transfer function is

$$H(f) = \frac{f_{3dB}^2}{f_{3dB}^2 - f^2 + j\sqrt{2}\, f f_{3dB}} \tag{12-37}$$

and the high-pass transfer function is

[5] The Z_0 filters discussed here are second-order Butterworth filters.

$$H(f) = \frac{f^2}{f_{3dB}^2 - f^2 + j\sqrt{2}\, f f_{3dB}} \tag{12-38}$$

The filter response is 3 dB down at f_{3dB}. After that, it rolls off at 40 dB per decade, due to the fact that it has two poles. If the filter characteristics must be sharper, the reader is encouraged to consult one of the many books devoted to filter design.[6]

Example 12.5

Determine the component values for a low-pass 50 Ω filter with a 3 dB frequency equal to 10 MHz. What is the approximate filter attenuation at 100 MHz?

$$f_{3dB} = 10 \text{ MHz}$$

$$L = \frac{\sqrt{2}\, Z_0}{2\pi f_{3dB}} = \frac{\sqrt{2}\, 50}{2\pi\,(10 \times 10^6)}$$

$$= 1.125\ \mu\text{H}$$

$$C = \frac{\sqrt{2}}{2\pi f_{3dB}\, Z_0} = \frac{\sqrt{2}}{2\pi(10 \times 10^6)\, 50}$$

$$= 450 \text{ pF}$$

At 100 MHz, which is one decade above the 3 dB frequency, the response will be attenuated by 40 dB. (The filter response rolls off at 40 dB/decade.)

REFERENCES

1. "Attenuating the Classical Attenuator Problem." *Hewlett-Packard Journal* (May 1975).
2. Hewlett-Packard Company. "High Frequency Swept Measurements," Application Note 183, December 1978. Palo Alto, CA.
3. Hewlett-Packard Company. "Coaxial and Waveguide Measurement Accessories Catalog," Publication number 5954-6401, November 1986. Palo Alto, CA.
4. Williams, Arthur B. *Electronic Filter Design Handbook*. New York: McGraw-Hill Book Company, 1981.
5. Witte, Robert A. *Electronic Test Instruments: A User's Sourcebook*. Indianapolis, IN.: Howard W. Sams & Company, 1987.

[6] See Williams (1981).

13

Two-Port Networks

Two-port network theory provides the theoretical basis for making network measurements. Two-port network theory can be expanded to N-port theory for networks having more than two ports while one-port measurements are essentially a subset of two-port measurements. The simplest of two-port measurements is the gain or transfer function of the device. This assumes a fairly simple model of the device under test. More complete two-port models such as impedance parameters provide a better view of device behavior, while scattering parameters present a two-port model which is consistent with transmission line theory and measurements.

13.1 SINUSOIDAL SIGNALS

The standard forcing function for network analysis is the sinusoid, either the sine or cosine function. This stimulus is appropriate if we make the assumption that the network being measured is a linear, time-invariant system. Applying a sinusoid to the network's input and measuring the amplitude and phase of the network's output (both as a function of frequency) adequately characterizes the network. Selecting the cosine representation, the input forcing function (or stimulus) is

$$v(t) = V_M \cos(2\pi ft + \theta) \tag{13-1}$$

which is equal to the real part of an exponential function:

$$v(t) = Re[V_M\, e^{j(2\pi ft+\theta)}] \tag{13-2}$$

This can be verified easily by use of Euler's identity;

$$e^{jx} = \cos(x) + j\sin(x) \tag{13-3}$$

Splitting the exponential gives

$$v(t) = Re[V_M e^{j2\pi ft} e^{j\theta}] \tag{13-4}$$

In a linear system, the output signal will always be the same frequency as the input signal (with no other frequencies present). The *Re*[ ] and the $e^{j2\pi ft}$ term is dropped to produce the *vector* or *phasor form* of the previous equation:

$$\overline{V} = V_M e^{j\theta} \tag{13-5}$$

This is the *polar form* of the vector. Expanding the exponential using Euler's identity gives the *rectangular form*

$$\begin{aligned}\overline{V} &= V_M[\cos\theta + j\sin\theta] \\ &= V_M\cos\theta + j\,V_M\sin\theta\end{aligned} \tag{13-6}$$

$$V_{\text{RE}} = Re[\overline{V}] = V_M\cos\theta \tag{13-7}$$

$$V_{\text{IM}} = Im[\overline{V}] = V_M\sin\theta \tag{13-8}$$

Other useful conversion formulas are

$$\theta = \tan^{-1}(V_{\text{IM}}/V_{\text{RE}}) \quad \text{for } V_{\text{RE}} \geq 0 \tag{13-9}$$

$$\theta = \tan^{-1}(V_{\text{IM}}/V_{\text{RE}}) + 180 \quad \text{for } V_{\text{RE}} < 0,\ V_{\text{IM}} > 0 \tag{13-10}$$

$$\theta = \tan^{-1}(V_{\text{IM}}/V_{\text{RE}}) - 180 \quad \text{for } V_{\text{RE}} < 0,\ V_{\text{IM}} < 0 \tag{13-11}$$

$$V_M = \sqrt{V_{\text{RE}}^2 + V_{\text{IM}}^2} \tag{13-12}$$

A further change in notation produces a yet more concise vector form:

$$\overline{V} = V_M \angle\theta \tag{13-13}$$

To summarize, the following three expressions all represent the same signal:

$$v(t) = V_M\cos(2\pi ft + \theta) \tag{13-14}$$

$$\overline{V} = V_M e^{j\theta} \tag{13-15}$$

$$\overline{V} = V_M \angle\theta \tag{13-16}$$

The vector form is a first step toward a true frequency domain representation of the signal. Note that the vector forms do not explicitly state the frequency of the input and output signals. So far, we have only considered the vector form to be valid at one specific frequency. Later we will expand on this concept and consider vectors which are a function of frequency.

Example 13.1

Express the following signal in polar and rectangular vector forms: $v(t) = 5\cos(20t + 30°)$.

The amplitude of the waveform is 5 and the phase is 30 degrees. Therefore, the polar vector form is $5\ e^{j(30°)}$ or $5\ \angle 30°$.

The rectangular form is found by

$$V_R = 5\cos(30°) = 4.33$$

$$V_I = 5\sin(30°) = 2.5$$

$$\overline{V} = 4.33 + j2.5$$

Example 13.2

Express the following 20 MHz vector signal as a time domain function:

$$\overline{V} = 8 + j4$$

First, convert to polar form.

$$V_M = \sqrt{8^2 + 4^2} = 8.94$$

$$\theta = \tan^{-1}(8/4) = 63.43°$$

$$\overline{V} = 8.94\ \angle 63.43°$$

$$v(t) = 8.94\cos[2\pi(20 \times 10^6)t + 63.43°]$$

13.2 THE TRANSFER FUNCTION

Engineering circuit theory textbooks introduce the concept of the transfer function (or system function), which is the output voltage of a network divided by its input voltage, both in vector form and both a function of frequency (Figure 13-1).

$$H(f) = \frac{V_2(f)}{V_1(f)} \tag{13-17}$$

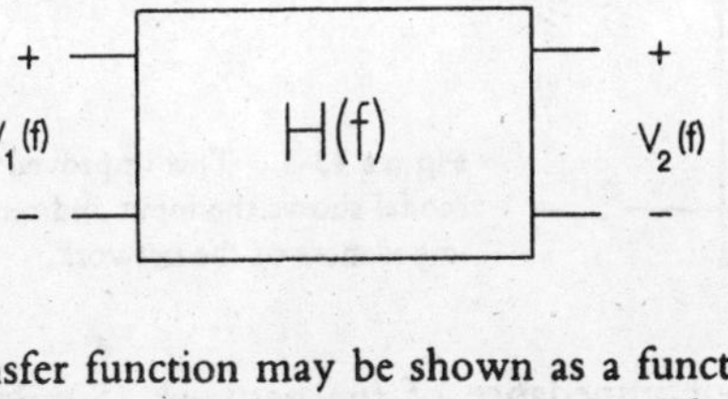

Figure 13-1. The transfer function of a network, $H(f)$, relates the input voltage to the output voltage.

The transfer function may be shown as a function of scalar frequency (ω or f) or complex frequency (s), where $s = \sigma + j\omega$. The former is compatible with Fourier theory and the latter is oriented toward Laplace transforms. The complex frequency approach supports transient analysis, which for traditional network measurements is not used. Thus, the transfer function is shown here as a function of f, with units of hertz.

The transfer function can be reduced in complexity by ignoring the phase information contained in $H(f)$. Taking the magnitude of the voltages results in voltage gain as a function of frequency.

$$G_V(f) = |H(f)| = \frac{|V_2(f)|}{|V_1(f)|} \tag{13-18}$$

As shown in Chapter 2, the voltage gain can be expressed in decibel form as

$$G_{V(\text{dB})}(f) = 20 \log[G_V(f)] \tag{13-19}$$

A circuit model for the transfer function is shown in Figure 13-2. Since V_1, V_2, and H are all functions of frequency, we will drop the explicit designation at this point. This simple model ignores any loading considerations and is an appropriate model for studying network theory, but needs to be improved on for practical use.

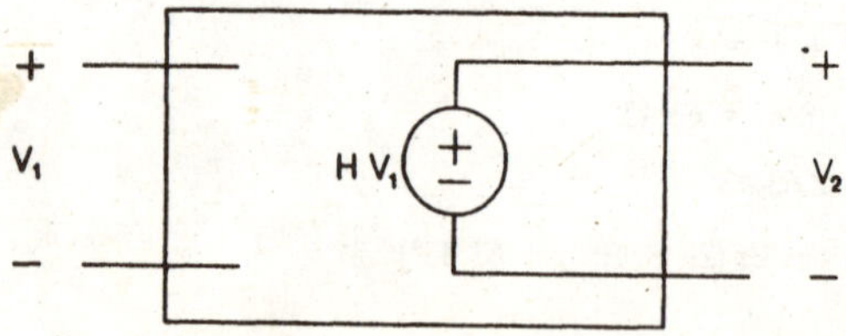

Figure 13-2. This circuit models the behavior of a network, ignoring loading conditions.

13.3 IMPROVED TWO-PORT MODEL

The previous circuit model implies that the impedance looking into the input terminals is infinite and the impedance looking into the output terminals is zero. This model is inaccurate for circuits having finite input and output impedances. A better model is shown in Figure 13-3. The circuit providing the input voltage to the two-port network is loaded by the input impedance of the network. Similarly, the output of the network is loaded by the impedance connected across its output (Figure 13-4). The voltage source driving the network is shown as having a finite output impedance, Z_S. The relationship of the source and output voltages is now

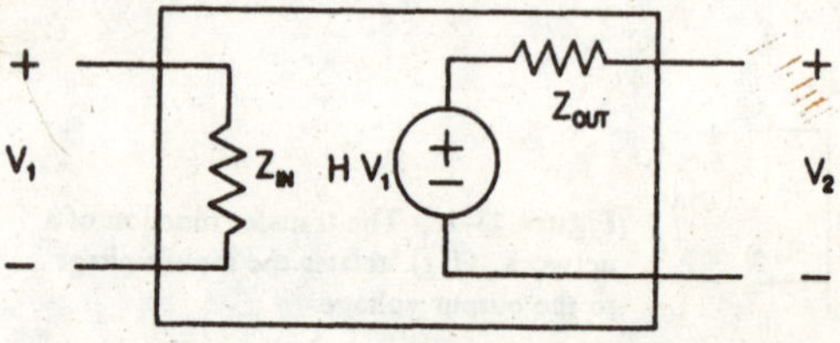

Figure 13-3. This improved circuit model shows the input and output impedances of the network.

$$\frac{V_2}{V_S} = \frac{Z_{IN}\, Z_L\, H}{(Z_S + Z_{IN})(Z_L + Z_{OUT})} \tag{13-20}$$

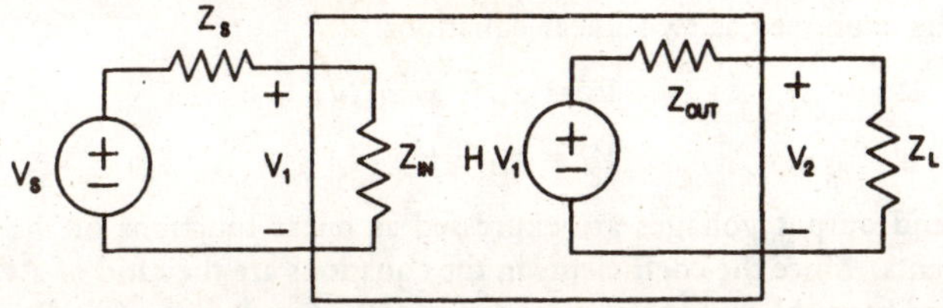

Figure 13-4. When driven by a nonzero source impedance and/or loaded with a finite load impedance, loading effects will occur.

The ratio of the output voltage to source voltage is clearly affected by the source impedance and load impedance. If the source impedance is low relative to the network input impedance, the effect may be negligible. Similarly, if the load impedance is large relative to the output impedance of the network, that effect may be ignored. In many practical measurement situations, the source impedance and load impedance are specified. That is, the measurement procedure calls for a source with a particular output impedance and for a particular load impedance to be installed at the output. If only the forward transfer characteristics (V_2/V_S) of the network are required, specifying the terminating impedances may be sufficient to ensure an accurate characterization of the network.

13.4 IMPEDANCE PARAMETERS

So far in this chapter, we have discussed two-port network models which are incomplete in that they do not provide for both forward and reverse transfer characteristics. What is needed is a model which takes into account the influence of the output port on the input port. There is an entire class of linear two-port models which completely characterize a two-port network. The first of these which we will discuss is the use of *impedance parameters*.

The impedance parameter model is shown in Figure 13-5. Note that in the circuit are two impedances and two dependent voltage sources. Alternatively, the

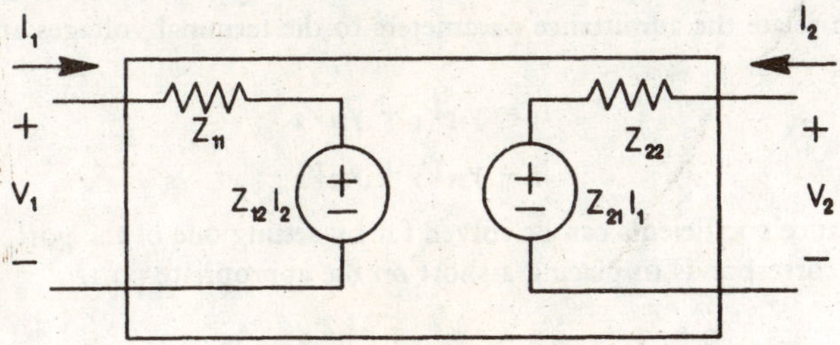

Figure 13-5. The circuit model associated with two-port impedance parameters.

model can be expressed as two linear equations:

$$V_1 = z_{11}I_1 + z_{12}I_2 \tag{13-21}$$

$$V_2 = z_{21}I_1 + z_{22}I_2 \tag{13-22}$$

The input and output voltages are expressed as linear functions of the input and output currents. Since the coefficients in the equations are the ratio of a voltage and a current, and have the units of ohms, they are called impedance (or Z) parameters. The first subscript of the particular Z parameter indicates the port at which the effect occurs, while the second subscript indicates the source of the effect. For example, z_{21} represents the effect on the voltage at port 2 due to the current at port 1. In general, Z parameters are complex numbers which vary as a function of frequency. To solve for the particular impedance parameter, the currents I_1 and I_2 are selectively set to zero. In an actual measurement situation this means that the appropriate port is left open.

$$z_{11} = \left.\frac{V_1}{I_1}\right|_{I_2=0} \tag{13-23}$$

So z_{11} is the input impedance of the network, with the stipulation that the output is left open. Therefore, z_{11} is called the *open-circuit input impedance*. The other parameters can be determined in a similar manner and have appropriate names.

$$z_{12} = \left.\frac{V_1}{I_2}\right|_{I_1=0} \quad \text{open-circuit reverse transfer impedance} \tag{13-24}$$

$$z_{21} = \left.\frac{V_2}{I_1}\right|_{I_2=0} \quad \text{open-circuit forward transfer impedance} \tag{13-25}$$

$$z_{22} = \left.\frac{V_2}{I_2}\right|_{I_1=0} \quad \text{open-circuit output impedance} \tag{13-26}$$

13.5 ADMITTANCE PARAMETERS

Another set of two-port parameters is the *admittance parameters*. The linear equations which relate the admittance parameters to the terminal voltages and currents are

$$I_1 = y_{11}V_1 + y_{12}V_2 \tag{13-27}$$

$$I_2 = y_{21}V_1 + y_{22}V_2 \tag{13-28}$$

The admittance coefficients can be solved for by setting one of the port voltages to zero. This corresponds to placing a short on the appropriate port.

$$y_{11} = \left.\frac{I_1}{V_1}\right|_{V_2=0} \quad \text{short-circuit input admittance} \tag{13-29}$$

$$y_{12} = \left.\frac{I_1}{V_2}\right|_{V_1=0} \quad \text{short-circuit reverse transfer admittance} \tag{13-30}$$

$$y_{21} = \left.\frac{I_2}{V_1}\right|_{V_2=0} \quad \text{short-circuit forward transfer admittance} \tag{13-31}$$

$$y_{22} = \left.\frac{I_2}{V_2}\right|_{V_1=0} \quad \text{short-circuit output admittance} \tag{13-32}$$

13.6 HYBRID PARAMETERS

The *hybrid parameters* (or *h* parameters) are often used to describe transistor characteristics. The coefficients of the linear equations are not consistently impedances nor admittances, hence the name hybrid parameters.

The parameters are defined by the following equations.

$$V_1 = h_{11}I_1 + h_{12}V_2 \tag{13-33}$$

$$I_2 = h_{21}I_1 + h_{22}V_2 \tag{13-34}$$

Solving for the h parameters

$$h_{11} = \left.\frac{V_1}{I_1}\right|_{V_2=0} \quad \text{short-circuit input impedance} \tag{13-35}$$

$$h_{12} = \left.\frac{V_1}{V_2}\right|_{I_1=0} \quad \text{open-circuit reverse voltage gain} \tag{13-36}$$

$$h_{21} = \left.\frac{I_2}{I_1}\right|_{V_2=0} \quad \text{short-circuit forward current gain} \tag{13-37}$$

$$h_{22} = \left.\frac{I_2}{V_2}\right|_{I_1=0} \quad \text{open-circuit output admittance} \tag{13-38}$$

13.7 TRANSMISSION PARAMETERS

Yet another variation on the basic concept of two-port parameters are *transmission parameters* (also called *ABCD parameters*). These parameters are defined by

$$V_1 = AV_2 - BI_2 \tag{13-39}$$

$$I_1 = CV_2 - DI_2 \tag{13-40}$$

The parameters are given by the following equations.

$$A = \left.\frac{V_1}{V_2}\right|_{I_2=0} \quad \text{open-circuit voltage ratio} \tag{13-41}$$

$$B = \left. \frac{V_1}{-I_2} \right|_{V_2=0} \quad \text{negative short-circuit transfer impedance} \tag{13-42}$$

$$C = \left. \frac{I_1}{V_2} \right|_{I_2=0} \quad \text{open-circuit transfer admittance} \tag{13-43}$$

$$D = \left. \frac{I_1}{-I_2} \right|_{V_2=0} \quad \text{negative short-circuit current ratio} \tag{13-44}$$

13.8 SCATTERING PARAMETERS

Scattering parameters (or *s* parameters) are most commonly used at high frequencies and are the most important set of two-port parameters relating to network measurements. Unlike the previous sets of two-port parameters, *s* parameters use a traveling wave approach to describe the activity at each port. A traveling wave is incident on the port; a portion of it is reflected, and the remainder is transmitted to the network (Figure 13-6). The approach is consistent with transmission line theory, hence its common usage in high frequency measurements. *s* parameter measurements are referenced to a characteristic impedance (Z_0), which is also usually the nominal input and output impedance of the network.

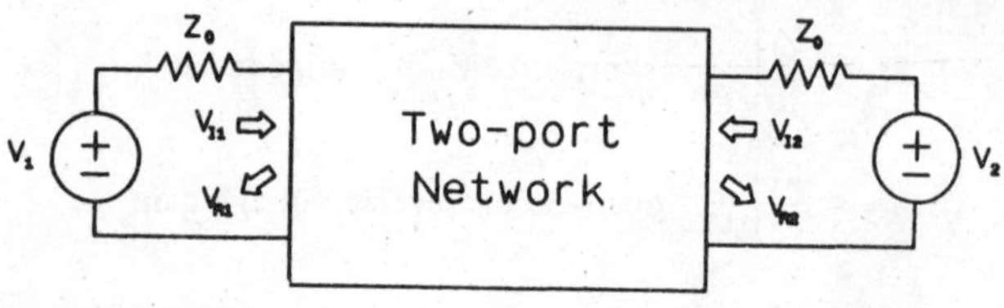

Figure 13-6. A two-port network can be characterized using *s* parameters, which describe the behavior of the incident and reflected voltages at each port.

The defining equations for *s* parameters are

$$V_{R1} = s_{11} V_{I1} + s_{12} V_{I2} \tag{13-45}$$

$$V_{R2} = s_{21} V_{I1} + s_{22} V_{I2} \tag{13-46}$$

Notice that the equations are made up only of incident and reflected voltages (and not currents). Again we will solve for the individual coefficients, the *s* parameters.

To solve for s_{11} it is necessary to set V_{I2} to zero. One might be tempted to attach mentally a short circuit to port 2 in order to accomplish this. But remembering that the *s* parameter model of the network deals with incident and reflected voltages, we will attach a Z_0 load to port 2 (Figure 13-7). This sets V_{I2} to zero while maintaining a nonreflecting load at port 2.

$$s_{11} = \left. \frac{V_{R1}}{V_{I1}} \right|_{Z_0 \text{ load on port 2}} \tag{13-47}$$

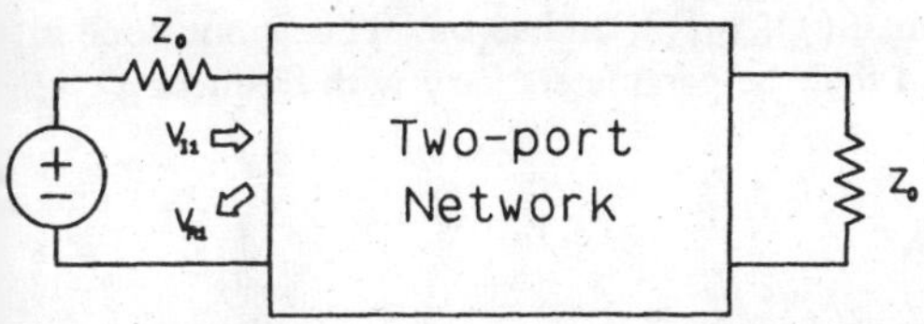

Figure 13-7. s_{11} is found by determining the amount of reflection at the input port while applying a Z_0 load to the output port.

So s_{11} is the same as the reflection coefficient at port 1 when port 2 is terminated with a Z_0 load. s_{11} is called the *input reflection coefficient.*

Solving for s_{21} with a Z_0 load still connected to port 2

$$s_{21} = \frac{V_{R2}}{V_{I1}}\bigg|_{Z_0 \text{ load on port 2}} \tag{13-48}$$

s_{21} is called the *forward transmission coefficient,* which loosely corresponds to the transfer function of the device.[1]

Continuing on

$$s_{22} = \frac{V_{R2}}{V_{I2}}\bigg|_{Z_0 \text{ load on port 1}} \tag{13-49}$$

So s_{22} is the reflection coefficient at port 2 (with port 1 terminated in Z_0) and is called the *output reflection coefficient.*

$$s_{12} = \frac{V_{R1}}{V_{I2}}\bigg|_{Z_0 \text{ load on port 1}} \tag{13-50}$$

s_{12} is called the *reverse transmission coefficient.*

The s parameter equations are often shown with the incident and reflected voltages normalized by the square root of Z_0. The defining equations are then

$$b_1 = s_{11}a_1 + s_{12}a_2 \tag{13-51}$$

$$b_2 = s_{21}a_1 + s_{22}a_2 \tag{13-52}$$

where

$$a_1 = \frac{V_{I1}}{\sqrt{Z_0}}, \quad a_2 = \frac{V_{I2}}{\sqrt{Z_0}}, \quad b_1 = \frac{V_{R1}}{\sqrt{Z_0}}, \quad b_2 = \frac{V_{R2}}{\sqrt{Z_0}}$$

This notation is introduced to provide continuity with other literature that the reader may encounter.

It is worth mentioning again that, in general, s parameters (like all two-port parameters) are a function of frequency. We could emphasize this point by always

[1] We will examine the subtleties in this statement later on in the chapter.

writing the s parameters in the form $s_{11}(f)$, $s_{21}(f)$, and so on. This is not done in practice, but instead it is understood that the parameters vary with frequency.

13.9 TRANSFER FUNCTION AND s_{21}

As previously stated, s_{21}, the forward transmission coefficient is similar to the conventional notion of transfer function. The differences will now be examined. Normally, the concept of a V_{OUT}/V_{IN} type of transfer function means that the voltages at the input and output ports of the network are measured directly, perhaps with some specified source and load impedances. Given that the source and load impedance constraints are observed, the transfer function measurement degenerates to a vector voltage measurement (preserving the phase information).

Now consider s_{21} which is equal to the reflected voltage at the output port, V_{R2}, divided by the incident voltage at the input port, V_{I1} (with a Z_0 load on the output port). The reflected voltage at the output port is somewhat of a misnomer, since it is really a traveling wave leaving the output port, due to activity on the input port. Since there is no incident wave on the output port and since V_{R2} will be totally absorbed by the Z_0 load, V_{R2} is the same as the V_{OUT} of the transfer function measurement.

Things are not quite the same on the input side. The incident wave, V_{I1} is the voltage that would be delivered by the Z_0 voltage source to a Z_0 load. If the input to the network is a perfect Z_0, then none of the incident voltage will be reflected from the input port. In that case, V_{I1} is the same as V_{IN} of the transfer function measurement. However, if the input impedance of the network is not Z_0, some of the incident voltage will be reflected, making V_{IN} different from V_{I1}. Another way of saying this is that s_{21} expresses the output voltage relative to the voltage available from the source when it is driving Z_0. For devices that have input impedances near Z_0, s_{21} is basically the same as the notion of a transfer function. When the input impedance is not close to Z_0, then the two measurements will be different.

13.10 WHY *s* PARAMETERS?

So why use s parameters at all? The answer is multifaceted. s parameters are closely related to and are an extension of transmission line theory, in that the input and output voltages are treated as incident and reflected traveling waves. For low-frequency design and measurement, this may not be a factor, but at higher frequencies transmission line concepts are unavoidable due to the shorter wavelength.

s parameters are measured with the network ports terminated in Z_0 impedances. Other two-port parameters require open or short circuits to be connected at the input and output ports. At higher frequencies, open and short circuits can be difficult to implement. Stray capacitance and inductance as well as transmission line effects get in the way. More importantly, many circuits will not behave well when presented with an open or short termination. Severe distortion or oscillation may occur.

Directional couplers and return loss bridges (which are discussed in Chapter 16) provide a means of separating incident and reflected waves, while maintaining a Z_0 match in the system. Thus, the individual traveling waves can be measured without significantly disturbing the device under test.

A wide body of design methodology has been developed which relates directly to and relies upon *s* parameters. Reflection coefficients (s_{11} and s_{22}) are often plotted on the Smith chart to design impedance matching and other networks. Flow graphs can be used to analyze systems which are characterized by *s* parameters. Entire textbooks have been written on these design techniques, which is beyond the scope of this book.[2]

REFERENCES

1. Carson, Ralph S. *High Frequency Amplifiers*. New York: John Wiley & Sons, Inc., 1975.
2. Gonzalez, Guillermo. *Microwave Transistor Amplifiers*. Englewood Cliffs, NJ: Prentice-Hall, Inc., 1984.
3. Hayt, William H., and Jack E. Kemmerly. *Engineering Circuit Analysis*, 2nd ed. New York: McGraw-Hill Book Company, 1971.
4. Hewlett-Packard Company. "S-Parameter Techniques for Faster, More Accurate Network Design," Application Note 95-1, February 1967. Palo Alto, CA.
5. Hewlett-Packard Company. "S-Parameter Design," Application Note 154, April 1972. Palo Alto, CA.
6. Irwin, J. David. *Basic Engineering Circuit Analysis*. New York: Macmillan Publishing Company, 1984.

[2] See Carson (1975) and Gonzalez (1984).

14

Network Analyzers

Network measurements can be divided into two types—transmission through the network and reflection at the network's input or output port. Full two-port network analysis normally requires the use of a multichannel network analyzer and an s parameter test set. Simpler measurements, such as transmission-only measurements, can be performed with less sophisticated equipment.

14.1 BASIC NETWORK MEASUREMENTS

The transmission through a two-port network is measured by applying a signal to one port and measure the response at the other port (Figure 14-1). The forward transmission characteristics of a network are measured by connecting the signal source to the input port and measuring the response at the output port. The reverse transmission characteristics of the network can be measured by driving the network at the output port and measuring the response at the input.

The reflection at the input port of a network is measured by applying a signal to the port and measuring the traveling wave reflected by the input. The reflection at the output port can be measured in a similar manner, while driving the output port.

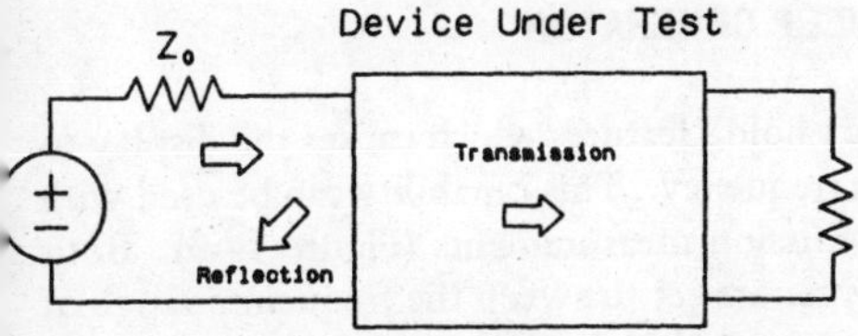

Figure 14-1. Transmission and reflection in a two-port network.

14.2 OSCILLOSCOPE AND SWEEP GENERATOR

An oscilloscope and a sweep generator can be used to implement a simple scalar network analyzer (Figure 14-2). This measurement setup is useful for making transmission measurements when only the magnitude (and not the phase) of the transfer function is required. The oscilloscope is operated in *X-Y* (channel versus channel) mode with the *X* (horizontal) input being the sweep voltage from the generator. The *Y* (vertical) input is the output of the device under test. The sweep voltage of the generator is proportional to the generator frequency. As the generator sweeps, the output of the device under test is plotted across the oscilloscope display.

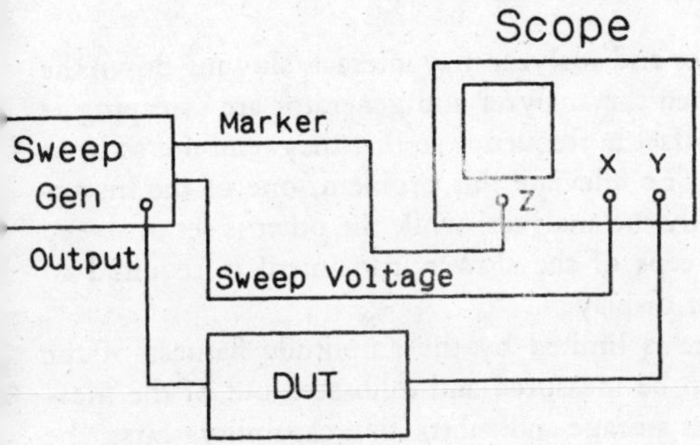

Figure 14-2. A sweep generator and an oscilloscope can be configured to perform basic transmission measurements.

Marker outputs from the sweep generator may be used to identify accurately frequencies on the oscilloscope display. These marker outputs are used to drive the intensity (*Z*-axis) input of the oscilloscope. When the marker frequency is swept through, the intensity on the scope changes. Some sweep generators can increase the output level slightly as the generator passes through the marker frequency, causing a "blip" on the oscilloscope display.

This method relies upon the sweep generator output to be constant with changes in frequency. Any imperfection in the flatness of the sweep generator will show up as an error in the measurement. Another major disadvantage of this network measurement technique is that the oscilloscope display is linear and has a limited dynamic range.

14.3 SPECTRUM ANALYZER WITH SWEEP GENERATOR

Most spectrum analyzers provide a "max hold" feature, which causes the display to retain the largest measured value at each frequency. This capability can be used with a sweep generator to make scalar transmission measurements (Figure 14-3). Both the sweep generator and spectrum analyzer are set to sweep the frequency range of interest. The spectrum analyzer is set to max hold. As the sweep generator excites the device under test, the spectrum analyzer measures the device's output. Gradually, the spectrum analyzer accumulates the entire response of the device.

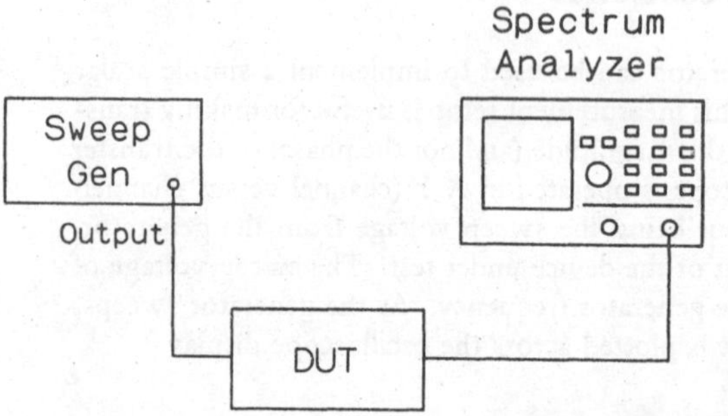

Figure 14-3. A sweep generator and spectrum analyzer can be used to make transmission measurements.

The sweep times of the generator and analyzer may interact, slowing down the measurement. The worst case is when the analyzer and generator are sweeping at approximately the same speed, but offset in frequency so that they tend not to be at the same frequency simultaneously. To alleviate this problem, one of the instruments is set to sweep very fast (usually the analyzer) while the other is set to sweep slowly. In this case, just a few sweeps of the slower instrument is required to produce a useful plot on the analyzer display.

The accuracy of this technique is limited by the amplitude flatness of the generator. However, the flatness can be measured and calibrated out of the measurement if the spectrum analyzer has storage and subtraction capability. First, the device under test (DUT) is removed and the generator is connected directly to the analyzer so that the generator's amplitude response is measured and stored. The DUT is reinstalled and the spectrum analyzer is set to subtract the generator response from the measured DUT response.[1] Of course, error due to the spectrum analyzer is still present in the measurement.

If the sweep generator and spectrum analyzer sweeps are synchronized, the measurement can be completed in one sweep and max hold is unnecessary. The spectrum analyzer may have an external trigger which can be driven by a trigger

---

[1] The desired mathematical operation is actually division, not subtraction, but the spectrum analyzer normally displays the response in logarithmic (decibel) form. Since the subtraction is done after the log, it is equivalent to division before the log function: $\log(a/b) = \log a - \log b$.

signal from the generator. Or perhaps both instruments can be triggered simultaneously by an external signal. Synchronizing the two instruments is usually more difficult than it first appears, due to latency between the trigger signal and the start of sweep and the difference in sweep rates between the two instruments.

This technique is particularly useful when the input and output frequency ranges of the device under test are not the same. Most network analyzers measure the output at the same frequency as the input. Therefore, a conventional network analyzer may not be able to measure the transfer characteristics of a frequency translating device such as a mixer.

14.4 SPECTRUM ANALYZER WITH TRACKING GENERATOR

The problem of synchronizing the generator and analyzer sweeps can be circumvented by designing the generator as part of the spectrum analyzer. Since the generator tracks the analyzer's receiver frequency, it is called a *tracking generator*. This combination is essentially a simple network analyzer, capable of making magnitude-only transmission measurements (Figure 14-4).

The flatness of the generator is still a source of error in the measurement, but again it can be calibrated out by measuring it and subtracting it from the DUT response.

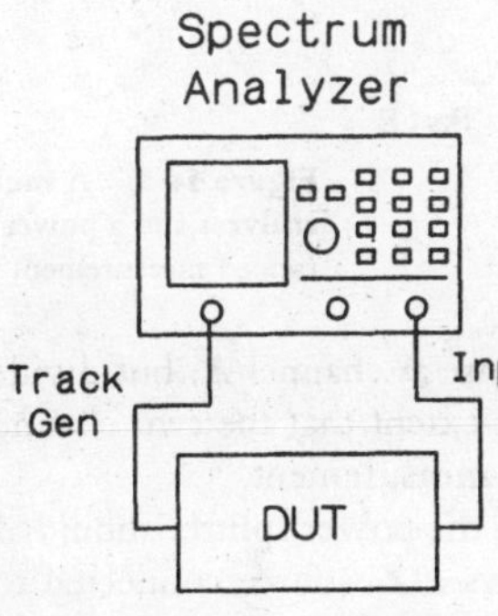

Figure 14-4. Some spectrum analyzers include a tracking generator which is useful in making basic network measurements.

14.5 RATIOING

Network analyzers normally have multiple receiver channels with closely matched gain and phase characteristics. Figure 14-5a shows a typical two-channel network analyzer. The generator (or source) is connected to a two-way power splitter, with one splitter output going to the DUT and the other output connected directly to channel R (reference channel) of the analyzer. The output of the DUT is connected to channel A of the analyzer. The network analyzer uses the split-off source signal to correct for frequency response imperfections in the source. The network analyzer

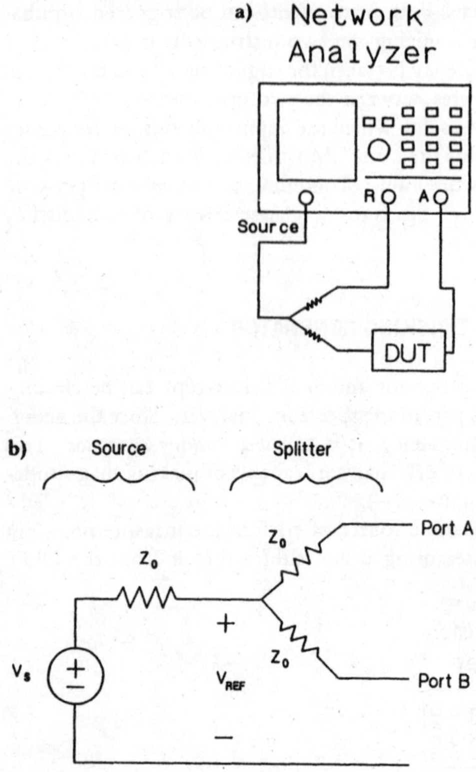

Figure 14-5. A multichannel network analyzer and a power splitter allow a ratioed measurement to be performed.

is configured to display the DUT response at channel A, but divided by the signal present at the reference channel. To the extent that the two channels are matched, the source flatness is removed from the measurement.

When making ratio measurements, the power splitter should always be a two-resistor type splitter. Figure 14-5b shows a Z_0 source connected to a two-resistor power splitter. V_{REF} is considered a virtual voltage source, with both port A and port B seeing V_{REF} through a Z_0 impedance. This implies that both ports always receive the same incident voltage. V_{REF} and the incident voltage may change with frequency, but this effect will be removed by the ratioing of the two channels.

14.6 DIRECTIONAL COUPLER

A *directional bridge* or *directional coupler* has the ability to sense the energy traveling in one direction along a transmission line. It is used to separate the incident voltage from the reflected voltage on the line when performing network measurements.

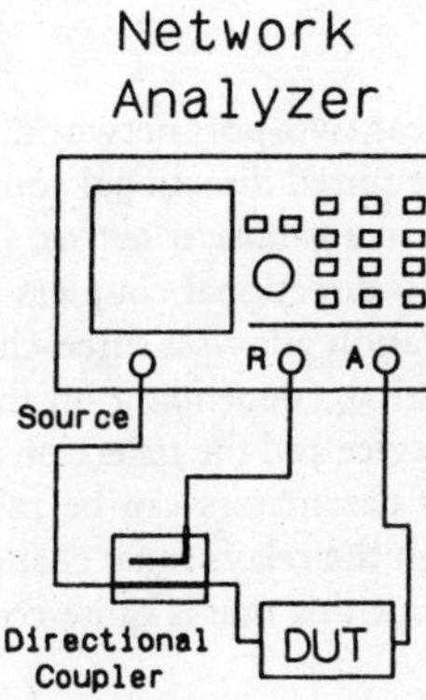

Figure 14-6. A directional coupler can also be used to produce a radioed measurement.

As configured in Figure 14-6, the directional coupler senses the incident wave at the input port of the device under test, by diverting a small amount of the incident power to the auxiliary port. The diverted incident wave is connected to the reference channel of the network analyzer and is used to obtain a ratioed measurement. Reversing the directional coupler allows the reflected wave at the DUT input port to be measured. Figure 14-7 shows the use of a directional coupler, along with a power splitter, to produce a reflection measurement. The power splitter is used to produce the reference channel signal (for ratioing), while the directional coupler measures the reflected wave from the DUT. Here the DUT is shown as a one-port device, since this is a one-port measurement. When multiport devices are measured in this manner, the unused ports are normally terminated with a Z_0 load.

Directional couplers are discussed further in Chapter 16.

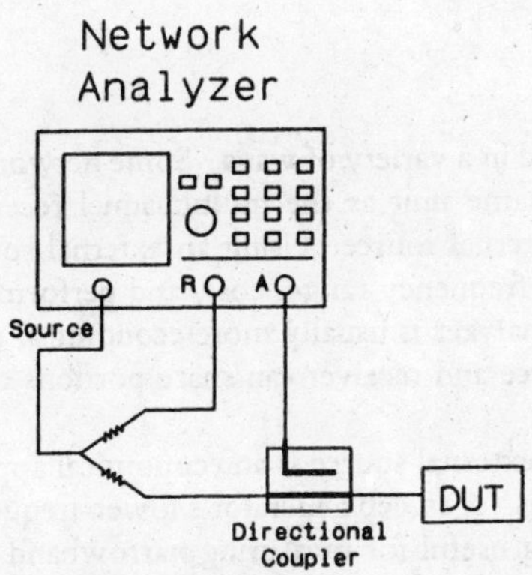

Figure 14-7. A directional coupler and a power splitter are shown here configured to measure the reflection from the device under test.

14.7 *s* PARAMETER TEST SET

For complete characterization of a linear two-port network, all four of the two-port parameters must be measured. The required directional couplers and power splitter are assembled into one device called an *s parameter test set*. A test set is commonly configured with a power splitter, two directional couplers and switching relays as shown in Figure 14-8. This configuration allows a three-channel network analyzer to measure either s_{11} and s_{21} or s_{12} and s_{22}, depending on the position of the relay. Thus, the transmission through the device and the reflection at the driven port can be displayed simultaneously. All four *s* parameters can be measured without reconnecting the device under test (although the relays must change position). Other test set configurations are also possible, but this one is quite common.

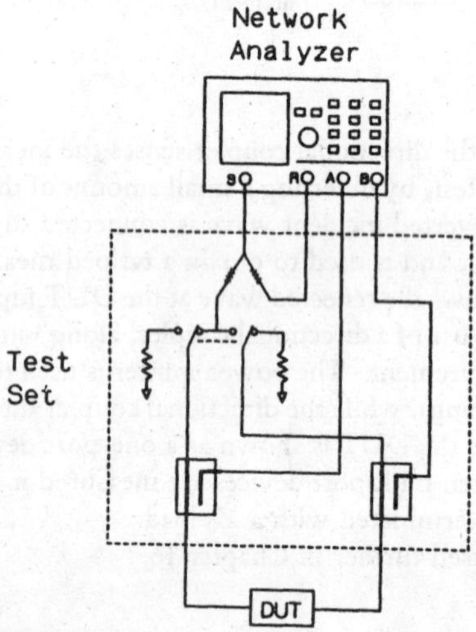

Figure 14-8. An *s* parameter test set allows all four *s* parameters to be measured without reconnecting the device under test.

14.8 SOURCES

Network analyzers can be configured in a variety of ways. Some network analyzers have signal sources built into the same unit as the multichannel receiver. Other network analyzers use a separate external source. Using an external source allows the user to optimize the system for frequency range, cost, and performance tradeoffs. Building the source into the analyzer is usually more economical (for a given level of performance) since the source and receiver can share portions of circuitry, but it reduces flexibility.

A sweep generator used as an external source is an economical approach suitable for measuring broadband devices. A sweep generator's lower-frequency stability and accuracy keeps it from being useful for measuring narrowband devices. A synthesized source provides excellent frequency stability, making it ideal for measuring narrowband devices, but at a higher cost.

14.9 SWEEP LIMITATIONS

A network analyzer has sweep rate (Hz/sec) limitations just as the spectrum analyzer does. Neither analyzer can be swept arbitrarily fast. The sweep rate limitation of a spectrum analyzer is proportional to the square of the resolution bandwidth, as discussed in Chapter 5. The sweep rate limitation in a network measurement is less obvious since it depends on both the IF bandwidth of the analyzer and the response of the device under test.

Since the source and receiver of a network analyzer are tuned to the same frequency, the signal out of the source lands in the center of the receiver passband.[2] Compare this with the spectrum analyzer case, where the signal is generated external to the analyzer, usually at some fixed frequency. The analyzer's receiver is swept past the stationary signal such that the signal is seen passing across the passband of the receiver. In the spectrum analyzer case, the signal is assumed to be a pure spectral line and the characteristics of the resolution bandwidth filter determines the maximum sweep rate.

In the network analyzer case, the signal is not sweeping past the receiver since the source and receiver are moving together. However, the signal amplitude at the output of the DUT will change as the sweep progresses. Consider a device under test which has the filter shape shown in Figure 14-9. As the sweep starts, the receiver will see a small signal due to the finite stop band of the filter. For the first part of the sweep, the signal amplitude will change slowly. When the sweep progresses to the filter skirt, the signal amplitude will begin to increase. The rate of amplitude increase will depend on the sweep rate and also the steepness of the filter response. The steeper the filter, the faster the amplitude will change. During this time, the receiver must respond quickly enough to track the increasing signal amplitude; otherwise, the measured response will be smeared as the receiver is unable to keep up. The wider the receiver bandwidth, the quicker it will respond.

To provide a little more perspective, consider a device which has a very flat amplitude response—a cable. If the cable were perfectly flat, the analyzer could sweep extremely fast because there are no amplitude variations for the analyzer to track.

Because sweep rate is essentially determined by the device under test, network analyzers usually place the entire burden for determining the sweep rate on the user. The sweep rate is rarely automatically chosen (as in the case of a spectrum analyzer).

---

[2] One exception is when the device under test has a large amount of delay relative to the sweep rate such that the receiver moves in frequency by a significant amount before the source's signal propagates through the device. When this happens, the sweep rate must be decreased to minimize the effect.

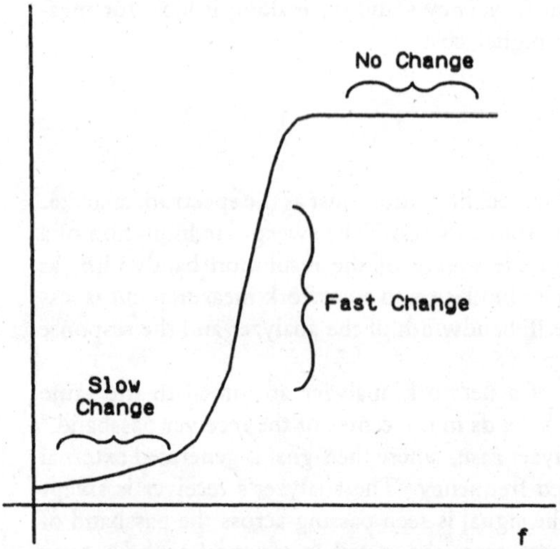

Figure 14-9. The rate of amplitude change seen by the network analyzer depends on the shape of the DUT's transfer function and the sweep speed of the analyzer.

So how does one determine the optimum (fastest) sweep rate? Unfortunately, the most common way of setting the sweep rate is by trial and error. First, a starting sweep rate is chosen (probably by past experience). The frequency response of the device is noted, and the sweep rate is decreased. If the response does not change, the device is not being swept too fast. If the response does change, the device is being swept too fast and the sweep rate must be decreased. This process is repeated until the response is stabilized, implying that the sweep rate is adequate.

14.10 AMPLITUDE SWEEP

Some network analyzers provide an *amplitude sweep* or *power sweep* capability which allows the user to measure network parameters as a function of amplitude level. For instance, the gain of an amplifier might be measured as a function of its input power in order to determine the gain compression point.

Unlike most other network measurements, amplitude sweep measurements imply that the network is nonlinear. Normal *s* parameter measurements assume that the network is either linear or nearly linear. The purpose of an amplitude sweep is to uncover and measure network parameters which change with varying amplitude and such changes are inherently nonlinear.

14.11 FREQUENCY OFFSET MEASUREMENTS

Earlier in the chapter, we stated that most network analyzers always measure devices with the analyzer source and receiver set to the same frequency. This limitation means that devices which have differing input and output frequencies cannot be characterized with this type of network analyzer. However, there are network analyzers available which allow the source and receiver to be offset in frequency. Frequency offset capability provides two additional types of measurements—swept harmonic measurements and mixer measurements.

Swept harmonic measurements characterize the harmonic distortion performance of a device over a range of frequencies. For swept harmonic measurements, the source is set to the fundamental frequency, f and the receiver is tuned to nf, where n is the number of the harmonic to be measured. As the network analyzer sweeps in frequency, the receiver automatically tracks and measures the chosen harmonic. Since the entire frequency range can be measured in one sweep, it represents a large productivity increase over other methods where the user must individually measure the harmonic level at each fundamental frequency with a spectrum analyzer.

Mixers translate an input frequency (called the RF frequency) to a different output frequency (called the IF frequency). Conventional network analyzers cannot measure this device, but with frequency offset capability, this measurement can be automated. The test configuration shown in Figure 14-10 is used to characterize the conversion loss of a mixer. The network analyzer source provides a swept signal to the RF port of the mixer, while a signal generator provides the LO (local oscillator) signal. The IF signal out of the mixer is fed into a network analyzer receiver. The network analyzer sweeps its source and receiver offset in frequency by the LO frequency. This produces a swept response of the mixer's conversion loss. Note the measurement instruments are connected to the mixer ports with attenuators, to minimize errors due to impedance mismatch at the mixer ports. Also, a low-pass filter is included at the mixer output to remove the unwanted mixer products.

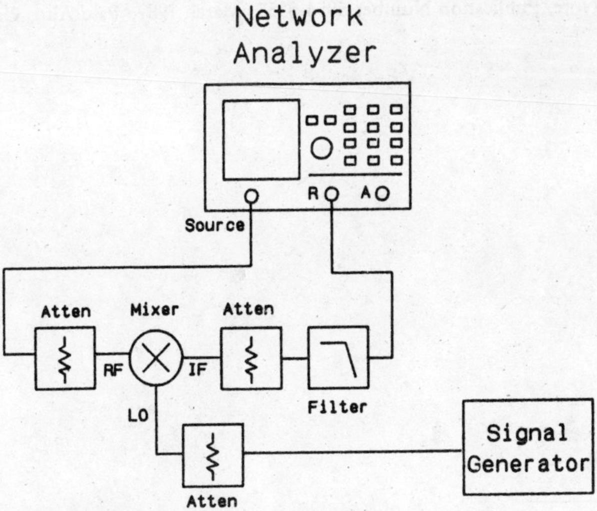

Figure 14-10. A network analyzer with frequency offset capability can be configured to characterize the conversion loss of a mixer.

14.12 TIME DOMAIN MEASUREMENTS USING THE INVERSE FFT

The inverse fast Fourier transform (IFFT) can be used to change a vector frequency domain network measurement back into the time domain response of the network. This time domain response is usually displayed as either the impulse response or the step response of the network. This feature can be of great use if the time domain characteristics of the network are of interest.

Even if the time domain characteristics of the network are not desired, the time domain response can be used to remove certain imperfections in the measurement. For example, the reflections due to a poor connector may not be easily visible in the frequency domain, but when transformed to the time domain, the connector's reflection will be apparent. This reflection can be removed by use of a gating function in the time domain. Then, the gated time domain response can be transformed back into the frequency domain using an FFT. The result is a frequency domain measurement which no longer contains the errors due to the reflections of an imperfect connector.[3]

REFERENCES

1. Curran, Jim. "Simplify Your Amplifier and Mixer Testing." *RF Design* (April 1988).
2. Hewlett-Packard Company. "High Frequency Swept Measurements," Application Note 183, Publication Number 5952-9200, December 1978. Palo Alto, CA.
3. Hewlett-Packard Company. "Vector Measurements of High Frequency Networks," Application Note, Publication Number 5954-8355, March 1987. Palo Alto, CA.

[3] For more information on time domain network analysis, see Hewlett-Packard (1987).

15

Transmission Measurements

The most common network measurement is characterizing the transmission through a device. In many electronic systems, the response at the output of a system block due to a signal at the input is a critical parameter. For distortionless transmission through a device, the output signal must be identical to the input signal, perhaps delayed in time and scaled in amplitude. This implies the device must have a flat amplitude response and a linear phase response. These criteria are not usually completely met, but can be approached in practice.

Measurement error is introduced into transmission measurements via a variety of mechanisms. These error mechanisms can be quantified so that the quality of the measurement is known.

15.1 DISTORTIONLESS TRANSMISSION

A system or network is called *distortionless* if its output is an exact replica of its input, except for amplitude scaling and time delay. Put mathematically,

$$y(t) = kx(t - t_0) \tag{15-1}$$

where

$y(t)$ = output signal

$x(t)$ = input signal

k = amplitude scale factor

t_0 = time delay in the system

Note that k and t_0 are constants and are not allowed to be a function of frequency. Figure 15-1 shows an example of input and output signals of a linear system. The input pulse has an amplitude of 1 and a pulse width of T. The output is the same shape as the input but is delayed by t_0 and the amplitude of the output has been changed by the amplitude scale factor, k.

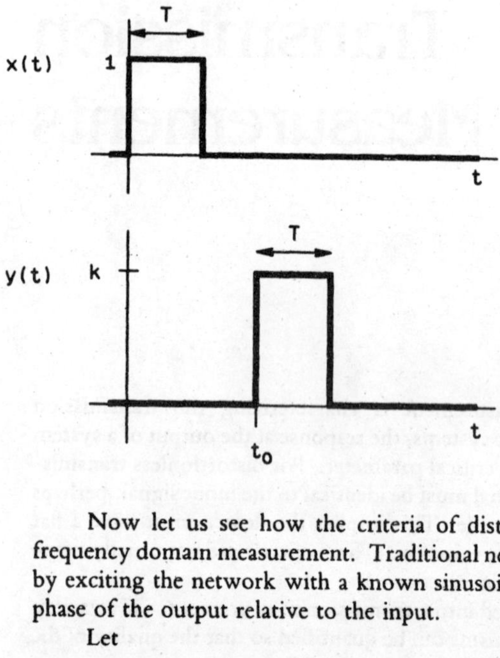

Figure 15-1. For distortionless transmission, the output of a system, $y(t)$, must be the same as the input, $x(t)$, except for time delay and amplitude scaling.

Now let us see how the criteria of distortionless transmission relates to a frequency domain measurement. Traditional network measurements are performed by exciting the network with a known sinusoid and measuring the amplitude and phase of the output relative to the input.

Let

$$x(t) = A \cos \omega t \tag{15-2}$$

For distortionless transmission,

$$y(t) = kA \cos[\omega(t - t_0)] \tag{15-3}$$

$$y(t) = kA \cos[\omega t - \omega t_0] \tag{15-4}$$

$$y(t) = kA \cos[\omega t - \theta(\omega)] \tag{15-5}$$

where $\theta(\omega)$ is the phase response of the system

$$\theta(\omega) = \omega t_0 = 2\pi f t_0.$$

Thus, for distortionless transmission, the amplitude response of the system is a constant (flat with frequency) and the phase response is a linear function of frequency (Figure 15-2). Phase measurements are usually limited to ±180 degrees and the phase plot in Figure 15-2 is shown wrapping around in order to stay within this

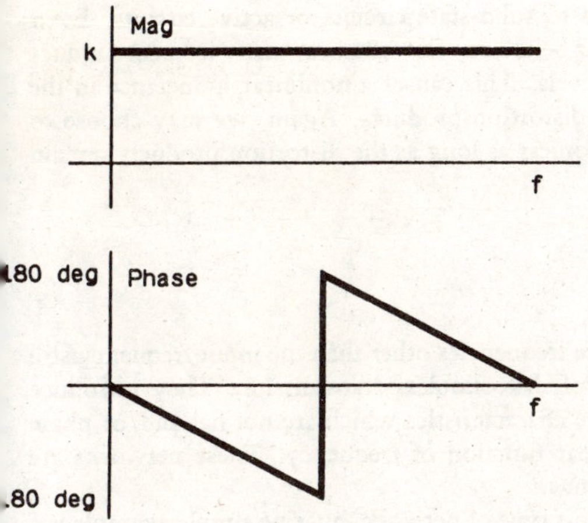

Figure 15-2. In the frequency domain, distortionless transmission implies a constant amplitude response and a linear phase response.

range.[1] Note that the output sinusoid has the same frequency as the input sinusoid and that no other frequencies are present.

Having introduced a strict definition of distortionless transmission, we will note that only devices which have infinite bandwidth, a completely flat magnitude response and a linear phase response will meet this definition of distortionless. Often, this definition is too strict and can be relaxed without jeopardizing the usefulness of the device.

15.2 NONLINEARITY

Many networks exhibit nonlinearities of the form

$$y(t) = k_0 + k_1x(t) + k_2x^2(t) + k_3x^3(t) + k_4x^4(t) + \cdots \tag{15-6}$$

As shown in Chapter 7, a sinusoidal input into this type of network produces output frequencies at the harmonics of the input. Obviously, this violates the distortionless transmission criteria since the output will not have the same shape as the input.

Many networks which are considered linear will exhibit this type of response under some operating conditions. For instance, a typical "linear" amplifier will have some finite level of harmonics at its output, due to distortion introduced in the amplifier. As long as these harmonics are small enough (as determined by the system requirements), we may choose to ignore them and consider the amplifier to be linear. However, if the amplifier is overdriven, the harmonic distortion may become severe, in which case we may choose to consider it a nonlinear system.

[1] It certainly can be argued that the phase response continues on in a straight line, but the usual measuring techniques limit it to a principal angle covering one 360 degree range.

Nonlinearities are not limited to solid-state circuits or active circuits. Even passive circuits can exhibit nonlinear behavior. For instance, many iron core inductors will saturate at high current levels. This causes a nonlinear inductance in the circuit, which can cause the usual distortion products. Again, we may choose to consider such a network as distortionless as long as the distortion products remain below a certain level.

15.3 LINEAR DISTORTION

Many networks that do not produce frequencies other than the input frequency still do not meet the strict definition of distortionless transmission. They introduce distortion by introducing amplitude characteristics which are not flat and/or phase characteristics which are not a linear function of frequency. These networks are sometimes said to have *linear distortion*.

There are many examples of this type of network, but one simple example is a single-pole low-pass filter (Figure 15-3). At low frequencies, both the amplitude and the phase are constant and introduce no distortion. As the frequency increases, the amplitude rolls off and the phase changes. The amplitude rolling off is a clear violation of distortionless transmission. The phase is allowed to change, but it must change in a linear manner over the entire frequency range. In a single-pole filter, this is not the case, so the phase characteristic also introduces distortion.

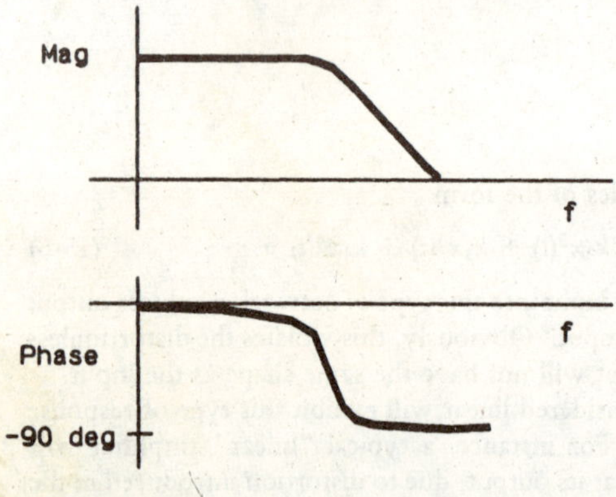

Figure 15-3. The amplitude and phase response of a single-pole low-pass filter.

But now consider the purpose of such a network. It is to remove (apparently) undesirable high frequencies while retaining the low frequencies. At low frequencies, little or no amplitude or phase distortion is introduced so the frequencies that appear at the output are not distorted. At higher frequencies, distortion is introduced, but these frequencies tend to be removed from the system anyway. So for frequencies of interest, we may choose to think of this as a distortionless network even though it does not meet the rigorous definition.

Many other examples exist. Bandpass networks are common in radio receivers and their amplitude response may be flat over some limited frequency range, but not over all frequencies. Still we may choose to consider them distortionless over that range. Even broadband amplifiers, which are usually considered flat in amplitude response, introduce linear distortion because they do not exhibit infinite bandwidth. Amplifiers that are AC coupled do not pass arbitrarily low frequencies (and certainly not DC) and the amplifier's response rolls off on the high-frequency side. Over its frequency range, we may wish to consider an amplifier as linear.

In summary, "linearity" and "distortionless transmission" are assumed with many networks that do not meet the strict definition. In practice, this is not a problem as long as it is understood what is meant by "distortionless" for a particular measurement situation.

15.4 INSERTION GAIN AND LOSS

Measurement of insertion gain or loss is shown in Figure 15-4. The output level of a signal generator is measured by a power meter. The device under test (DUT) is then inserted between the generator and the power meter. The gain or loss of the device is determined by taking the ratio of the output power to the generator power or, in decibels, the input power is subtracted from the output power.

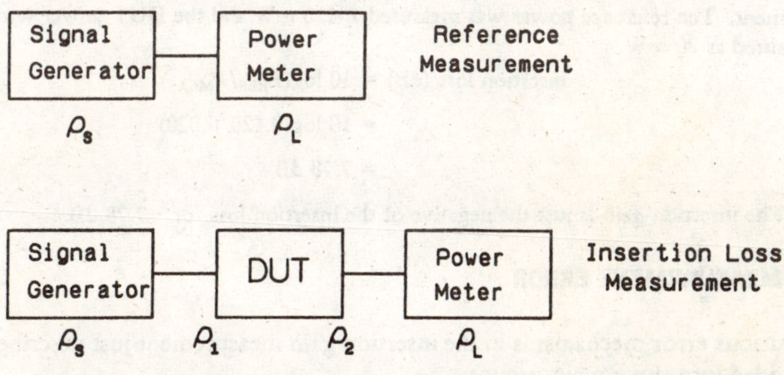

Figure 15-4. Measurement of insertion gain (or loss) can be accomplished with a signal generator and a power meter.

$$\text{insertion loss (dB)} = 10 \log(P_{REF}/P_{MEAS}) \qquad (15\text{-}7)$$

where

P_{REF} = reference power

P_{MEAS} = power measured at the DUT output

The insertion loss equation can be inverted to produce insertion gain:

$$\text{insertion gain (dB)} = 10 \log(P_{MEAS}/P_{REF}) \tag{15-8}$$

The reference power is determined by

$$P_{REF} = K_M P_S \tag{15-9}$$

where K_M is a constant which corresponds to the mismatch error in the power meter and P_S is the power out of the source.

$$P_{MEAS} = K_M P_S K_{DUT} \tag{15-10}$$

where K_{DUT} is the power transfer through the device under test.

The insertion gain can be expressed in terms of these two equations:

$$\begin{aligned} \text{insertion gain (dB)} &= 10 \log(K_M P_S K_{DUT}/K_M P_S) \\ &= 10 \log(K_{DUT}) \end{aligned} \tag{15-11}$$

The result is that the insertion gain measurement depends only on the DUT's transfer characteristics and not on the mismatch error at the meter or the source's power level. This is an idealized view of the insertion gain/loss measurement. Many of the error mechanisms have been ignored which, in practice, need to be considered in determining the accuracy of the measurement.

Example 15.1

Determine the insertion loss and insertion gain in decibels for the following measurement. The reference power was measured at 120 mW and the DUT power was measured as 20 mW.

$$\begin{aligned} \text{insertion loss (dB)} &= 10 \log(P_{REF}/P_{MEAS}) \\ &= 10 \log(0.120/0.020) \\ &= 7.78 \text{ dB} \end{aligned}$$

The insertion gain is just the negative of the insertion loss, or −7.78 dB.

15.5 MEASUREMENT ERROR

The various error mechanisms in the insertion gain measurement just described can be divided into three main groups:

Mismatch Error—errors due to the imperfect impedances of the source, power meter, and device under test.

Instrument Accuracy—errors due to imperfections internal to the power meter (not including mismatch at its input). Instrument accuracy includes both *absolute accuracy* and *relative accuracy*. Absolute accuracy indicates how accurately the meter can measure the power in a signal while relative accuracy pertains to how well the meter can measure *changes* in signal power.

Instrument Drift—errors due to drift in the test instruments, particularly the signal generator.

Error mechanisms which occur consistently in both the reference measurement and the insertion gain measurement will tend to be removed from the insertion gain calculation. On the other hand, error mechanisms which are not the same in the two measurements will introduce an error or uncertainty in the final insertion gain result.

Mismatch error, which was already discussed in Chapter 11, is caused by reflections at the various connections in the system. First, consider the reference measurement. A mismatch loss occurs at the output of the signal generator and the input to the power meter, both due to their imperfect Z_0 impedances. The mismatch loss at the signal generator is consistent between the reference and insertion loss measurement and is considered to be included in P_S, which has been shown to be removed from the insertion loss calculation. Similarly, the mismatch loss at the power meter is included in K_M and has already been shown to be removed from the insertion loss calculation.

Another mismatch error occurs, namely, the mismatch uncertainty which is due to the reflection off of the power meter being reflected again at the signal generator. The doubly reflected wave ends up again at the power meter and introduces an error into the reference measurement. Since this same error mechanism does *not* occur during the insertion gain measurement, it will introduce an error into the insertion gain calculation. Including this error in the reference measurement gives a more accurate equation for P_{REF}. (Note that the mismatch uncertainty has the same form as equation 11-33 in Chapter 11, except that it is squared to be consistent with the use of power instead of voltage.)

$$P_{REF} = \frac{K_M P_S}{(1 \pm \rho_S \rho_L)^2} \tag{15-12}$$

The insertion gain measurement has a different set of reflections due to the imperfect Z_0 impedances. There are double reflections between the signal generator and the DUT (ρ_S, ρ_1) and between the DUT and the power meter (ρ_2, ρ_L). These two sets of reflections introduce two more sources of measurement uncertainty, and the resulting measured power is

$$P_{MEAS} = \frac{K_M P_S K_{DUT}}{(1 \pm \rho_S \rho_1)^2 (1 \pm \rho_2 \rho_L)^2} \tag{15-13}$$

There is one more source of mismatch uncertainty. The reflected wave from the power meter can pass through the DUT and be reflected back from the signal generator. This reflection passes through the DUT again and is finally incident on the power meter, introducing a mismatch uncertainty. This mechanism will be ignored here—if the roundtrip loss through the DUT is greater than 10 dB, its effect is negligible.[2]

Now that the mismatch uncertainties have been included in the power measurements, consider again the insertion gain measurement:

---

[2] See Adam (1969) and Hewlett-Packard (1978) for a discussion of this effect.

$$\text{insertion gain (dB)} = 10\log(P_{\text{MEAS}}/P_{\text{REF}})$$

$$= 10\log\left[\frac{K_M P_S K_{\text{DUT}}(1 \pm \rho_S\rho_L)^2}{K_M P_S(1 \pm \rho_S\rho_1)^2(1 \pm \rho_2\rho_L)^2}\right] \tag{15-14}$$

$$= 10\log(K_{\text{DUT}}) + 20\log(1 \pm \rho_S\rho_L) - 20\log(1 \pm \rho_S\rho_1) - 20\log(1 \pm \rho_2\rho_L) \tag{15-15}$$

The first term of the equation is the ideal result for insertion gain, and the remaining terms represent the mismatch uncertainty in the measurement.

$$\text{mismatch uncertainty} = 20\log(1 \pm \rho_S\rho_L) - 20\log(1 \pm \rho_S\rho_1) - 20\log(1 \pm \rho_2\rho_L) \tag{15-16}$$

Example 15.2

The insertion loss of an attenuator is measured using a signal generator and power meter. Determine the mismatch uncertainty in measuring the 10 dB attenuator with an *SWR* at each port of 1.5. The signal generator and power meter have return losses of 20 dB and 30 dB, respectively.

First, compute the reflection coefficient of each device.

$$\rho = \frac{SWR - 1}{SWR + 1}$$

$$\rho_1 = \rho_2 = \frac{1.5 - 1}{1.5 + 1} = 0.20$$

$$\rho_S = 10^{(-RL/20)} = 10^{(-20/20)} = 0.10$$

$$\rho_L = 10^{(-RL/20)} = 10^{(-30/20)} = 0.0316$$

Now compute the mismatch uncertainty:

$$\begin{aligned}\text{mismatch uncertainty} &= 20\log(1 \pm \rho_S\rho_L) \\ &\quad -20\log(1 \pm \rho_S\rho_1) \\ &\quad -20\log(1 \pm \rho_2\rho_L) \\ &= 20\log[1 \pm (0.1)(0.0316)] \\ &\quad -20\log[1 \pm (0.10)(0.20)] \\ &\quad -20\log[1 \pm (0.20)(0.0316)] \\ &= (+0.0274, -0.0275) + (+0.175, -0.172) \\ &\quad +(+0.0551, -0.0547)\end{aligned}$$

$$\text{mismatch uncertainty} = +0.258\text{ dB}, -0.254\text{ dB}$$

Therefore, the errors due to mismatch uncertainty are bounded by +0.258 dB and −0.254 dB, for a total uncertainty of 0.521 dB. Note that the mismatch uncertainty is not the same in both directions, but with small mismatch uncertainty the two limits have approximately the same magnitude.

Now consider the errors internal to the power meter (instrument accuracy). The *absolute* accuracy of the power meter is not important as long as the *relative* accuracy of the meter is good. In other words, the meter does not need to be able to determine whether the signal power is exactly a certain power level, but it does need to measure accurately *changes* in power level. In the insertion gain measurement, the change in power level that must be measured is the change caused by the insertion of the device under test into the measurement system. The measurement is accurate to the extent that the meter correctly measures this change.

The signal generator may introduce an error due to instrument drift. If the signal generator is slightly off in signal level, it will not affect the measurement accuracy since the power meter measures the signal level anyway. However, the signal generator's output power must be stable so that it does not change between the time the reference measurement and the insertion loss measurement take place. Such a change in output power would introduce an error classified as instrument drift.

15.6 NORMALIZATION

The insertion gain/loss measurement technique just described measures a single frequency. The concept can be expanded to multiple frequencies by using a swept network analyzer and is referred to as *normalization*. Figure 15-5 shows a network analyzer with test set used to measure the transmission characteristics of a device. First, a through connection is used to connect the two ports of the network analyzer system. The response of the network analyzer, test set, and cabling is measured and stored in digital memory inside the network analyzer. Figure 15-6 shows the typical amplitude response of the measurement system. Besides the amplitude unflatness there is often significant linear phase response due to delays in the system. After the test system responses are stored, the measurement is made relative to them.[3] After normalization is performed but with the through still connected, the analyzer display will show a flat 0 dB magnitude response and a flat 0 degree phase response. The device under test is inserted into the measurement path and its characteristics are measured. The normalization data may be valid only with the particular analyzer setting so if the analyzer setup is changed, the test system may need to be renormalized.

---

[3] The measurement system responses are usually stored in digital memory as an array of complex (vector) numbers. The normalized measurement is computed by dividing the device response by the system response.

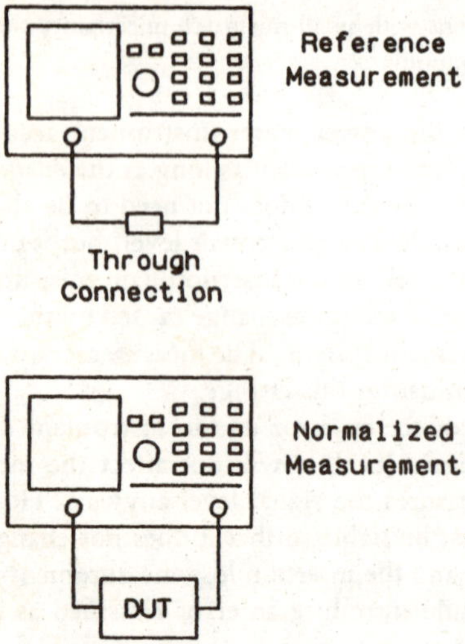

Figure 15-5. During normalization, a *through* connection is substituted for the device under test and the system response is measured and stored. Then the device under test is reinserted and the measurement is made relative to the stored normalization data.

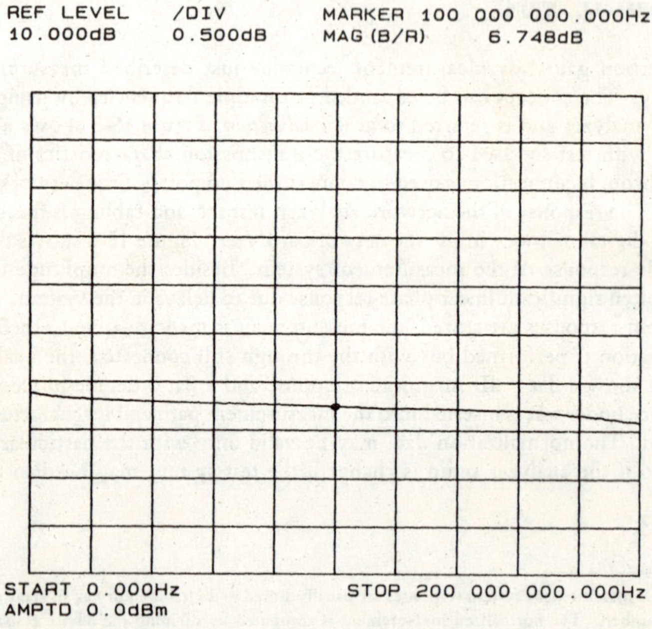

Figure 15-6. A typical system response, measured during normalization. After normalization, the measured response will be perfectly flat.

Normalization is a simple, but effective technique for removing error from the measurement. It requires that the network analyzer is stable with time. If the analyzer's response drifts significantly, it will need to be normalized often. Assuming that the network analyzer is stable from one measurement to another, its absolute accuracy is no longer an issue. Any absolute measurement error is removed by the normalization. Take the example of a perfect 0 dB insertion loss device. During normalization, the output power of the source is measured by the receiver. This measurement may have some absolute error in it, but when the perfect 0 dB device is inserted, the network analyzer will read the exact same value. Thus, at 0 dB, which is where the normalization occurred, we can achieve extremely good accuracy. The only requirement is that the network analyzer is stable. Now suppose that we insert an attenuator that has a loss of 10 dB as our device under test. The power level at the network analyzer's receiver drops by 10 dB. We are not particularly interested in whether the network analyzer can measure the exact power level at the DUT output, but we do need the analyzer to measure accurately the DUT's output *relative to the power level during normalization*. The network analyzer's absolute accuracy is not important, but its *relative* accuracy is critical. In this example, the output power seen by the analyzer will drop by 10 dB. We are interested in how accurately the analyzer can measure this 10 dB change, which is normally specified by the instrument manufacturer as *dynamic accuracy*.

Example 15.3

A network analyzer is used to make a normalized measurement of the insertion loss through a filter with a nominal loss of 3 dB. The network analyzer specifications include

Source level accuracy: ±1 dB

Source flatness: 1.5 dB peak to peak over the frequency range of the analyzer

Receiver absolute accuracy: ±0.15 dB

Receiver dynamic accuracy: ±0.02 dB

How much error may be introduced into the measurement due to these instrument errors?

Since normalization is used in this measurement, most of these instrument errors are removed. The effects of source level accuracy, source flatness, and receiver absolute accuracy are all removed during normalization. This leaves only the receiver dynamic accuracy contributing to error in the measurement. The total error introduced by these effects is ±0.02 dB. Note that this is much better than the other specifications would imply.

15.7 IMPORTANCE OF LINEAR PHASE

The importance of a linear phase response is often overlooked in considering distortion. Distortion is often discussed in terms of amplitude distortion of the waveform, particularly in the form of harmonics. But a device could have a perfectly flat amplitude response and still severely distort the waveform if the device's phase is not linear with frequency.

Consider the square wave as an example. Figure 15-7a shows the square waveform that results from only the fundamental and third harmonic of a true square wave. Recall that it would take an infinite number of harmonics to recreate

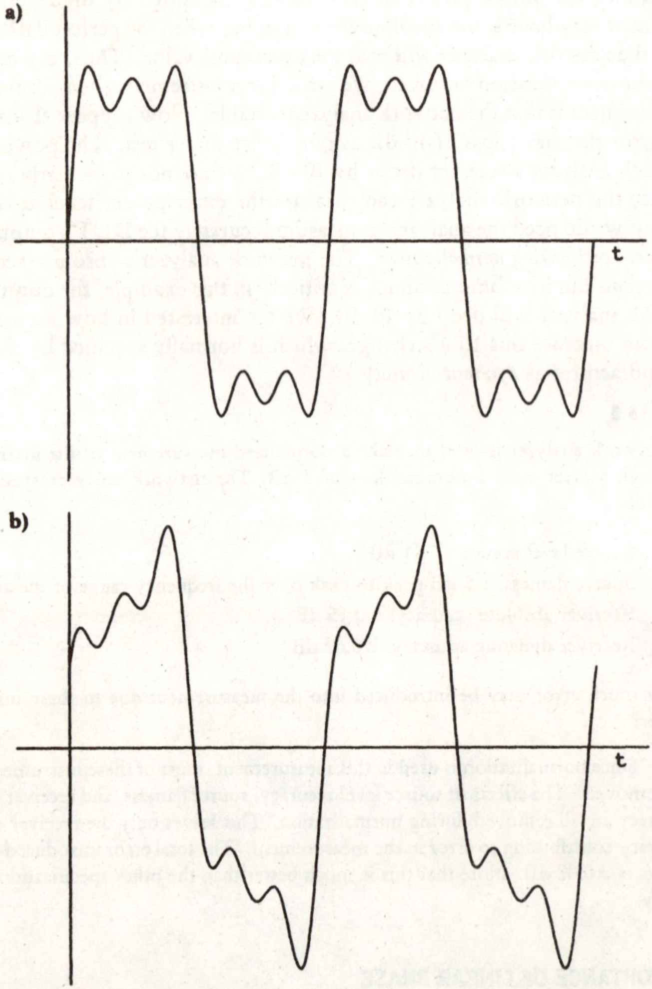

Figure 15-7. (a) The waveform that results from the first fundamental and third harmonic of a square wave. (b) The same waveform after passing through a system with nonlinear phase.

the square wave exactly. However, the two sine waves combine to produce a waveform that is recognizable as a square wave. Notice how the fundamental is lined up in phase with the square wave and that the third harmonic adds to the fundamental in just the right places to make the combined waveform more square.

Suppose this signal is passed through a device which alters the phase relationship between the two sine waves (Figure 15-7b). Now the two sine waves add together in such a way that produces a new waveform, one that does not approximate the ideal square wave nearly as well. The amplitudes of the sine waves are still the same, but the phase relationship has been altered.

If both sine waves are delayed by an equal amount of time, the waveform appears at the output undistorted, but delayed. A constant delay implies linear phase since a degree of phase at a low frequency is a much longer time than a degree of phase at high frequency. Thus, for the delay to be constant, the amount of phase shift required will increase with frequency.

Linear phase response is becoming more important in electronic systems as the use of digital data and pulsed signals increases. If a single sine wave is passed through a system, it may not be important that the phase response is well controlled. However, when a pulsed or digital signal is transmitted, the phase response of the system must be linear so that the fidelity of the pulse is retained.

15.8 DEVIATION FROM LINEAR PHASE

As stated previously, the phase response of a system must be linear with frequency if distortion is to be avoided. Often, this definition is modified to require linear phase over only the frequency range of interest, such as the passband of a filter.

Many devices such as high-order narrowband filters have a large amount of phase change over a small frequency range. It is difficult to determine how linear the phase response of a device is from a plot as shown in Figure 15-8a. Most network analyzers provide a feature to introduce or remove a user selectable amount of linear phase. The user can adjust the amount of linear phase introduced in order to flatten the measured phase response of the device under test. After the phase is made as flat as possible, the deviation from linear phase can be measured (Figure 15-8b).

15.9 PHASE ERROR

So far the error analysis has concentrated on the magnitude error introduced by various mechanisms. Phase error is also important and can be derived from the magnitude error. Consider the vector diagram shown in Figure 15-9. In general, a magnitude error ΔV may have an arbitrary phase relationship with the signal (V) and introduces some corresponding phase error, $\Delta\theta$. The worse case for the phase

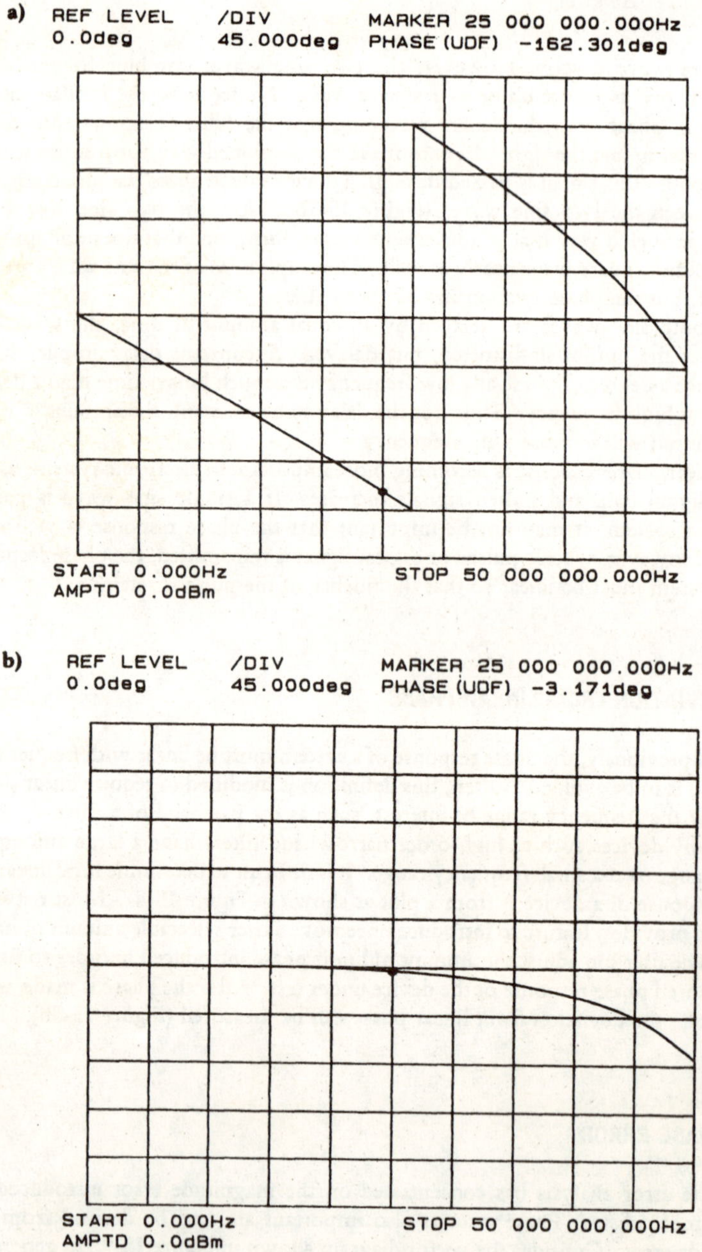

Figure 15-8. (a) A typical measurement of a high-order filter with a large amount of delay in the passband. (b) The same measurement after a large amount of linear phase is removed.

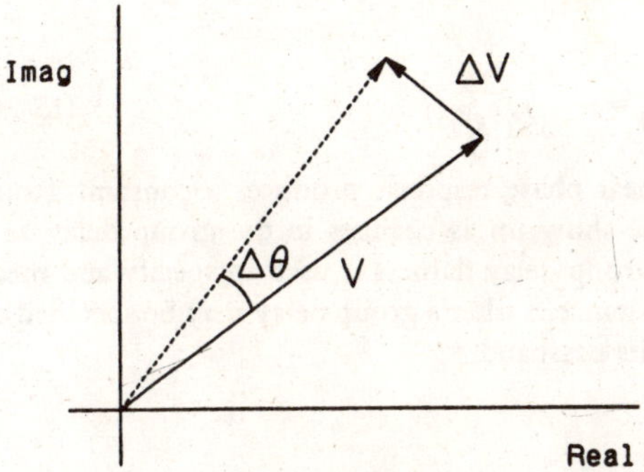

Figure 15-9. Phase error or uncertainty can be related to magnitude error using this vector diagram.

error is when the new vector is perpendicular to the error vector. Under such conditions the phase error can be determined from the following equation:

$$\sin(\Delta\theta) = \Delta V/V \tag{15-17}$$

$$\Delta\theta = \sin^{-1}(\Delta V/V) \tag{15-18}$$

Example 15.4

A particular measurement has a worst case magnitude error of ±1.0 dB. What is the corresponding phase error?

A magnitude error of ±1.0 dB implies that

$$1 \text{ dB} = 20 \log(1 + \Delta V/V)$$

$$\Delta V/V = 10^{(1/20)} - 1 = 0.122$$

The phase error is

$$\Delta\theta = \sin^{-1}(\Delta V/V)$$

$$= \sin^{-1}(0.122) = \pm 7.01°$$

15.10 GROUP DELAY

The group delay through a device is defined as the negative of the derivative of its phase response.

$$t_g = -\frac{d\phi}{d\omega} \tag{15-19}$$

where

ϕ = phase response in radians

ω = frequency in radians/sec

If degrees and hertz are used,

$$t_g = -\frac{1}{360}\frac{d\phi}{df} \tag{15-20}$$

Because of the differentiation, a linear phase response produces a constant group delay. Deviations from linear phase show up as changes in the group delay as a function of frequency. Therefore, group delay flatness is used to specify and measure phase-related distortion. For instance, a filter's group delay may be specified to be flat within some tolerance over its passband.

15.11 DELAY APERTURE

In modern network analyzers, group delay is usually derived from the phase response by calculating an approximation to the derivative.[4] The derivative of the phase is approximated by taking a small Δf in frequency and determining the corresponding phase change, $\Delta\phi$, as shown in Figure 15-10. The group delay is computed as

$$t_g = -\frac{\Delta\phi}{360\,\Delta f} \tag{15-21}$$

The term Δf is called the *delay aperture,* since it is the frequency aperture over which the delay measurement is computed. The delay aperture is usually selectable by the instrument user, in order to optimize the measured results. Ignoring noise consider-

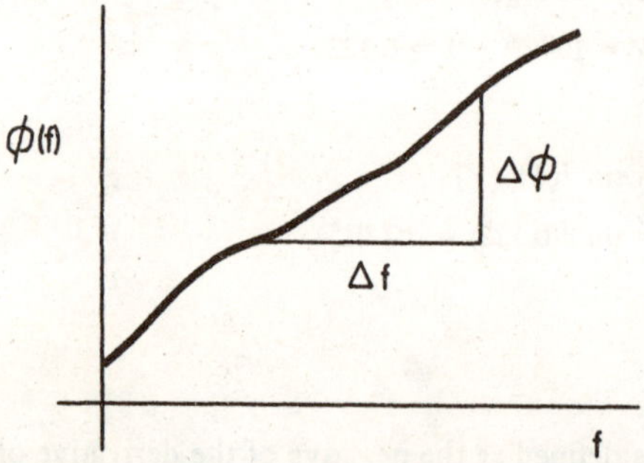

Figure 15-10. Group delay can be calculated by approximating the slope of the phase response.

ations, a small delay aperture seems appropriate since it more closely approximates the true derivative operation. However, the derivative operation tends to exaggerate any noise that happens to be present in the measurement. Since Δf appears in the denominator of the group delay calculation, making Δf smaller makes the noise in

[4] In some network analyzers, group delay is measured using a modulated source. However, the concept of delay aperture is still valid.

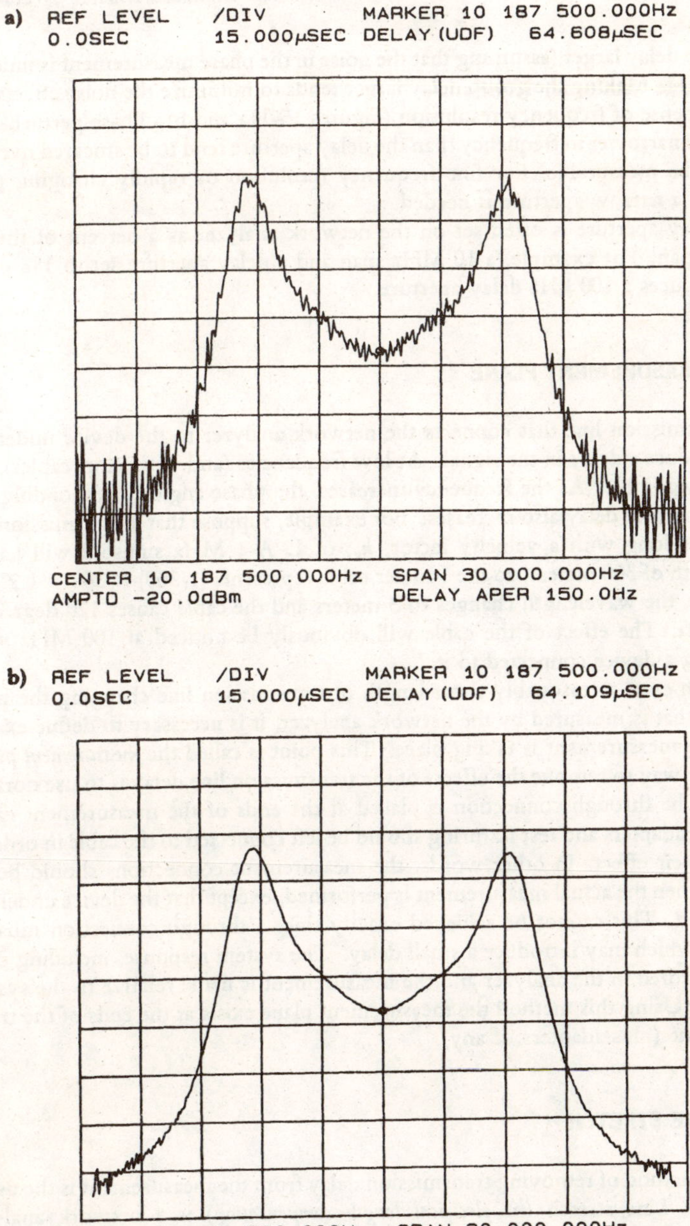

Figure 15-11. (a) A group delay measurement of a crystal filter using a narrow delay aperture. (b) The same group delay measurement with a wider delay aperture.

the group delay larger (assuming that the noise in the phase measurement is independent of Δf). Making the group delay larger tends to minimize the noise effects, but at the expense of frequency resolution (Figures 15-11a and b). Phase perturbations which are narrower in frequency than the delay aperture tend to be smeared over and will not be measurable. For fine frequency resolution or rapidly changing phase response, a narrow aperture is needed.

Delay aperture is often set on the network analyzer as a percent of the frequency span. For example, a 10 MHz span and a delay aperture set to 1% of the span produces a 100 kHz delay aperture.

15.12 MEASUREMENT PLANE

The transmission line that connects the network analyzer to the device under test will introduce a delay in the signal. At low frequencies (and with short cables), this delay is negligible. As the frequency increases, the phase angle corresponding to a fixed amount of delay also increases. For example, suppose that a transmission line is 1 meter long with a velocity factor, k_v, of 1. A 1 MHz sinusoid will have a wavelength of 300 meters, so the 1 meter cable represents $(1/300) \times 360 = 1.2°$. At 100 MHz, the wavelength changes to 3 meters and the cable causes 120 degrees of phase shift. The effect of the cable will obviously be noticed at 100 MHz when measuring a device connected to it.

With even a reasonably short length of transmission line changing the phase response that is measured by the network analyzer, it is necessary to define exactly where the measurement is taking place. This point is called the *measurement plane*.

One way to remove the effects of the transmission line delay is to use normalization. The through connection is placed at the ends of the measurement cables while any adapters and test fixturing should be left connected to the cable in order to remove their effect. In other words, the measurement connections should be the same as when the actual measurement is performed, except that the device under test is removed. This cannot be achieved exactly since a through connection must be inserted, which may introduce a small delay. The system response, including cable delay, is stored in the analyzer and the measurement is made relative to the system response. Using this method the measurement plane exists at the ends of the transmission line (plus adapters, if any).

15.13 LINE STRETCH

Another method of removing transmission delay from the measurement is the use of *line stretch*. Line stretch (or *electrical length compensation*) is a network analyzer feature which adds or subtracts the effect of an ideal transmission line whose length is specified by the user. Usually, this is done mathematically by adding linear phase into the measured response. The user may be able to specify the actual length of the line or the amount of delay in seconds. Depending on the network analyzer, the propagation velocity of the line may have to be taken into account. The user may choose to measure the physical length of the line or may simply adjust the amount of line strength until the phase response flattens.

REFERENCES

1. Adam, Stephen F. *Microwave Theory and Applications.* Englewood Cliffs, N.J.: Prentice-Hall, Inc., 1969.
2. Hewlett-Packard Company. "High Frequency Swept Measurements," Application Note 183, Publication Number 5952-9200, December 1978. Palo Alto, CA.
3. Hewlett-Packard Company. "Vector Measurements of High Frequency Networks," Application Note, Publication Number 5954-8355, March 1987. Palo Alto, CA.
4. Oliver, Bernard M., and John M. Cage. *Electronic Measurements and Instrumentation.* New York: McGraw-Hill Book Company, 1971.
5. Ziemer, R. E., and W. H. Tranter. *Principles of Communications.* Boston: Houghton Mifflin Company, 1976.

16

Reflection Measurements

Reflection measurements characterize a two-terminal port of a device under test. The device may have only one port or may have multiple ports. The fundamental measurement is the complex reflection coefficient (as a function of frequency). Often, the magnitude of the reflection coefficient is displayed on a decibel scale, resulting in a return loss measurement. The reflection coefficient can also be converted to other forms such as standing wave ratio and impedance. Using *s* parameter notation, the reflection coefficient may be referred to as s_{11} or s_{22} of a two-port device.

16.1 RECTANGULAR DISPLAY FORMATS

The fundamental reflection measurement is the complex reflection coefficient, which is the ratio of the reflected voltage to the incident voltage.[1]

$$\Gamma = V_R/V_I \tag{16-1}$$

The magnitude of the reflection coefficient may be displayed on a linear scale but is more commonly shown in decibel form as return loss.

$$\text{RL} = -20\log(|\Gamma|) = -20\log(\rho) \tag{16-2}$$

[1] The reflection coefficient measurement will be magnitude only (scalar) if a scalar network analyzer is used.

The return loss calculation includes a minus sign which causes the return loss values to be positive. When measured on a network analyzer, the minus sign is often omitted, producing measured values which are negative. For example, a network analyzer may read −40 dB, corresponding to a return loss of 40 dB. Figure 16-1 shows a typical return loss measurement of a bandpass filter.

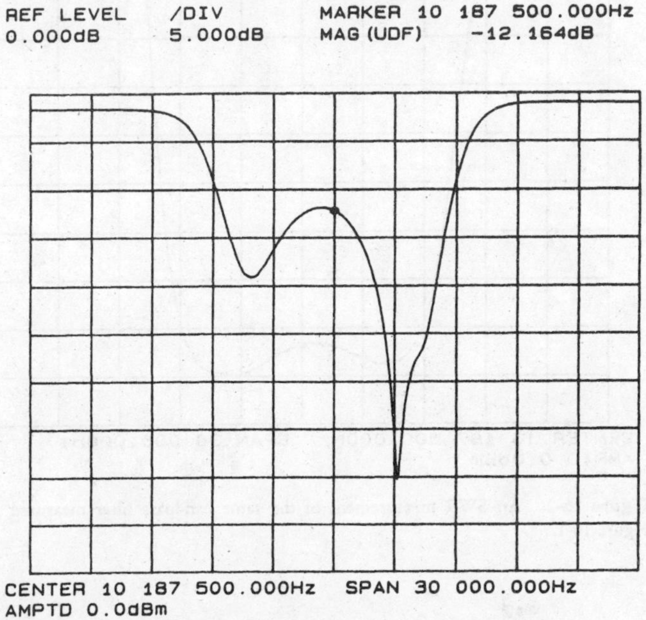

Figure 16-1. A return loss measurement of the input port of a bandpass filter. The return loss exceeds 10 dB in the center of the passband (the lower the trace, the better the match.)

SWR and Impedance

Other parameters can be derived from the complex reflection coefficient and displayed on a rectangular graticule. Equations for standing wave ratio and complex impedance have already been covered in Chapter 11. An example of an SWR plot is shown in Figure 16-2.

16.2 POLAR DISPLAY FORMATS

The reflection coefficient is a complex number and is often displayed in polar format, on a complex plane with a horizontal real axis and a vertical imaginary axis (Figure 16-3). In a polar display, the frequency axis is lost, but the network analyzer's marker or cursor may provide the user with the frequency of any particular data

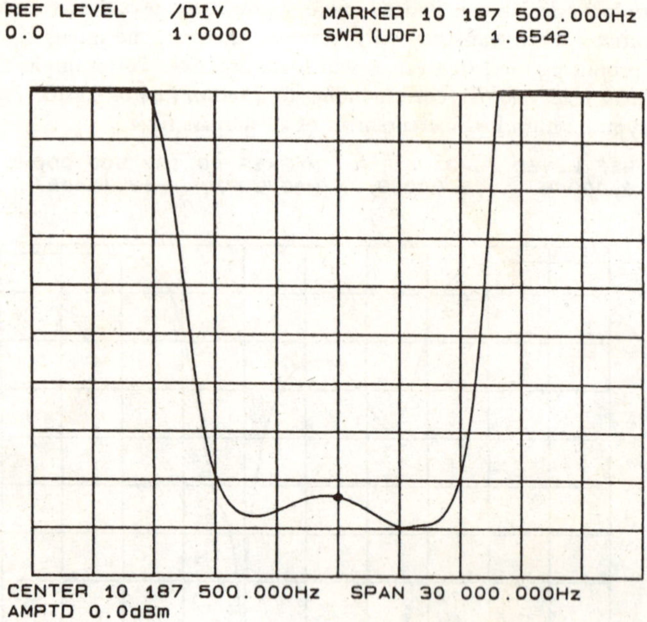

Figure 16-2. An SWR measurement of the same bandpass filter measured in Figure 16-1.

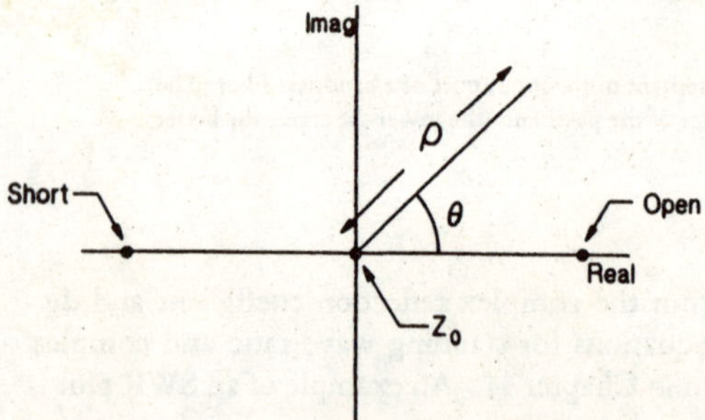

Figure 16-3. The complex reflection coefficient is often plotted in polar format.

point. Any particular frequency point is plotted according to the magnitude (ρ) and phase (θ) of the reflection coefficient. The point is plotted at the end of a vector which starts from the center of the polar plot and extends outward a distance equal to ρ and at an angle of θ. The angle is determined relative to the right hand real axis, which is defined as zero degrees.

If a perfect Z_0 load is connected to the test port, there will be no reflection and the reflection coefficient will be $0\angle 0°$, or $0 + j0$ which is the center of the polar plot. An open circuit produces complete reflection of the incident wave with a reflection coefficient of $1\angle 0°$, which is plotted at the right-hand side of the polar display. A short on the test port causes complete reflection of the incident wave and the reflection coefficient is -1 or in polar format, $1\angle 180°$. This point is plotted on the left-hand side of the polar plot. All three of these points are shown plotted on the complex plane in Figure 16-3.

The Smith Chart

A variation on the polar plot of the complex reflection coefficient is the *Smith chart*. The reflection coefficient is still plotted in polar form, but with a different graticule, called the Smith chart (Figure 16-4). The Smith chart is a familiar tool to radio frequency engineers and is used extensively in design work. As a network analyzer graticule, the Smith chart converts the complex reflection coefficient to normalized impedance. (Many other conversions are possible with the Smith chart.)

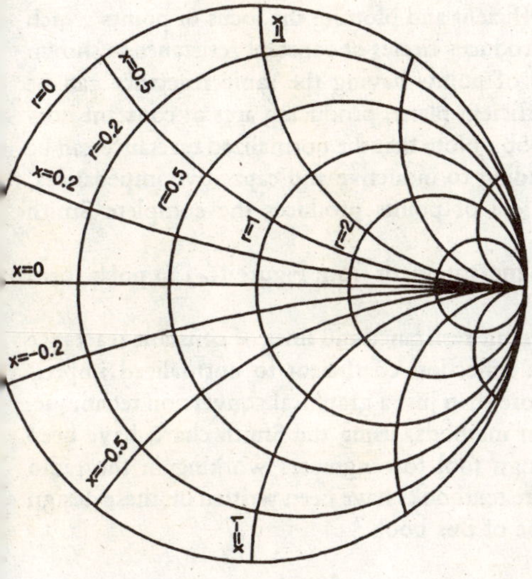

Figure 16-4. The Smith chart is a graphical mapping of the complex reflection coefficient into normalized complex impedance.

The normalized impedance, z is given by

$$z = Z/Z_0 \tag{16-3}$$

where

Z = unnormalized impedance

Z_0 = characteristic impedance of the system

The normalized and unnormalized impedances can also be expressed in terms of their resistive and reactive components.

$$Z = R + jX \tag{16-4}$$

$$z = r + jx \tag{16-5}$$

and

$$r = R/Z_0 \tag{16-6}$$

$$x = X/Z_0 \tag{16-7}$$

Example 16.1

In a 50 Ω system, a particular value of complex reflection coefficient is plotted on the Smith chart and the normalized impedance is 0.3 − j2. What is the impedance (unnormalized) for this value of reflection coefficient?

$$z = 0.3 - j2$$
$$z = Z/Z_0$$
$$Z = Z_0 z = 50(0.3 - j2) = 15 - j100\ \Omega$$

Evaluating the complex reflection coefficient and plotting the locus of points which have the same normalized resistance produces circles of constant resistance as shown in Figure 16-5a. Similarly, the locus of points having the same reactance can be plotted on the complex reflection coefficient plane, producing arcs of constant normalized reactance shown in Figure 16-5b. Note that the normalized reactance can be either positive or negative, corresponding to inductive and capacitive impedances, respectively. Combining these two loci of points produces the complete Smith chart.

Figure 16-6 shows the reflection measurement from Figure 16-1 in polar form with a Smith chart graticule.

The Smith chart's circles of constant resistance and lines of constant reactance provide a graphical conversion from reflection coefficient to normalized impedance.[2] But the Smith chart is much more than just a graphical conversion technique. A wide variety of analysis and design methods, using the Smith chart, have been developed, making it a standard design tool for engineers working in the radio frequency and microwave areas. Entire textbooks have been written on these design techniques, which is beyond the scope of this book.[3]

16.3 DIRECTIONAL BRIDGES AND COUPLERS

A directional bridge or a directional coupler can be used to extract the incident or reflected voltage along a transmission line or at a port of a device under test. Ideally,

---

[2] See Hayt (1974) for a more complete derivation of the Smith chart.

[3] See Gonzalez (1984).

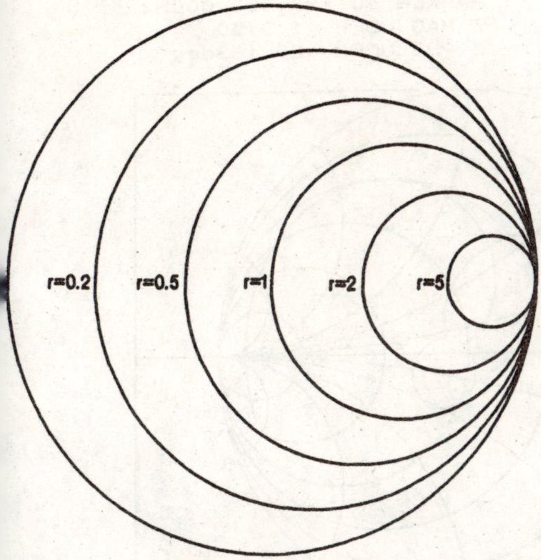

Figure 16-5a. Circles of constant resistance on the Smith chart.

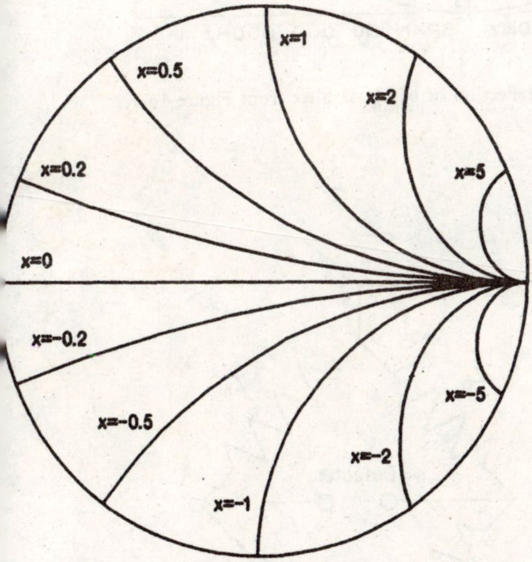

Figure 16-5b. Lines of constant reactance on the Smith chart.

the bridge or coupler measures only the wave traveling in the desired direction and ignores any traveling waves going the other way. However, bridges and couplers have practical limitations which will be discussed in detail later in the chapter.

The directional bridge and directional coupler perform the same basic function, but with different techniques. First, let's examine the operation of the directional bridge. A simplified circuit diagram of a directional bridge is shown in Figure

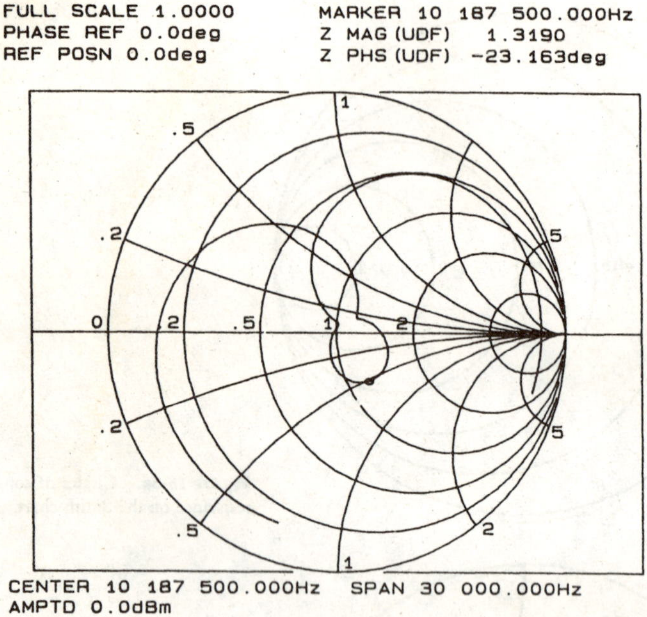

Figure 16-6. Polar plot of input reflection of bandpass filter from Figure 16-1, with Smith chart graticule.

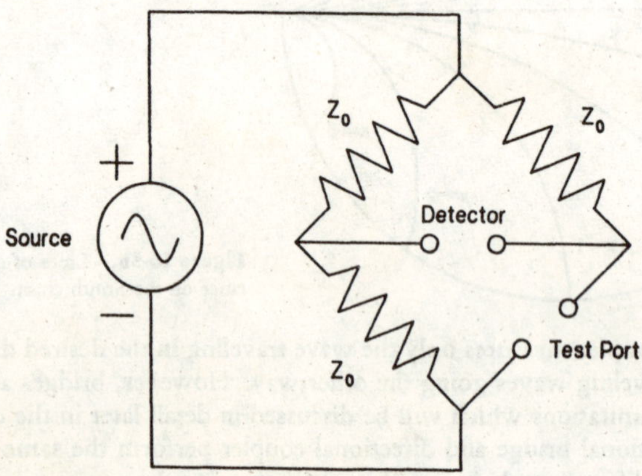

Figure 16-7. A simplified circuit of a directional bridge.

16-7. When the test port is terminated in a perfect Z_0, the bridge is balanced and the detector will measure zero volts, indicating that no reflected wave is present. Now suppose the test port were left open. The detector port would receive half of the source voltage, indicating a large reflection. On the other hand, if the test port were shorted, the detector port would again receive half of the source voltage, but with opposite polarity. Thus, this directional bridge produces a detector voltage which is proportional to the amount of reflection. Just as important, the detector voltage phase or polarity indicates the phase of the reflection.

Notice that the detector port has a voltage on it which is a replica of the source voltage. More specifically, this means that the detector port signal is at the source frequency and must be detected. When used with a network analyzer, this presents no problem since the analyzer will perform the detection and measurement of the detector signal. Directional bridges are also used in meters designed to display *SWR*, called *reflectometers* or simply *SWR meters*. In that case, a diode detector circuit converts the detector voltage to a DC level, which drives a conventional voltage meter.

One problem that must be dealt with is the fact that the detector port is balanced precariously on the directional bridge and is not referenced to ground. Network analyzers usually have one side of their receiver inputs connected to ground. The network analyzer's source is also usually grounded, so driving such a network analyzer directly with the bridge circuit shown in Figure 16-7 would cause the bridge circuit to be unbalanced. To sidestep this problem, the ground connection to the source or detector port must be broken. Directional bridges usually have a transformer (or balun) which either creates a balanced source (floating with respect to ground) or converts the balanced detector output to a single-ended output with one side connected to ground. While this transformer solves the problem, it does not operate at or near DC so the frequency response of the bridge is forced to roll off at some low frequency, typically 10 kHz to 100 kHz.[4]

At microwave frequencies, a directional coupler is most often used to separate traveling waves. A directional coupler provides the same basic function as the directional bridge, but using waveguide techniques to separate the traveling waves. For our purposes, directional bridges and directional couplers perform the same basic function. Therefore, the term "directional device" will be used to loosely refer to both bridges and couplers.[5]

16.4 INSERTION LOSS

Figure 16-8 shows a representation of a directional device, with the three ports labeled. The signal applied to the input will appear at the output with some amount of loss. The insertion loss of a directional device is the ratio of the input power to

[4] For a more detailed discussion of a practical directional bridge, see Spaulding (1984).

[5] See Laverghetta (1984) and Adam (1969) for more information on directional couplers.

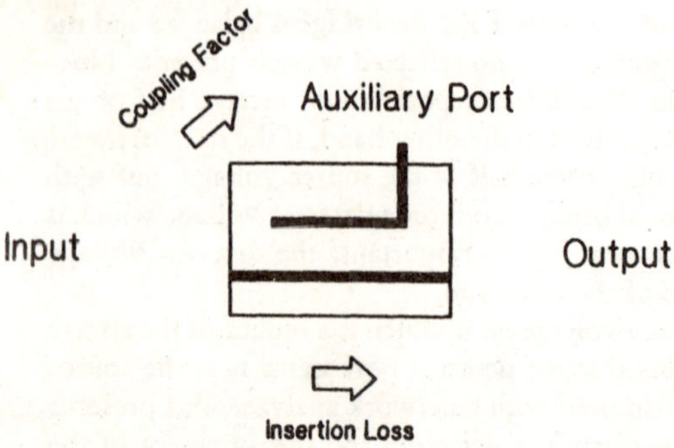

Figure 16-8. The representation of a three-port directional device.

the output power, expressed in decibels. A high-quality directional coupler will exhibit very little insertion loss, usually between 0.1 and 1 dB, while a directional bridge will have a significant insertion loss, typically 6 dB. For a given source power level, large insertion loss means less power delivered to the test port.

16.5 COUPLING FACTOR

The coupling factor of a directional coupler is a measure of how much signal appears at the auxiliary port for a given signal level at the input port (Figure 16-8). The coupling factor is defined as the ratio of the input power to the auxiliary port power, expressed in decibels. A typical coupling factor for a directional bridge is 6 dB, while a directional coupler may have a coupling factor closer to 20 dB.

The power present at the auxiliary port of a directional device is proportional to the directional traveling wave that is being measured. The terminology used here and shown in Figure 16-8 assumes that the forward traveling wave is being measured (i.e., a wave traveling from left to right in the figure).

16.6 DIRECTIVITY

The most important figure of merit for a directional device is *directivity*, which indicates how well a directional device can separate opposite traveling waves. Ideally, the directional device can completely separate the forward and reverse waves, but in reality, some of the forward wave is present in a measurement of the reverse wave (and vice versa). Directivity is defined as the ratio of the power present at the auxiliary port when the signal is traveling in the forward direction to the power present at the auxiliary port when the same signal is traveling in the reverse direction. This ratio is expressed in decibel form, is ideally infinite but typically is 30 to 40 dB. Finite directivity can be thought of as being caused by a leakage path which lets the undesired traveling wave couple into the auxiliary port.

Directivity is important because it limits the maximum return loss that can be measured using a particular directional device.

Example 16.2

A directional bridge has an insertion loss of 6 dB, a coupling factor of 6 dB and a directivity of 40 dB. Configured as shown in Figure 16-8 and with a source power of 0 dBm, determine the power levels at the auxiliary port and the output port with the output terminated in Z_0.

The insertion loss of the bridge is 6 dB; therefore, the output level is 0 dBm − 6 dB = −6 dBm. The coupling factor is also 6 dB, so the power at the auxiliary port is 0 dBm − 6 dB = −6 dBm. Since there is no reflected signal from the output port (due to the Z_0 termination), the directivity does not affect the signal level at the auxiliary port. Had there been a reflection, a portion of that signal would have also appeared at the auxiliary port.

16.7 REFLECTION CONFIGURATION

Now let's reverse the orientation of the directional device and rename the ports to be more appropriate for a reflection measurement (Figure 16-9). A signal will be applied to the input port (formerly the output port) which will excite the DUT at the test port. If the DUT is a perfect Z_0 match, no reflection will occur and no signal will be present at the auxiliary port. (Actually, the auxiliary port will have a small signal present due to the finite directivity of the directional device.) If the test port has a non-Z_0 load connected to it, a reflection will occur and the signal level at the auxiliary port will be proportional to the size of the reflected wave. A directional bridge or coupler is used in this manner to perform a reflection measurement.

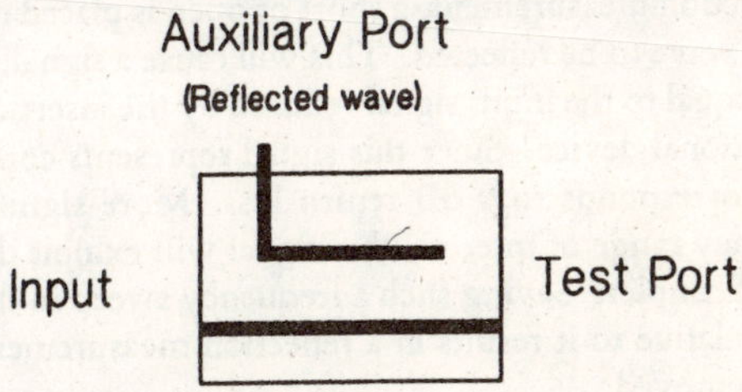

Figure 16-9. The representation of a directional device labeled with reflection measurement terminology.

Example 16.3

The bridge described in Example 16.2 is configured for a reflection measurement as shown in Figure 16-9. The input signal level is 0 dBm. Determine the signal level at the auxiliary port for the following loads at the test port: (a) short, (b) Z_0 load.

a. Load = short. The wave incident on the test port is 0 dBm − 6 dB (insertion loss) = −6 dBm. Since the load is a short circuit, all the incident wave is reflected and appears at the auxiliary port, but reduced by the coupling factor. Thus, the power at the auxiliary port is −6 dBm − 6 dB = −12 dBm. The directivity of the device will also contribute to the power at the auxiliary port. The auxiliary port signal due to the directivity is the input signal level reduced by the insertion loss, the coupling factor, and the directivity. This signal level is 0 dBm − 6 dB − 6 dB − 40 dB = 52 dBm. This signal will introduce an error at the auxiliary port, but since −52 dBm is much smaller than −12 dBm, its effect will be slight.

b. Load = Z_0. The incident wave (which is fully absorbed at the test port) is still −6 dBm. Since none of the wave is reflected, the auxiliary port would have (ideally) no signal. However, the finite directivity of the device will cause a signal to be present. As previously calculated the power in this signal will be −52 dBm. Therefore, even with a perfect Z_0 load, this bridge will indicate a reflected power of −52 dBm + 6 dB = −46 dBm, referred to the test port. Since −6 dB was incident on the test port, the reflected power of −46 dBm corresponds to a device with a return loss of 40 dB. Therefore, a directional device cannot directly measure a return loss value greater than the device's directivity.

16.8 NORMALIZATION

Normalization of transmission measurements was already discussed in Chapter 15. The same type of technique can be used to improve reflection measurements.

The coupling factor of the directional coupler is a nuisance when determining the return loss measured with a directional device. Since return loss is defined to be the ratio (in dB) of the reflected power to the incident power, the coupling factor and the insertion loss of the directional device must be accounted for. Also, the coupling factor and insertion loss will not be constant with frequency and will introduce a frequency response error. Both of these problems can be eliminated by the use of normalization.

During normalization of a reflection measurement, a short or open is placed on the test port, causing all the incident wave to be reflected. This will cause a signal to appear at the auxiliary arm which is equal to the input signal reduced by the insertion loss and coupling factor of the directional device. Since this signal represents complete reflection at the test port, it corresponds to 0 dB return loss. More significantly, when swept over the frequency range of interest, this signal will exhibit the frequency response of the directional coupler. Saving such a frequency sweep in the analyzer's memory and measuring relative to it results in a reflection measurement with the frequency response error removed.

We briefly mentioned that a short or open can be used to provide a totally reflected signal. At frequencies below a few hundred megahertz, either one will suffice so the choice is not critical. At higher frequencies, the short circuit is often preferred, since it provides a more reliable and repeatable termination than the open circuit, which suffers from stray capacitance effects. The short circuit, having a reflection coefficient of −1, introduces a 180 degree phase change at the test port. (The magnitude will be unaffected.) Some network analyzers provide a special normalization feature which removes the effect of this phase change. If not, the user must remember to invert the phase of the measured data.

16.9 ERROR IN REFLECTION MEASUREMENTS

The standard error model for a directional coupler is described by the following equation.[6]

$$\Gamma_M = D + \frac{(1 + T_R)}{(1 - M_S\Gamma_A)}\Gamma_A \tag{16-8}$$

where

Γ_A = actual reflection coefficient

Γ_M = measured reflection coefficient

D = directivity error

T_R = frequency response error

M_S = source match error

(All these variables are complex and are a function of frequency.)

Notice that the directivity error term appears as a constant in the equation. This implies that the absolute error introduced by the finite directivity is independent of the load. (Recall that directivity can be thought of as a leakage from the input port to the auxiliary port.) The $(1 + T_R)$ term in the equation represents the frequency response of the system. Any source, receiver or directional device unflatness will be accounted for here.

The last term in the equation is $1/(1 - M_S\Gamma_A)$, which itself depends on the reflection coefficient. This term accounts for source match errors, that is, error introduced due to the lack of a perfect Z_0 impedance looking back into the directional device. This type of mismatch will rereflect a portion of the signal which is reflected from the test port. The double reflection will show up at the test port and introduce an error.

The uncertainty in the reflection coefficient measurement is defined by

$$\Delta\Gamma = \Gamma_M - \Gamma_A \tag{16-9}$$

$$\Delta\Gamma = D + \frac{(1 + T_R)}{(1 - M_S\Gamma_A)}\Gamma_A - \Gamma_A \tag{16-10}$$

$$\Delta\Gamma = D + \frac{T_R\Gamma_A + M_S\Gamma_A^2}{(1 - M_S\Gamma_A)} \tag{16-11}$$

---

[6] This error model was originally described in Ely (1967).

For most measurement situations, the product of the source match coefficient, M_S and the actual reflection coefficient, Γ_A is much smaller than 1. The equation can be simplified to

$$\Delta\Gamma = D + T_R\Gamma_A + M_S\Gamma_A^2 \tag{16-12}$$

which is the classical result for predicting errors in reflection measurements. Since the phase of the complex reflection coefficient is not usually known, Γ is replaced by ρ as we consider the worst case.

$$\Delta\rho = D + T_R\rho_A + M_S\rho_A^2 \tag{16-13}$$

Example 16.4

Determine the measurement uncertainty in the following reflection measurement. The actual return loss is 10 dB, the directivity of the directional bridge is 40 dB, the effective source match is 20 dB, and the frequency response error is ±1 dB.

First, convert each of the measurement parameters from decibels into linear values.

$$\rho_A = 10^{-(10/20)} = 0.316$$

$$D = 10^{-(40/20)} = 0.01$$

$$M_S = 10^{-(20/20)} = 0.1$$

$$\text{freq. response (db)} = 20\log(1 + T_R)$$

$$T_R = 10^{(1/20)} - 1 = 0.122$$

$$\Delta\rho = D + T_R\rho_A + M_S\rho_A^2$$

$$= 0.01 + (0.122)(0.316) + 0.1(0.316)^2$$

$$\Delta\rho = \pm 0.585$$

ρ could actually be anywhere between 0.316 + 0.0585 = 0.375 and 0.316 − 0.0585 = 0.258. Or, expressed as return loss, 8.52 dB to 11.77 dB.

16.10 ERROR CORRECTION

The three main error mechanisms just discussed can be characterized and eliminated from the measurement through *error correction*, also known as *accuracy enhancement*. It was already mentioned that normalization provides a means of removing frequency response errors from a reflection measurement, but it does not improve the directivity of the directional device. Recall that finite directivity means that the traveling wave going in the opposite (undesired) direction will contribute to the measured level of the desired traveling wave. These two signals will add vectorally and may add constructively, destructively or anywhere in between, showing up as a ripple in the response of the reflection measurement. When the two signals add destructively (totally canceling), the error approaches infinity. When the two signals add together in phase, the measured value may be off by as much as a factor of 2 (or 6 dB).

Error correction involves the measurement of certain parameters of the directional device over the frequency range of interest, storing them in digital form and correcting the measured values to produce a more accurate measurement. In the past, the complexity of the error correction computation and the amount of digital storage needed required the use of an external computer. With decreasing memory cost and increasing microprocessor power, the error correction calculations can be performed internal to the instrument. Typically, the user is prompted to attach the appropriate terminations (Z_0, open and short) during error correction calibration, and the instrument does the rest. The calibration may be good only for the selected frequency range and changing the frequency range may require recalibration by the user.

16.11 NORMALIZATION REVISITED

Let us reexamine the concept of normalization, given our error model for directional devices. Normalization of a reflection measurement required the placement of an open or short at the test port of the directional device. This causes all of the signal to be reflected back to the auxiliary port. For a device with good directivity (30 or 40 dB), this reflected signal will be much larger than the error signal caused by finite directivity. Also, for a directional device and a source producing a good Z_0 impedance, the source match error can be ignored. Thus, reflection normalization measures the frequency response error term of the error model while setting $D = 0$ and $M_S = 0$.

$$\Gamma_M = (1 + T_R)\Gamma_A \quad \text{(normalization)} \tag{16-14}$$

The error correction or normalization term, E_N, is

$$E_N = 1 + T_R \tag{16-15}$$

which is a complex function of frequency stored away in digital memory. To produce the corrected measurement, the error model equation is reversed.

$$\Gamma_A = \Gamma_M/E_N \tag{16-16}$$

16.12 TWO-TERM ERROR CORRECTION

A two-term error correction can be performed by ignoring the source match error and correcting the directivity and frequency response errors. The advantage of this technique is shorter calibration time (compared with three-term error correction which will be described shortly), since only two terminations (open and Z_0 load) must be measured. Ignoring the source match error is not too much of a compromise in many cases, particularly if the impedance looking back into the directional coupler is very close to Z_0 and/or the input of the device under test is close to Z_0. The equation for the two-term error model is

$$\Gamma_M = D + (1 + T_R)\Gamma_A \quad \text{(two-term error correction)} \tag{16-17}$$

Two calibration measurements are required, the first one with a Z_0 load. This sets the reflection coefficient equal to zero, so that the directivity term can be measured. On the second calibration measurement, an open (or sometimes a short) is placed on the test port, causing the reflection coefficient to be equal to one. Thus, the frequency response term can be measured. The directivity term is still present in the second measurement, but may be subtracted off by the analyzer error correction algorithm.

16.13 THREE-TERM ERROR CORRECTION

Using the full three-term error correction model necessitates the use of three calibration measurements and three unique terminations—a Z_0 load, an open and a short. With the Z_0 load attached, the reflection coefficient is zero and only the directivity appears in the measurement. When a short is attached, the reflection coefficient is -1, and the error model reduces to

$$\Gamma_M = D + \frac{(-1)(1 + T_R)}{(1 + M_S)} \tag{16-18}$$

Then an open circuit is attached, making the reflection coefficient equal to one. This gives

$$\Gamma_M = D + \frac{(1 + T_R)}{(1 - M_S)} \tag{16-19}$$

With two equations and two unknowns (T_R and M_S), the two unknowns can be calculated, resulting in a complete three-term error correction. Again, the correction factors are stored in memory and the error model is used to enhance the accuracy of the measurement.

Figure 16-10 shows the return loss measurement of a Z_0 load, with and without three-term error correction. The difference between the two traces represents the improvement in directivity that the error correction provides. The directivity errors have been removed from the corrected measurement, effectively leaving no detected auxiliary port voltage. Thus, the analyzer measures its own input noise and the measurement appears noisier than the uncorrected trace. Even though the trace is noisier with a Z_0 load connected, the substantial directivity improvement provides a much more accurate measurement.

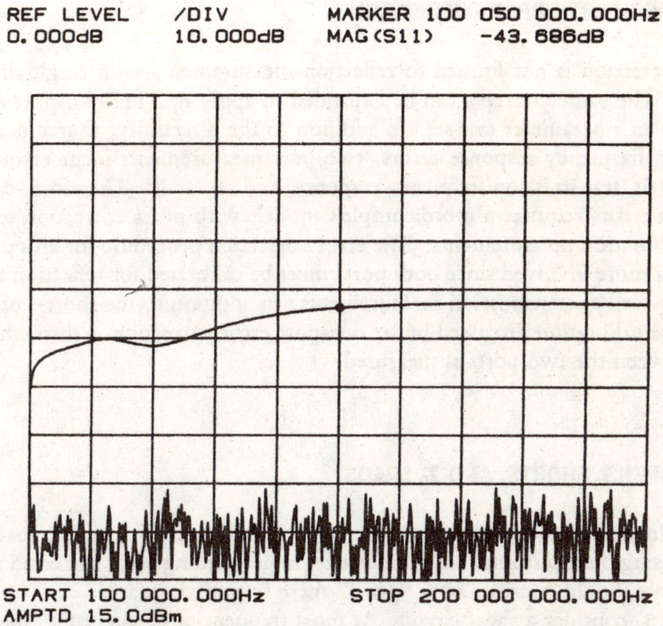

Figure 16-10. A return loss measurement of a Z_0 load without error correction (upper trace) and with three-term error correction (lower trace). The directivity is improved from approximately 43 dB to over 80 dB.

16.14 RESISTIVE BRIDGE

One limitation of conventional directional couplers and directional bridges is that they do not work for arbitrarily low frequencies. Directional couplers operate over a particular range of microwave frequencies, while directional bridges will operate down to around 10 kHz. Directional bridges require a transformer or balun in the signal path to convert the balanced bridge voltage to a voltage referenced to ground. Since this transformer operates only for AC signals, the bridge will not function at or near DC.

A resistive power splitter can function as a directional device and will operate at frequencies all the way down to DC. Unfortunately, it has very poor inherent directivity, around 6 dB. However, error correction can be used to improve the directivity to a useful level, providing reflection measurements over a wide frequency range, including DC.

16.15 TWO-PORT ERROR CORRECTION

Error correction is not limited to reflection measurements with single directional devices. The same concepts can be expanded to apply to a full two-port measurement on an *s* parameter test set. In addition to the directivity, source match, and reflection frequency response errors, two-port measurements incur errors due to load match, transmission frequency response, and crosstalk. Therefore, correction for these errors requires a more complex model, with more correction terms and more calibration measurements. The error correction procedure for an *s* parameter test set is more involved since both ports must be calibrated for reflection measurements as well as transmission measurements. In addition to the short-, open-, and load-type calibration discussed under one-port error correction, a through connection between the two ports is measured.

16.16 OPENS, SHORTS, AND Z_0 LOADS

In this chapter, we have referred to open circuits, short circuits, and Z_0 loads. Since we are using these terminations as references in performing error corrected measurements, we should consider how "good" these terminations are.

First, consider a short circuit. At most frequencies of interest a good-quality short circuit can be obtained. At very low frequencies, this might be as simple as a BNC connecter with a wire shorted across it. As the frequency increases, a higher-quality connector (such as N or SMA) will be required with a low-inductance short designed into it.

An open circuit is not quite as simple. At frequencies below a few hundred megahertz, an open circuit is just that—leaving the test connector open. At microwave frequencies, an open connector begins to have significant stray capacitance and we must instead attach a special termination specifically designed to be an open.[7] This termination is designed to control the stray capacitance between the inner conductor and the outer shield, such that the error correction algorithm can account for it.

A good quality Z_0 load may have a return loss between 40 and 50 dB. Since this is the reference load for the measurement, the corrected measurement cannot reliably exceed this return loss. The device under test is simply being compared with the reference load used during the error correction calibration. So an alternative view is that a corrected measurement which indicates that the device under test is a perfect Z_0 load, really means that the device under test has the same impedance as the reference load.

[7] This only seems like a contradiction in terms.

REFERENCES

1. Adam, Stephen F. *Microwave Theory and Applications*. Englewood Cliffs, NJ: Prentice-Hall, Inc., 1969.
2. Ely, Paul C., Jr. "Swept Frequency Techniques." *Proceedings of the IEEE* (June 1967).
3. Gonzalez, Guillermo. *Microwave Transistor Amplifiers*, Englewood Cliffs, N.J.: Prentice-Hall, Inc., 1984.
4. Hayt, William H., Jr. *Engineering Electromagnetics*, 3rd ed. New York: McGraw-Hill Book Company, 1974.
5. Hewlett-Packard Company. "Coaxial and Waveguide Measurement Accessories Catalog," Publication number 5954-6401, November 1986. Palo Alto, CA.
6. Hewlett-Packard Company. "Vector Measurements of High Frequency Networks," Publication Number 5954-8355, March 1987. Palo Alto, CA.
7. Laverghetta, Thomas S. *Practical Microwaves*. Indianapolis, IN: Howard W. Sams & Company, Inc., 1984.
8. Spaulding, William M. "A Broadband Two-Port S-Parameter Test Set." *Hewlett-Packard Journal* (November 1984).

17 Analyzer Performance and Specifications

Spectrum and network analyzer specifications are the instrument manufacturer's way of communicating to the user the level of performance which the user can expect from a particular instrument. Understanding and interpreting instrument specifications enables the instrument user to predict how the instrument will perform in a specific measurement situation. More specifically, the user can determine the overall accuracy of a measurement.

The form and style of the specifications are usually related somewhat to the block diagram and measurement techniques internal to the instrument. These specifications will often appear to be more complex than necessary. Oversimplifying an instrument data sheet may force the manufacturer to understate the performance level of an instrument in order to cover all possible cases in a single specification. For instance, if the high-impedance input of an instrument exhibits poorer amplitude accuracy than the 50 Ω input, it makes sense to specify accuracy for the two inputs separately, rather than compromising the overall accuracy specification.

17.1 SOURCE/TRACKING GENERATOR SPECIFICATIONS

The signal source on a network analyzer or a tracking generator on a spectrum analyzer may have a certain amount of error in its frequency and its amplitude. The manufacturer's specifications delineate these imperfections. In cases where the frequency of the source and receiver are both derived from a common reference oscillator, the source frequency specifications may also apply to the receiver.

Ideally, the source output produces only one frequency—the desired, fundamental frequency. In practice, a source also produces other frequencies, both harmonic and nonharmonic frequencies. A typical specification for a network analyzer source harmonic content might be −25 or −30 dBc, which may seem like rather poor distortion performance. However, if the source and receiver are tracking in frequency and the device under test is reasonably linear, the harmonic content will simply fall outside the passband of the receiver. Nonharmonic spurious signals are not necessarily so well behaved and may show up at what appears to be random frequencies. The analyzer manufacturer must make sure that the level of these spurious signals remains small enough and/or that these spurs do not show up near the source frequency so that they do not fall within the receiver passband.

Frequency Resolution—The frequency resolution specification indicates the smallest change possible when setting the source frequency. For example, a source with a 0.1 Hz frequency resolution may be set to 1000.1 Hz, 1000.2 Hz and 1000.3 Hz, but not 1000.05 Hz. This specification does *not* define how accurate the source frequency is, but only how finely it may be set.

Frequency Stability—A source's frequency varies over time due to thermal, aging, and other effects. The frequency stability specification describes this long term frequency drift, usually in terms of parts per million per day, often with a temperature range specified. (Short-term frequency fluctuation is specified in terms of phase noise.) A typical frequency stability specification might be $\pm 5 \times 10^{-8}$/day. With this spec, a 100 MHz source frequency could vary as much as $\pm(5 \times 10^{-8})$ (100 MHz) = ±5 Hz per day.

Level Accuracy—This specification indicates how much error there can be in the output or power level of the source. This is often specified at only one frequency and power level, with a linearity specification to describe the accuracy at other output levels and a flatness specification to describe how the level varies with frequency. A typical specification is ±0.5 dB at 50 MHz and +10 dBm output power.

Level Linearity—Level linearity describes how the level accuracy changes with changing output level. Often it is specified in table form with an accuracy specified for a particular range of output power. For example,

ERROR	OUTPUT LEVEL
±0.2 dB	−5 dBm to +15 dBm
±0.5 dB	+15 dBm to +20 dBm

Flatness—The flatness specification represents the frequency response of the source power level. The flatness spec alone does not indicate anything about the absolute accuracy of the source power, but instead indicates how much it varies over frequency. For example, a flatness specification of ±1 dB means that for a given amplitude setting, the actual source power level may vary over a 2 dB range when swept in frequency, usually measured relative to a low-frequency amplitude value.

Impedance—The nominal output impedance of the source is important in that a Z_0 (e.g., 50 Ω) system needs to be driven by a source having an output impedance of Z_0. In addition, the quality of this Z_0 source impedance will usually be specified in terms of return loss or *SWR*. This is important in predicting measurement error due to imperfect source match. Typical specification: >20 dB return loss.

Phase Noise—Very-short-term variations in frequency are specified in terms of phase noise. The phase noise is specified in dBc (dB relative to the "carrier" or source frequency) at some frequency offset away from the source frequency and normalized to a 1 Hz bandwidth. A typical specification is stated as <−90 dBc (1 Hz BW) at a 10 kHz offset.

Harmonics—The harmonic content present in the output signal is specified in terms of dBc (decibels relative to the carrier, in this case the fundamental frequency). Typical specification: <−30 dBc.

Nonharmonic Spurious Signals—The source may produce other spurious signals which are not harmonically related to the source frequency. These spurious signals are also usually specified in terms of dBc. Typical specification: <−50 dBc.

17.2 RECEIVER CHARACTERISTICS

Input Impedance—The nominal input impedance of the receiver is important if the input is required to properly terminate a port of the device under test. A Z_0-type input (e.g., 50 Ω) will usually have a return loss or *SWR* specification associated with it so that mismatch errors can be estimated. Typical specification: >25 dB return loss.

Analyzers operating in the under 50 MHz range may also supply a high impedance input, whose nominal resistive and capacitive components are specified (e.g., 1 megohm and 30 pF).

Sensitivity/Noise Level—Sensitivity (or noise level) specifies the noise floor of the analyzer (usually in dBm), either in a particular resolution bandwidth or normalized to a 1 Hz bandwidth. Either way, the sensitivity spec represents the reading of the analyzer due to noise with no signal present. Signals obscured by this noise level cannot be detected, while signals right at this level will be measurable, but with some error (see Chapter 8). Typical specification: −130 dBm (1 Hz BW).

Residual Responses—Residual responses are signals which appear on the analyzer display due to imperfections internal to the analyzer with no input signal connected. Typical specification: >100 dB below maximum input level.

Input-Related Spurious Responses—Input-related spurious responses are signals which appear on the analyzer display due to imperfections internal to the analyzer when an input signal is connected. When the input is disconnected, these responses disappear. These responses are artifacts of the analyzer's internal block diagram (in particular, local oscillator impurities), and they differ from distortion products in that they occur at frequencies not directly related to the input frequency.

DC Response/LO Feedthrough—Most analyzers that operate near 0 Hz generate a response at DC. In a swept analyzer, this is due mainly to the local oscillator feedthrough. In an FFT analyzer, this response is due to DC offsets in the signal path. The level of this response at 0 Hz is usually specified in decibels relative to a full-scale response. This specification may be omitted on analyzers whose low-frequency limit is significantly above 0 Hz (e.g., 100 kHz). Typical specification: >33 dB below full-scale input level.

Multichannel network analyzer amplitude characteristics are usually specified for the single-channel case as well as the dual-channel (or ratio) case. The accuracy of the ratio case is usually better since the channels are designed and built to be matched in magnitude and phase characteristics. Thus, when the ratio of these channels is used, their amplitude errors are reduced to the extent that the channels are matched. Phase and delay characteristics are normally specified only for ratio measurements.

Amplitude Accuracy—The absolute amplitude accuracy of the receiver is usually specified for a full-scale signal and may be restricted to center of screen. A dynamic accuracy specification is added to the absolute accuracy spec to determine the accuracy at lower amplitudes. Alternatively, amplitude accuracy may be composed of several different specifications such as *IF gain uncertainty*, *RF gain uncertainty*, *amplitude temperature drift*, and so forth.

Amplitude Dynamic Accuracy—Also known as *incremental accuracy* or *log scale fidelity*, this specification describes how accurate the analyzer is in a relative sense. That is, if a signal changes by 1 dB at the input, what change is shown on the analyzer display? A typical spec is stated as ± 0.05 dB/dB, meaning that for a 1 dB change in signal level, an error of ± 0.05 dB may be introduced. Alternatively, it may be specified in table form, with an error limit for each measurement range. This specification is important because it represents the main error remaining after a normalization is performed.

Amplitude Resolution—The smallest change in amplitude that can be detected by the analyzer, often related to the marker or cursor readout. The resolution should be significantly smaller than the typical amplitude accuracy, so that the resolution does not limit accuracy.

Amplitude Frequency Response or Flatness—This specification describes the variation in amplitude response due to changing frequency. In cases where the absolute accuracy of the receiver is specified at only one point, the amplitude flatness must be added in to determine the error at other frequencies. The amplitude flatness is also important in cases where a network measurement is performed without the use of normalization.

Phase specifications usually only apply when two receiver channels are used together in a relative or ratio phase measurement.

Phase Accuracy—The absolute phase accuracy of the receiver is usually specified for a full-scale signal and may be restricted to center of screen. A dynamic accuracy specification is added to the absolute accuracy spec to determine the phase accuracy at lower amplitudes.

Phase Dynamic Accuracy—This specification describes how accurate the phase response of the analyzer is with changes in signal amplitude. This specification is important because it represents the main error remaining after a normalization is performed.

Phase Resolution—The smallest change in phase that can be detected by the analyzer, often related to the marker or cursor. The resolution should be significantly smaller than the typical phase accuracy, so that the resolution does not limit accuracy.

Phase Frequency Response—This specification describes the variation in phase response due to changing frequency. In cases where the absolute phase accuracy of the receiver is specified at only one point, the phase frequency response must be added in to determine the error at other frequencies. The phase frequency response is also important in cases where a network measurement is performed without the use of normalization.

Delay Specifications—Since most analyzers calculate the delay measurement from the phase measurement, delay specifications are obtained by translating the phase specs into delay specs. For example, delay accuracy might be given as

$$\text{delay accuracy} = (\text{phase accuracy})/(360 \times \text{delay aperture})$$

with phase accuracy in degrees and delay aperture in hertz.

17.3 DYNAMIC RANGE

Dynamic range is a receiver specification (or, more precisely, a set of receiver specifications) and is important enough to be treated separately. Dynamic range describes the range of signal levels that can be reliably measured simultaneously. In particular, it describes the analyzer's ability to measure small signals in the presence of large signals. This ability is critical to the function of an analyzer, since its main function is to measure the individual frequency components of a signal or the frequency response of a network.

Spectrum Analyzer

First, we will examine dynamic range as it applies to the spectrum analyzer case. Then, we will explore how those concepts relate to the network analyzer case.

Dynamic range is defined as the maximum ratio of two signal levels simultaneously present at the input which can be measured to a specified accuracy.[1] We can imagine connecting two signals to the analyzer input—one which is the maximum allowable level for the analyzer's input range and the other which is much smaller (Figure 17-1). The smaller one is reduced in amplitude until it is no longer detectable by the analyzer. When the smaller signal is just measurable, the ratio of the two signal levels (in dB) defines the dynamic range of the analyzer.

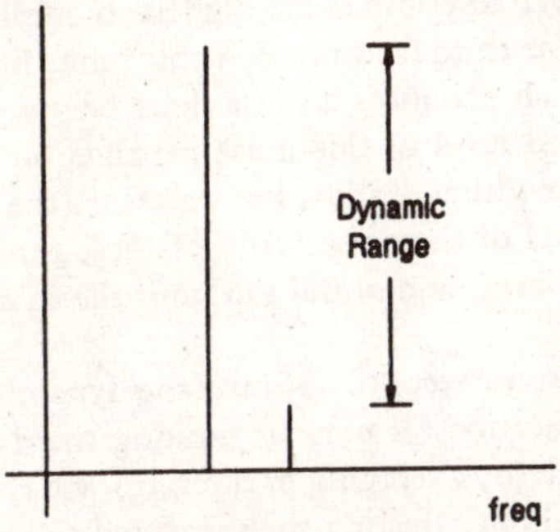

Figure 17-1. The dynamic range of a spectrum analyzer is the ratio (expressed in dB) of the largest and smallest signals that can be reliably measured at the same time.

What effects might make it undetectable? Such things as residual responses of the analyzer, harmonic distortion of the large signal (due to analyzer imperfections), and the internal noise of the analyzer will be large enough to cover up the smaller signal as we decrease its amplitude. The smaller signal might not fall on top of, for instance, a harmonic of the larger signal, but when considering the general case, we must assume that it might. Alternatively, we could note that we can't tell the difference between the smaller signal and an imperfection of the analyzer such as a distortion product or residual response. Thus, the dynamic range of the instrument determines the amplitude range over which we can reliably make measurements.

Figure 17-2 shows the error mechanisms that can limit the dynamic range of the analyzer. A single frequency is at the input of the analyzer, but its harmonics are generated internal to the analyzer due to harmonic distortion. For simplicity, we have only shown one input frequency. Had there been more than one frequency, we would also have intermodulation distortion products (in addition to the harmonic distortion products). Other sources of error are the residual and input-related spurious responses in the analyzer. If the harmonics, residual or spurious responses are, for example, 80 dB down from the fundamental, the dynamic range of the analyzer

[1] From Institute of Electrical and Electronics Engineers (1979).

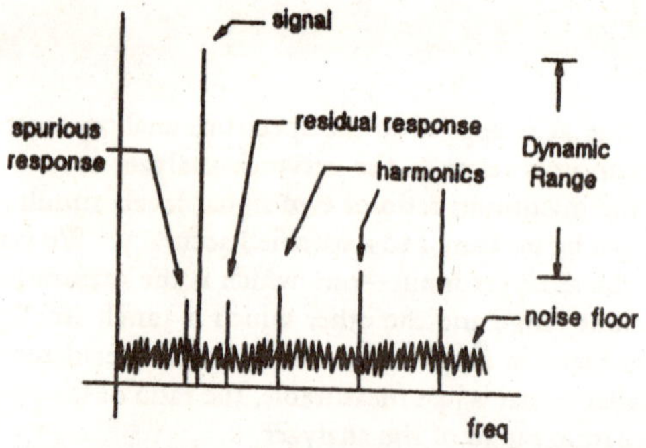

Figure 17-2. Dynamic range is limited by the analyzer's harmonic distortion, internal noise, residual responses, and input-related spurious responses.

will be limited to 80 dB. Whichever response is the highest in amplitude will ultimately limit the dynamic range. The third factor in dynamic range limitations is the internal noise of the analyzer, which produces a noise floor below which a signal cannot be measured. The measured level of this noise depends on the resolution bandwidth used. A narrower bandwidth will allow less noise into the measurement, thereby reducing the measured level of the noise. Any of these three mechanisms (distortion, residual/spurious responses, and noise) can limit the dynamic range of the instrument.

The instrument user can take some steps to optimize the dynamic range for the user's particular measurement application. If noise is limiting the dynamic range, reducing the predetection bandwidth (by averaging or filtering) will reduce the noise level measured by the analyzer without affecting the measured signal level. If the distortion products are the limiting factor, they can be reduced by reducing the signal level. As discussed in Chapter 7, the distortion products will drop in amplitude by a larger amount than the signal level, causing an increase in dynamic range. An external attenuator supplied by the user or the analyzer's internal attenuator can reduce the signal level. Of course, as the signal level is reduced, the dynamic range may be limited by the noise floor.

Network Analyzer

The dynamic range considerations are somewhat different for network analyzers. The spectral characteristics of both the source and receiver must be considered, since either can limit the dynamic range. In the case of the spectrum measurement, very little is known about the signal, but in the case of a network measurement, the source and receiver are both tuned to the same frequency. Therefore, harmonic distortion (in either the source or receiver) is not usually a problem since the harmonics fall outside the receiver passband. Intermodulation distortion is usually negligible since network measurements are performed with only one frequency stimulating the DUT, which is linear or near linear.

Source spurious responses can potentially cause measurement error but don't usually limit dynamic range. Since the source is always at the measurement frequency, its amplitude tends to dominate over small spurious frequencies. The device under test attenuates these spurious responses along with the desired source frequency.

The remaining analyzer imperfections which normally limit the dynamic range of a network measurement are receiver residual responses and the receiver noise floor. As in the spectrum analyzer case, the noise floor can be reduced by narrowing the predetection bandwidth, perhaps at the expense of increased measurement time.

REFERENCES

1. Hewlett-Packard Company. "HP 3577A Network Analyzer Technical Data," Publication Number 5953-5118. Palo Alto, CA, 1983.
2. Hewlett-Packard Company. "HP 8568B Spectrum Analyzer Technical Data," Publication Number 5952-9394. Palo Alto, CA, 1985.
3. Hewlett-Packard Company. "HP 8753A RF Network Analyzer Technical Data," Publication Number 5954-1550. Palo Alto, CA, 1986.
4. Hewlett-Packard Company. "HP 3588A Spectrum Analyzer Technical Data," Publication Number 5952-1439. Palo Alto, CA, 1989.
5. Institute of Electrical and Electronics Engineers, Inc. "I.E.E.E. Standard for Spectrum Analyzers," I.E.E.E. Standard 748-1979. New York: I.E.E.E., September 1979.

Index

Also Available from SciTech Publishing, Inc., USA
New Age International Pvt. Ltd.

Radio-Electronic Transmission Fundamentals
B. Whitfield Griffith, Jr.
ISBN: 978-81-224-3094-3, Original Edition Price: \$ 95.00, Indian Edition Price: Rs. 395.00

A Guide to MATLAB® Object-Oriented Programming
Andy H. Register
ISBN: 978-81-224-3095-0, Original Edition Price: \$ 83.95, Indian Edition Price: Rs. 395.00

Introduction to Electronic Defense Systems 2nd Ed.
Filippo Neri
ISBN: 978-81-224-3096-7, Original Edition Price: \$ 109.00, Indian Edition Price: Rs. 395.00

Spectrum and Network Measurements
Robert A. Witte
ISBN: 978-81-224-3097-4, Original Edition Price: \$ 75.00, Indian Edition Price: Rs. 495.00